BASIC ELECTRONICS AND LINEAR CIRCUITS

BASIC ELECTRONICS AND LINEAR CIRCUITS

Basic Electronics and Linear Circuits

N N BHARGAVA
Technical Teachers' Training Institute, Chandigarh

D C KULSHRESHTHA
Delhi College of Engineering, Delhi

S C GUPTA
Government Polytechnic, Gauchar, U.P.

Tata McGraw-Hill Publishing Company Limited
NEW DELHI

McGraw-Hill Offices

New Delhi New York St Louis San Francisco Auckland Bogotá Guatemala
Hamburg Lisbon London Madrid Mexico Milan Montreal Panama
Paris San Juan São Paulo Singapore Sydney Tokyo Toronto

© 1984, Technical Teachers' Training Institute, Chandigarh

38th reprint 2004
RYDARRAIRXZYQ

This edition can be exported from India only by the publishers,
Tata McGraw-Hill Publishing Company Limited

ISBN 0-07-451965-4

Published by Tata McGraw-Hill Publishing Company Limited,
7 West Patel Nagar, New Delhi 110 008, and printed at
Pushp Print Services, Delhi 110 053

The **McGraw·Hill** Companies

Foreword

The present textbook on "Basic Electronics and Linear Circuits" is yet another contribution by Technical Teachers' Training Institute, Chandigarh, in its efforts to develop basic instructional material for the polytechnics. The book has been written as a part of the programme of writing textbooks on basic subjects according to the revised curriculum developed by the Curriculum Development Centre of this Institute and accepted by various states in the Northern Zone.

Writing a textbook for technical education is indeed a very challenging task. The author has to possess not only expertise in the subject matter but also the technique of selecting appropriate material from the vast fund of knowledge in the subject matter and present it in a way which the students can easily understand. Judging from the remarks of the experts who reviewed this book before publishing and also on the basis of earlier field trials by the authors on the teachers and students of polytechnics, I have no doubt in my mind that the authors have done an excellent job.

The Institute will, therefore, feel amply rewarded if the teachers and students of the polytechnics accept this book as a basic textbook for the teaching and understanding of the subject. Any suggestions for the improvement of this book from all quarters will be most welcome.

P D KULKARNI
Principal
Technical Teachers' Training Institute
Chandigarh

Preface

The Ministry of Education, Government of India, established four regional Technical Teachers' Training Institutes in the country with a view to improve technical education, in general, and polytechnic education, in particular. Each of these institutes has a Curriculum Development Centre (CDC). After a detailed study of the technical education system and present-day technology, the CDC undertook the task of developing curricula for diploma courses in the various branches of engineering. It was found necessary to update the contents in the subject of electronics, because of the rapid developments that have taken place in this area in the last decade. Thus, a new curriculum was designed to answer the needs of the industry. All the modern devices and related circuits were included in the curriculum.

Effective implementation of any curriculum calls for appropriate support material for students as well as for teachers. Textbooks are amongst the most important resources for student-learning, provided they are written within the framework of the objectives laid down. This work is an honest effort in this direction, and is a result of a number of workshops conducted from time to time

The diploma engineer in the middle-level supervisory position forms an important link between the craftsmen of varying degrees of skill on the one end of the technical manpower spectrum and the engineers/scientists on the other. Whilst he must be conversant with the skills of the craftsman, he must also have a working knowledge of the principles of engineering and technology. This book aims at developing these principles without taking recourse to advanced mathematics. After the principles are developed the accent is on their applications in industry. Since this book deals with linear electronic circuits, it presupposes an elementary knowledge of electronic devices.

Following are the salient features of the book:

1. More emphasis has been given to semiconductor devices and their applications; only a passing reference is made to vacuum tube circuits as they have a shrinking future.
2. Each chapter starts with its *Objectives*. This enables the student to know what is expected of him after he goes through the chapter.
3. The subject is explained in an easy-to-understand language, without using advanced mathematics.
4. Profuse use of illustrations makes the apparently complex points easily understandable.

5. A large number of solved problems are included so as to help the student in applying the principles in practice.

6. A set of Review Questions and Objective-Type Questions provided at the end of each unit helps the student in self-evaluation.

7 Two to three tutorial sheets are provided at the end of each unit. Each tutorial sheet contains a set of three or four numerical problems (along with their answers); they can be solved by the students in about an hour while sitting in the classroom. The teachers will thus have a ready set of problems for tutorial work.

8. Experimental Exercises given at the end of each unit provide the student an opportunity to design experiments in the laboratory. They can then conduct the experiments so as to reinforce their theoretical knowledge.

We wish to express our deepest gratitude to Prof. T.K. Vaidyanathan, the Principal, and Prof. K.B. Raina, the Head of Electrical Engineering Department, TTTI, Chandigarh, for their constant encouragement. We are also grateful to Prof. T.R. Ramanna and Shri S.C. Laroiya of Curriculum Development Centre, TTTI, Chandigarh, for providing all the facilities for conducting the workshops. We also wish to express our apperciation for the accurate and neat typing of the manuscript done by Shri V.P. Chopra.

Despite the best efforts put in by us, it is possible that some unintentional errors might have eluded us. We shall acknowledge with gratitude any such errors, if pointed out. Any suggestions from students and colleagues for improvements in future editions of this book are most welcome.

N N BHARGAVA
D C KULSHRESHTHA
S C GUPTA

Contents

BASIC ELECTRONICS AND LINEAR CIRCUITS

Introduction to Electronics

OBJECTIVES: After completing this unit, you will be able to: ○ Define the scope of electronics. ○ State some of the applications of electronics in your day-to-day life. ○ State the latest trends in the field of electronics. ○ Draw the symbols, and state the main applications of some of the important *active devices* such as vacuum tubes, transistors, FETs, SCR, UJT, etc. ○ Recognize resistors, capacitors and inductors of various types from their physical appearance. ○ Read the values of resistors and capacitors from the code marks usually found on such components. ○ Draw the symbol of different passive components as per ISI specifications. ○ Explain the important specifications of resistors, capacitors and inductors. ○ Write SI units of various physical quantities used in electronics.

1.1 WHAT IS ELECTRONICS ?

The word 'electronics' is derived from *electron* mechanics which means the study of the behaviour of an electron under different conditions of externally applied fields.

The Institution of Radio Engineers (IRE) has given a standard definition of electronics in the *Proceedings of IRE*, Vol. 38, (1950) as "that field of science and engineering, which deals with electron devices and their utilization." Here, an electron device is "a device in which conduction takes place by the movement of electrons—through a vacuum, a gas, or a semiconductor."

Compared to the more established branches such as civil, mechanical, electrical, etc. electronics is a newcomer in the field of engineering. Until recently, it was considered an integral part of electrical engineering, but due to the tremendous advancement during the last few decades, it has now gained its rightful place.

We shall study, in the chapters that follow, how electronic devices function, and how they could be used to advantage in our daily life.

1.2 APPLICATIONS OF ELECTRONICS

Life today offers many conveniences which involve the use of electronic devices. As can be seen from Table 1.1, electronics plays a major role in almost every sphere of our life.

1.2.1 Communications and Entertainment

The progress of a country depends upon the availability of economical and rapid means of communication. During the earlier part of this century, the main application of electronics was in the field of telegraphy and telephony.

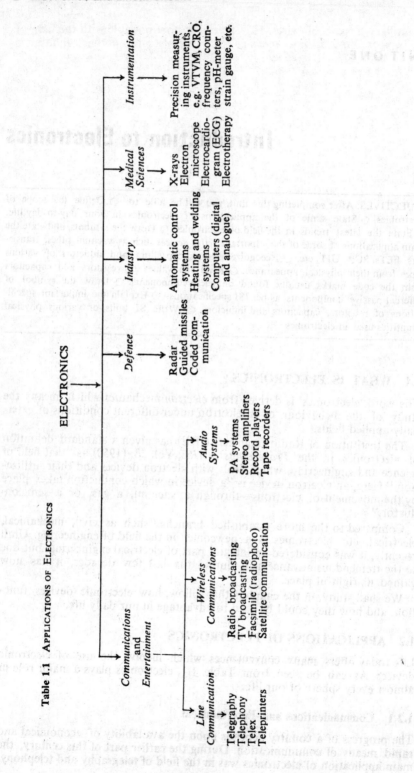

Table 1.1 APPLICATIONS OF ELECTRONICS

ELECTRONICS

Communications and Entertainment

Line Communication
→ Telegraphy
Telephony
Telex
Teleprinters

Wireless Communications
→ Radio broadcasting
TV broadcasting
Facsimile (radiophoto)
Satellite communication

Audio Systems
→ PA systems
Stereo amplifiers
Record players
Tape recorders

Defence
→ Radar
Guided missiles
Coded communication

Industry
→ Automatic control systems
Heating and welding systems
Computers (digital and analogue)

Medical Sciences
→ X-rays
Electron microscope
Electrocardiogram (ECG)
Electrotherapy

Instrumentation
→ Precision measuring instruments, e.g. VTVM, CRO, frequency counters, pH-meter strain gauge, etc.

This utilizes a pair of wires. However, it is now possible with the help of radio waves to transmit any message from one place to another, thousands of kilometres away, without any wires. With such *wireless* communication (radio broadcasting), people in any part of the world can know what is happening in other parts. With the help of a teleprinter, it is possible to type the message on a typewriter kept in another city. Photographs of events occurring somewhere, can be transmitted on facsimile (radiophoto). They can then be printed in the newspapers all over the world.

Radio and TV broadcasting provide a means of both communication as well as entertainment. With the help of satellites it has become possible to establish instant communication between places very far apart. Electronic gadgets like tape recorders, record players, stereo systems, public-address systems, etc. are widely used for entertainment.

1.2.2   Defence Applications

One of the most important developments during World War II was the RADAR (which is the short form for 'RAdio Detection And Ranging'). By using radar it is possible not only to detect, but also to find the exact location of the enemy aircraft. The anti-aircraft guns can then be accurately directed to shoot down the aircraft. In fact, the radar and anti-aircraft guns can be linked by an automatic control system to make a complete unit.

Guided missiles are completely controlled by electronic circuits. In a war, success or defeat for the nation depends on the reliability of its communication system. In modern warfare, communication is almost entirely electronic.

1.2.3   Industrial Applications

Use of automatic control systems in industries is increasing day by day. Electronic circuits are used in industrial applications like control of thickness, quality, weight and moisture content of a material. Electronic amplifier circuits are used to amplify signals and thus control the operations of automatic door-openers, lighting systems, power systems and safety devices, etc. Electronic circuits are used to produce stroboscopic lights of any desired frequency. When this is directed on a fast rotating object, it can be made to appear stationary or to be in slow motion by adjusting the frequency of light. This principle makes it possible to study the movement of various parts of a machine under normal running conditions.

For quick arithmetical calculations, desk calculators are commonly used in banks, departmental stores, etc. Some of you must be using calculators in your classroom while solving problems. Electronic computers, also called 'electronic brains', are used for automatic record keeping and solving of complicated problems.

Electronically-controlled systems, using suitable timers, are used for heating and welding in the industry. Even the power stations, which generate thousands of megawatts of electricity are controlled by tiny electronic devices and circuits.

1.2.4   Medical Sciences

Doctors and scientists are constantly finding new uses for electronic systems in the diagnosis and treatment of various diseases. Some of the instruments which have been in use are:

(i) *X-rays*, for taking pictures of internal bone structures and also for treatment of some diseases.
(ii) *Electrocardiographs* (*ECG*), to find the condition of the heart of a patient.
(iii) *Short-wave diathermy units*, for healing sprains and fractures.
(iv) *Oscillographs* for studying muscle action.

The use of electronics in medical science has expanded so enormously as to start a new branch of study, called 'bioelectronics'. Electronics is proving useful in saving mankind from a lot of suffering and pain.

1.2.5 Instrumentation

Instrumentation plays a very important role in any industry and research organization, for precise measurement of various quantities. It is only due to electronic instruments that an all-round development in every walk of life has been possible. VTVM, cathode-ray oscilloscopes, frequency counters, signal generators, pH-meters, strain-gauges, etc. are some of the electronic instruments without which no research laboratory is complete.

1.3 MODERN TRENDS IN ELECTRONICS

The real beginning in electronics was made in 1906, when Lee De Forest invented the *vacuum triode*. Without this device, the amplifier (which is the heart of all intricate and complex electronic gadgets) would not have been possible. Until the end of World War II, the vacuum tubes (valves) dominated the field of electronics.

In 1948, the invention of the *transistor* by the three Nobel laureates—John Bardeen, Walter Brattain and William Shockley at the Bell Laboratory, completely revolutionized the electronics industry. Transistors opened the floodgate to further developments in electronics. Within almost ten years of its discovery, the process of miniaturization of electronic equipments had gained momentum. The first integrated circuits (ICs) appeared in the market during the early sixties. Man's desire to conquer space accelerated this growth even further. The electronic age had truly begun. Now, during the eighties, this tremendous growth rate is not only continuing but is accelerating every year. The use of valves nearly became obsolete during the sixties. Due to the rapid developments in integrated circuit technology—starting from the small scale integration (SSI), then medium scale integration (MSI), large scale integration (LSI) and now with the most recent, very large scale integration (VLSI) technique—even the use of individual transistors is becoming unnecessary. The vast changes that have taken place during the last 20 years can best be understood by noting the reduction in size and price of modern digital computers. A small, modern minicomputer is more than 100 times smaller in size and 1/100th of the price of a computer designed 20 years ago to do similar jobs.

From an ordinary wrist watch to the control room of 400 000 tonne supertanker carrying cargo across the sea; from the telephone repeaters buried deep under the ocean to the spaceships far out in space; from a modern household to the gigantic steel mills and power houses, electronics has penetrated everywhere.

Electronics deals in the *micro* and *milli* range of voltage, current and power, but it is capable of controlling *kilo* and *mega* volts, amperes and watts. Therefore, it is not surprising to find the fundamentals of electronics as a core subject in all branches of engineering nowadays.

1.4 ELECTRONIC COMPONENTS

An electronic circuit may appear quite complicated and may be capable of performing fantastic functions. But, all electronic circuits, however complicated, contain a few basic components. Generally speaking, there are only five components—three passive and two active (see Table 1.2). An integrated circuit (for example, a microprocessor) may contain thousands of transistors, a few thousand resistors, etc. on a very small chip. The total number of components used in an electronic circuit may run into thousands—yet each component will be one of the above *five* types.

Table 1.2 TYPES OF ELECTRONIC COMPONENTS

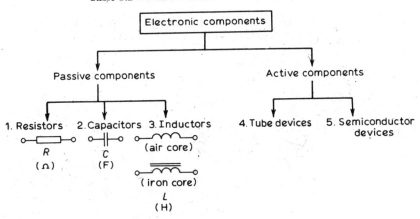

1.4.1 Passive Components

Resistors, capacitors and inductors are called *passive* components. These components by themselves are not capable of amplifying or processing an electrical signal. However, these components are as important, in an electronic circuit, as active (such as transistors) components are. Without the aid of these components a transistor cannot be made to amplify signals.

Resistors The flow of charge (or current) through any material, encounters an opposing force similar in many respects to mechanical friction. This 'opposing force' is called the *resistance* of the material. It is measured in *ohms*, for which the symbol is Ω (the greek capital letter omega). The circuit symbol for resistance (R) is shown in Table 1.2.

In some parts of an electronic circuit, resistance is deliberately introduced. The device or component to do this is called a *resistor*. Resistors are made in many forms. But all belong to either of two groups—*fixed* or *variable*.

Fixed resistors The most common of the low wattage, fixed-type resistors is the *moulded-carbon composition resistor*. The basic construction is shown in Fig. 1.1. The resistive material is of carbon-clay composition. The leads are made of tinned copper. Resistors of this type are readily available in values ranging from few ohms to about 22 MΩ, having a tolerance range of 5 to 20 %. They are quite inexpensive. A resistor may cost only fifty paisa.

The relative sizes of all fixed (and also variable) resistors change with the wattage (power) rating. The size increases for increased wattage rating in order to withstand the higher currents and dissipation losses. The relative sizes of moulded-carbon composition resistors for different wattage ratings are shown in Fig. 1.2.

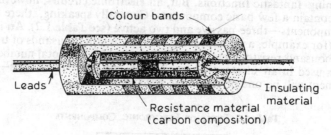

Fig. 1.1 The basic construction of a fixed, moulded-carbon composition resistor

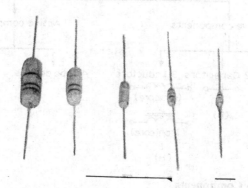

Fig. 1.2 Moulded-carbon composition resistors of different wattage ratings

Another variety of carbon composition resistors is the metalized type. Its basic structure is shown in Fig. 1.3. It is made by depositing a homogeneous film of pure carbon (or some metal) over a glass, ceramic or other insulating core. The carbon film can be deposited by pyrolysis of some hydrocarbon gas (e.g. benzyne) on the ceramic core. Only approximate values of resistance can be obtained by this method. Desired values are obtained by either trimming the layer thickness or by cutting helical grooves of suitable

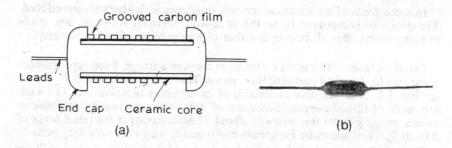

Fig. 1.3 Carbon-film resistor: (a) Construction; (b) A carbon-film resistor

pitch along its length. During this process, the value of resistance is monitored constantly. The cutting of grooves is stopped as soon as the desired value of resistance is obtained. Contact caps are fitted on both ends. The lead wires, made of tinned copper, are then welded to these end caps. This type of film-resistor is sometimes called *precision type*, since it can be obtained with an accuracy of $\pm 1\%$.

A *wire wound resistor* uses a length of resistance wire, such as nichrome. This wire is wound onto a round, hollow porcelain core. The ends of the winding are attached to metal pieces inserted in the core. Tinned copper wire leads are attached to these metal pieces. This assembly is coated with an enamel containing powdered glass. It is then heated to develop a coating known as *vitreous* enamel. This coating is very smooth and gives mechanical protection to the winding. It also helps in conducting heat away from the unit quickly. In other wire-wound resistors, a ceramic material is used for the inner core and the outer coating (see Fig. 1.4). Commonly available wire-wound resistors have resistance values ranging from 1 Ω to 100 kΩ, and wattage ratings up to about 200 W.

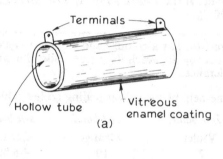

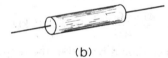

(b)

Fig. 1.4 Wire-wound fixed resistors:
(a) Vitreous enamel type
(b) Ceramic type

Colour coding and standard resistor values Some resistors are large enough in size to have their resistance (in Ω) printed on the body. However, there are some resistors that are too small in size to have numbers printed on them. Therefore, a system of *colour coding* is used to indicate their values. For the fixed, moulded composition resistor, four colour bands are printed on one end of the outer casing as shown in Fig. 1.5a.

The numerical value associated with each colour is indicated in Table 1.3. The colour bands are always read left to right from the end that has the bands closest to it, as shown in Fig. 1.5a.

The first and second bands represent the first and second significant digits, respectively, of the resistance value. The third band is for the number of zeros that follow the second digit. In case the third band is gold or silver, it represents a multiplying factor of 0.1 or 0.01. The fourth band represents

Table 1.3 COLOUR CODING

Colour	Digit	Multiplier	Tolerance
Black	0	$10^0 = 1$	
Brown	1	$10^1 = 10$	
Red	2	10^2	
Orange	3	10^3	
Yellow	4	10^4	
Green	5	10^5	
Blue	6	10^6	
Violet	7	10^7	
Gray	8	10^8	
White	9	10^9	
Gold	–	$0.1 = 10^{-1}$	$\pm 5\%$
Silver	–	$0.01 = 10^{-2}$	$\pm 10\%$
No colour	–	–	$\pm 20\%$

Mnemonics: As an aid to memory in remembering the sequence of colour codes given above, the student can remember the following sentence (all the capital letters stand for colours):

(a) Bill Brown Realized Only Yesterday Good Boys Value Good Work.
(b) Bye Bye Rosie Off You Go Bristol Via Great Western.

the manufacturer's tolerance. It is a measure of the precision with which the resistor was made. If the fourth band is not present, the tolerance is assumed to be $\pm 20\%$.

Example 1.1 A resistor has a colour band sequence: yellow, violet, orange and gold. Find the range in which its value must lie so as to satisfy the manufacturer's tolerance.

Solution: With the help of the colour coding table (Table 1.3), we find:

1st band	2nd band	3rd band	4th band
Yellow	Violet	Orange	Gold
4	7	10^3	$\pm 5\% = 47\ k\Omega \pm 5\%$

Now, 5% of $47\ k\Omega = \dfrac{47 \times 10^3 \times 5}{100}\ \Omega = 2.35\ k\Omega$

Therefore, the resistance should be within the range $47\ k\Omega \pm 2.35\ k\Omega$, or between **44.65 k$\Omega$** and **49.35 k$\Omega$**.

Example 1.2 A resistor has a colour band sequence: gray, blue, gold and gold. What is the range in which its value must lie so as to satisfy the manufacturer's tolerance?

Solution: The specification of the resistor can be found by using the colour coding table as follows:

1st band	2nd band	3rd band	4th band
Gray	Blue	Gold	Gold
8	6	10^{-1}	$\pm 5\% = 86 \times 0.1\ \Omega \pm 5\%$
			$= 8.6\ \Omega \pm 5\%$

5% of $8.6\ \Omega = \dfrac{8.6 \times 5}{100} = 0.43\ \Omega$

The resistance should lie somewhere between the values $(8.6 - 0.43)\ \Omega$ and $(8.6 + 0.43)\ \Omega$, or **8.17 Ω** and **9.03 Ω**.

The colour coding for wire-wound resistors, and composition resistors with radial leads is shown in Figs. 1.5b and c, respectively. Note that the first band in Fig. 1.5b is of double the width compared to the rest. The system of colour coding used for the moulded resistors with radial leads is called body-end-dot system. The numerical values associated with each colour is the same for all the three methods of colour coding.

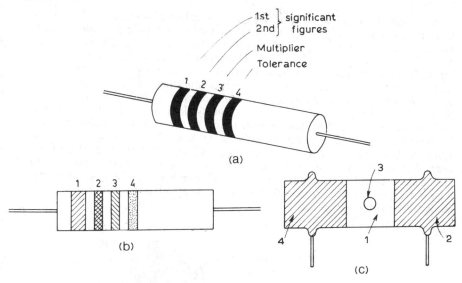

Fig. 1.5 Colour coding: (a) Moulded composition resistor; (b) Wire-wound resistor (c) Moulded composition resistor with radial leads (this system of colour coding is called body-end-dot system)

In practical electronic circuits, the values of the resistors required may lie within a very wide range (say, from a few ohms to about 20 MΩ). In most of the circuits, it is not necessary to use resistors of exact values. Even if a resistor in a circuit has a value which differs from the desired (designed) value by as much as 20 %, the circuit still works quite satisfactorily. There-fore, it is not necessary to manufacture resistors of all the possible values. A list of readily available *standard values* of resistors appears in Table 1.4.

Table 1.4 STANDARD VALUES OF COMMERCIALLY AVAILABLE RESISTORS (HAVING 10 % TOLERANCE)

Ohms (Ω)			Kilohms (kΩ)			Megohms (MΩ)	
1.0	10	100	1.0	10	100	1.0	10
1.2	12	120	1.2	12	120	1.2	12
1.5	15	150	1.5	15	150	1.5	15
1.8	18	180	1.8	18	180	1.8	18
2.2	22	220	2.2	22	220	2.2	22
2.7	27	270	2.7	27	270	2.7	
3.3	33	330	3.3	33	330	3.3	
3.9	39	390	3.9	39	390	3.9	
4.7	47	470	4.7	47	470	4.7	
5.6	56	560	5.6	56	560	5.6	
6.8	68	)	6.8	68	680	6.8	
8.2			8.2	82	820	8.2	

If resistors of very precise values are required for some specific application, special requests are to be made to the manufacturer.

Variable resistors In electronic circuits, sometimes it becomes necessary to adjust the values of currents and voltages. For example, it is often desired to change the volume (or loudness) of sound, the brightness of a television picture, etc. Such adjustments can be done by using variable resistors.

Although the variable resistors are usually called *rheostats* in other applications, the smaller variable resistors commonly used in electronic circuits are called *potentiometers* (usually abbreviated to 'pots'). The symbol for potentiometer is shown in Fig. 1.6a. The arrow in the symbol is a contact movable on a continuous resistive element. The moving contact will determine whether the resistance in the circuit is minimum $(0\,\Omega)$ or maximum value, R. The construction of all potentiometers is basically the same. Some have a wire-wound resistance as their primary element, while others have a carbon-film element. The basic construction of a wire-wound potentiometer is shown in Fig. 1.6b. The resistance wire is wound over a dough-shaped core of bakelite or ceramic. There is a rotating shaft at the centre of the core. The shaft moves an arm and a contact point from end to end of the resistance element. There are three terminals coming out of a potentiometer. The outer two are the end points of the resistance element, and the middle leads to the rotating contact.

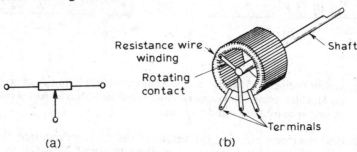

Resistance wire winding   Shaft

Rotating contact

Terminals

(a)   (b)

Fig. 1.6 Potentiometer: (a) Symbol; (b) Basic construction of a wire-wound potentiometer

A potentiometer can be either *linear* or *nonlinear*. Figure 1.7 shows the construction of both a linear and nonlinear (tapered type) potentiometer. In the linear type, the former (the part over which the wire is wound) is of uniform height and that is why the resistance varies linearly with the rotation of the contact. In a nonlinear potentiometer, the height of the former

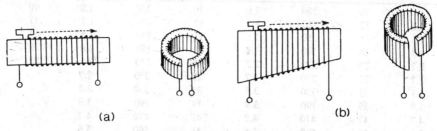

(a)   (b)

Fig. 1.7 Wire-wound potentiometer: (a) Linear type; (b) Nonlinear type

is not uniform. To make a potentiometer of this type, a tapered strip is taken and the resistance wire is wound over it, ensuring a uniform pitch. The strip is then bent into a round shape. The tapered strip gives a non-linear variation of resistance with the rotation of the moving contact. The strip can be tapered suitably so as to obtain a desired variation in resistance per unit rotation of the moving contact. The 'pots' used as volume control in sound equipment are generally of the nonlinear type (logarithmic variation).

Capacitors Capacitors of different kinds are found in nearly every electronic circuit. A capacitor is basically meant to store electrons (or electrical energy), and release them whenever desired. The circuit symbol of a capacitor is shown in Table 1.2. *Capacitance* is a measure of a capacitor's ability to store charge. It is measured in farads (F). However, the unit farad being too large, practical capacitors are specified in microfarads (μF), or picofarads (pF).

A capacitor offers low impedance to ac, but very high impedance to dc. So, capacitors are used when we want to couple alternating voltage from one circuit to another, while at the same time blocking the dc voltage from reaching the next circuit. It is also used as a *bypass capacitor*, where it by-passes the ac through it without letting the ac to go through the circuit across which it is connected. A capacitor forms a tuned circuit in series or parallel with an inductor.

A capacitor consists of two conducting plates, separated by an insulating material known as a *dielectric*. Since the two plates of a capacitor can be of many different conducting materials and the dielectric may be of many different insulating materials, there are many types of capacitors.

Capacitors, like resistors, can either be fixed or variable. Some of the most commonly used fixed capacitors are mica, ceramic, paper, and electrolytic. Variable capacitors are mostly air-gang capacitors.

Mica capacitors Mica capacitors are constructed from plates of aluminium foil separated by sheets of mica as shown in Fig. 1.8. The plates are connected to two electrodes. The mica capacitors have excellent characteristics under stress of temperature variations and high voltage applications. Available capacitances range from 5 to 10 000 pF. Mica capacitors are usually rated at 500 V. Its leakage current is very small ($R_{leakage}$ is about 1000 MΩ).

Ceramic capacitors Ceramic capacitors are made in many shapes and sizes. However, the basic construction is the same for each. A ceramic disc is coated on two sides with a metal, such as copper or silver. These coatings act as the two plates (see Fig. 1.9). During the manufacture of the capacitor, tinned wire leads are also attached to each plate. Then the entire unit is coated with plastic and marked with its capacitance value—either using numerals or a colour code. The colour coding is similar to that used for resistances. Figure 1.10 explains the colour code used for resistors and capacitors. Besides the value of the resistor (or capacitor), it also indicates the tolerance and temperature coefficients. Ceramic capacitors are very versatile. Their working voltage ranges from 3 V (for use in transistors) up to 6000 V. The capacitance ranges from 3 pF to about 2 μF. Ceramic capacitors have a very low leakage currents ($R_{leakage}$ is about 1000 MΩ) and can be used in both dc and ac circuits.

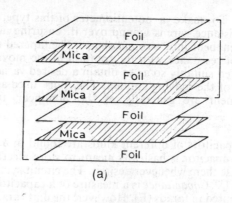

(a)

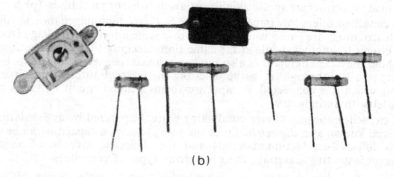

(b)

Fig. 1.8 Mica capacitors: (a) Construction; (b) Some mica capacitors

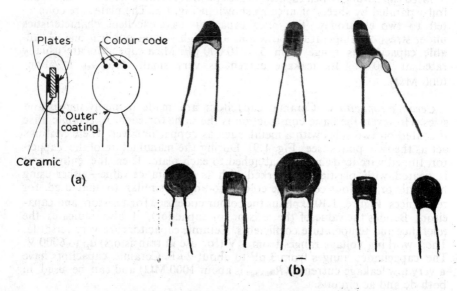

Fig. 1.9 Ceramic capacitor: (a) Construction; (b) Some ceramic capacitors

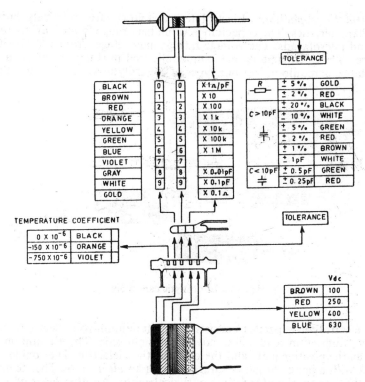

BLACK	0	0	X 1Ω/pF
BROWN	1	1	X 10
RED	2	2	X 100
ORANGE	3	3	X 1k
YELLOW	4	4	X 10k
GREEN	5	5	X 100k
BLUE	6	6	X 1M
VIOLET	7	7	
GRAY	8	8	X 0.01pF
WHITE	9	9	X 0.1pF
GOLD			X 0.1Ω

R	± 5 %	GOLD
	± 2 %	RED
C > 10pF	± 20 %	BLACK
	± 10 %	WHITE
	± 5 %	GREEN
	± 2 %	RED
	± 1 %	BROWN
	± 1 pF	WHITE
C < 10pF	± 0.5 pF	GREEN
	± 0.25pF	RED

TEMPERATURE COEFFICIENT

0 X 10⁻⁶	BLACK
-150 X 10⁻⁶	ORANGE
- 750 X 10⁻⁶	VIOLET

Vdc

BROWN	100
RED	250
YELLOW	400
BLUE	630

Fig. 1.10 Explanation of colour code used for resistors and capacitors

Paper capacitors The basic construction of a paper capacitor is shown in Fig. 1.11. Since paper can be rolled between two metals foils, it is possible to concentrate a large plate area in a small volume. The capacitor consists of two metal foils separated by strips of paper. This paper is impregnated with a dielectric material such as wax, plastic or oil.

Paper capacitors have capacitances ranging from 0.0005 µF to several µF, and are rated from about 100 V to several thousand volts. They can be used for both dc and ac circuits. Its leakage resistance is of the order of 100 MΩ.

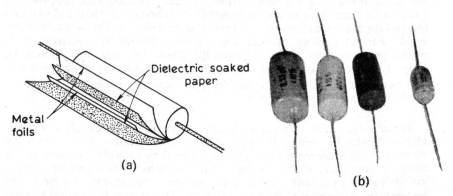

Dielectric soaked paper

Metal foils

(a)

(b)

Fig. 1.11 Tubular paper capacitors: (*a*) Construction; (*b*) Some tubular paper capacitors

Electrolytic capacitors Electrolytic capacitors are extremely varied in their characteristics. The capacitance value may range from 1 μF to several thousand microfarads. The voltage ratings may range from 1 V to 500 V, or more. These capacitors are commonly used in situations where a large capacitance is required. Various types of electrolytic capacitors are shown in Fig. 1.12.

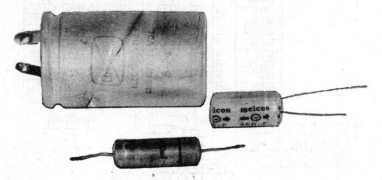

Fig. 1.12 Electrolytic capacitors

The electrolytic capacitor consists of an aluminium-foil electrode which has an aluminium-oxide film covering on one side. The aluminium plate serves as the positive plate and the oxide as the dielectric. The oxide is in contact with a paper or gauze saturated with an *electrolyte*. The electrolyte forms the second plate (negative) of the capacitor. Another layer of aluminium without the oxide coating is also provided for making electrical contact between one of the terminals and the electrolyte. In most cases, the negative plate is directly connected to the container of the capacitor. The container then serves as the negative terminal for external connections. The aluminium oxide layer is very thin. Therefore, the capacitor has a large capacitance in a small volume. It has high *capacitance-to-size ratio*. It is primarily designed for use in circuits where only dc voltages will be applied across the capacitor. Ordinary electrolytic capacitors cannot be used with alternating currents. However, there are capacitors available that can be used in ac circuits (for starting motors) and in cases where the polarity of the dc voltage reverses for short periods of time. The reason for the *polarized* (positive and negative electrodes) nature of the capacitor is that the aluminium foil and the aluminium oxide layer form a *semiconductor*. This semiconductor blocks current coming through the oxide film toward the electrode, but it readily passes current in the opposite direction. The capacitor should be properly connected so that the applied voltage encounters the high resistance.

A new type of electrolytic capacitor is the *tantalum* capacitor. It has an excellent capacitance-to-size ratio.

Variable capacitors In some circuits, such as a tuning circuit, it is desirable to be able to change the value of capacitance readily. This is done by means of a variable capacitor. The most common variable capacitor is the *air-gang capacitor*, shown in Fig. 1.13. The dielectric for this capacitor is air. By rotating the shaft at one end, we can change the common area between the movable and fixed set of plates. The greater the common area, the larger the capacitance.

Fig. 1.13 Air-gang capacitor (variable)

In some applications, the need for variation in the capacitance is not frequent. One setting is sufficient for all normal operations. In such situations we use a variable capacitor called a *trimmer* (sometimes called *padder*). Both mica and ceramic are used as the dielectric for trimmer capacitors. Figure. 1.14 shows the basic construction of a mica trimmer.

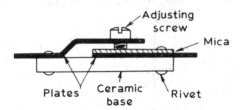

Fig. 1.14 Construction of variable capacitor

Inductors When current flows through a wire that has been coiled, it generates a magnetic field. This magnetic field reacts so as to oppose any change in the current. This reaction of the magnetic field, trying to keep the current flowing at a steady rate, is known as *inductance*; and the force it develops is called *induced emf*. The electronic component producing inductance is called an *inductor*. The symbols of an air-core and an iron-core inductor are shown in Table 1.2. The inductance is measured in henrys (H).

All inductors, like resistors and capacitors, can be listed under two general categories: fixed and variable. Different types of inductors are available for different applications.

Filter chokes are the inductors used in smoothing the pulsating current produced by rectifying ac into dc. A typical filter choke has many turns of wire wound on an iron core. To avoid power losses, the core is made of laminated sheets of E- and I-shapes (Fig. 1.15). Many power supplies use filter chokes of 5 to 20 H, capable of carrying current up to 0.3 A.

Audio-frequency chokes (AFCs) are used to provide high impedance to audio frequencies (say, 60 Hz to 5 kHz). Compared to filter chokes, they are smaller in size, and have lower inductance. Chokes having still smaller inductances are used to block the radio frequencies. Such chokes are called

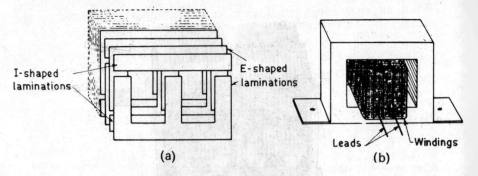

Fig. 1.15 Typical filter choke: (*a*) Exploded view of laminated core; (*b*) Assembly choke

radio-frequency chokes (RFCs). Variable inductors are used in tuning circuits for radio frequencies. The *permeability-tuned variable coil* has a ferromagnetic shaft. This shaft can be moved within the coil to vary the inductance, as shown in Fig. 1.16.

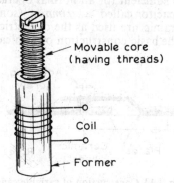

Fig. 1.16 Permeability-tuned variable coil

A transformer is quite similar in appearance to an inductor. It consists basically of two inductors having the same core (Fig. 1.17). One of these inductors, or windings, is called *primary*. The other is called *secondary*.

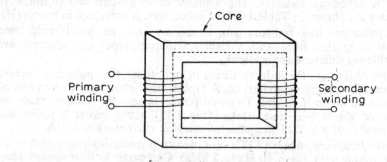

Fig. 1.17 Basic structure of a transformer

When an alternating current is applied at the primary, an induced voltage appears in the secondary. In a *step-up transformer*, the number of turns in the secondary is more than that in the primary. The secondary voltage is more than the primary. If the number of turns in the secondary is less than that in the primary, the voltage will be stepped-down. The transformer is then called a *step-down transformer*. A transformer of suitable *turns-ratio* is often used in electronic circuits for impedance matching.

1.4.2 Active Components

There are many active components used in electronic circuits. But all the active devices or components can be broadly classified into two categories: tube-type and semiconductor-type. Tube devices can again be of two types: vacuum tubes and gas tubes. These devices came prior to the semiconductor devices. Because of their advantages, the semiconductor devices are replacing the tube devices in almost all electronic applications. Table 1.5 gives brief information about commonly used active components, and Fig. 1.18 shows photographs of some active components.

Table 1.5 ACTIVE COMPONENTS

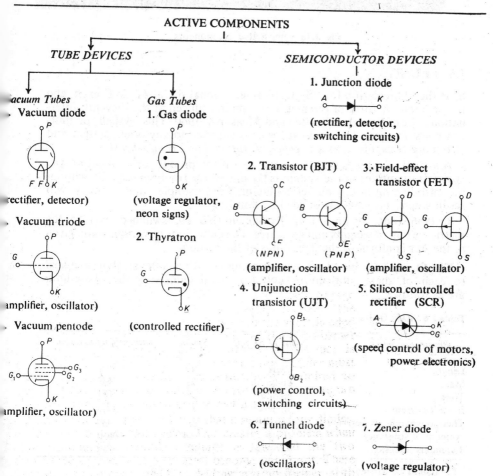

ACTIVE COMPONENTS

TUBE DEVICES

SEMICONDUCTOR DEVICES

Vacuum Tubes
- Vacuum diode

(rectifier, detector)

- Vacuum triode

(amplifier, oscillator)

- Vacuum pentode

(amplifier, oscillator)

Gas Tubes
1. Gas diode

(voltage regulator, neon signs)

2. Thyratron

(controlled rectifier)

1. Junction diode

(rectifier, detector, switching circuits)

2. Transistor (BJT)

(NPN)   (PNP)
(amplifier, oscillator)

3. Field-effect transistor (FET)

(amplifier, oscillator)

4. Unijunction transistor (UJT)

(power control, switching circuits)

5. Silicon controlled rectifier (SCR)

(speed control of motors, power electronics)

6. Tunnel diode

(oscillators)

7. Zener diode

(voltage regulator)

Fig. 1.18 Some active components

1.5 SI UNITS

SI is the abbreviation for "Systemé International d'Unités" in French, and is the modern form of the metric system introduced at the Eleventh International Conference of Weights and Measures, 1960. This system of units possesses features that make it logically superior to any other system and also more convenient, as it is *coherent, rational* and *comprehensive.*

A system of units is said to be coherent if the product or quotient of any two unit quantities in the system is the unit of the resultant quantity without the introduction of any numerical factor. For example, unit velocity will result when unit length is divided by unit time.

In 1956, India, by an Act of Parliament No. 89, switched over to the metric system of weights and measures. The definitions of various units given in the Act conform to the definitions of the SI units.

The SI units are based on seven base units with a unit symbol assigned to each of them as given in Table 1.6. The definitions of these base units are as follows:

Table 1.6 BASE UNITS

Physical quantity	Name of SI unit	Symbol	Dimensional notation
Length	metre	m	[L]
Mass	kilogram	kg	[M]
Time	second	s	[T]
Electric current	ampere	A	[I]
Thermodynamic temperature	kelvin*	K	[θ]
Amount of substance	mole	mol	[mol]
Luminous intensity	candela	cd	[ϕ]

*It should be written as kelvin only, and not degree kelvin or °K.

Length: The metre is the length equal to 1 650 760.73 wavelengths, in vacuum, of the radiation corresponding to the transition between the levels $2p^{10}$ and $5d^5$ of the krypton-86 atom.

Mass: The kilogram is equal to the mass of the international prototype kilogram stored at Sevrés, France.

Time: The second is the duration of 9 192 631 770 periods of the radiation corresponding to the transition between the two hyperfine levels of the ground state of the caesium-133 atom.

Electric Current: The ampere is that current which, if maintained in two straight parallel conductors of infinite length and of negligible circular cross-section and placed one metre apart in vacuum, would produce between these conductors a force equal to 2×10^{-7} newton per metre of length (N/m).

Thermodynamic Temperature: The kelvin is the 1/273.16 fraction of the thermodynamic temperature of the triple point of water*.

Amount of Substance: The mole is the amount of substance in a system which contains as many elementary entities as there are atoms in 0.012 kg of carbon-12.

Luminous Intensity: The candela is the luminous intensity, in the perpendicular direction, of a surface of 1/600 000 m² of a blackbody at the temperature of freezing platinum (2046 K), under a pressure of 101 325 newtons per square metre.

Table 1.7 SOME DERIVED UNITS

Physical quantity	Name of SI unit	Symbol
Frequency	hertz	Hz = cycles/s = 1/s
Force	newton	N = kg m/s²
Work, energy, quantity of heat	joule	J = N m
Power	watt	W = J/s
Electric charge	coulomb	C = A s
Electric potential	volt	V = W/A
Electric capacitance	farad	F = A s/V
Electric resistance	ohm	Ω = V/A
Electric conductance	siemens*	S = A/V
Magnetic flux	weber	Wb = V s
Magnetic flux density	tesla	T = Wb/m²
Inductance	henry	H = V s/A
Customary temperature	degree celsius	°C
Pressure	pascal	Pa = N/m

*The unit *siemens* is same as *mho* ($\mho$) which was used earlier.

The two dimensionless quantities, plane angle and solid angle, are treated as independent quantities with SI units radian (rad) and steradian (sr), respectively. These are known as *supplementary units*. The *radian* is the plane angle between two such radii of a circle which cuts off, on the circumference, an arc equal to the length of the radius. Thus,

$$\theta \text{ (in radians)} = \frac{\text{arc}}{\text{radius}}$$

---

*The temperature at which ice, water and water vapours coexist.

The *steradian* is the solid angle which, with its vertex at the centre of a sphere, cuts off an area of the surface of the sphere equal to that of a square having sides equal to the radius of the sphere. Thus, if S is the area cut off on the surface of a sphere of radius r, the solid angle at the centre of the sphere is

$$\Omega \text{ (in steradians)} = \frac{S}{r^2}$$

All other units are known as *compound* or *derived SI units*, some of which may have special names, as given in Table 1.7. The SI units cover all fields of physics and engineering.

1.5.1 Decimal Multiple and Submultiple Factors

Since all the coherent units are not of a convenient size for all applications, provision had to be made for multiples and submultiples of the coherent units. A complete list of such factors is given in Table 1.8. The guidelines for the application of these prefixes are as follows:

Table 1.8 SI PREFIXES

Factor	Prefix	Symbol	Factor	Prefix	Symbol
10^1	deca	da	10^{-1}	deci	d
10^2	hecto	h	10^{-2}	centi	c
10^3	kilo	k	10^{-3}	milli	m
10^6	mega	M	10^{-6}	micro	μ
10^9	giga*	G	10^{-9}	nano	n
10^{12}	tera	T	10^{-12}	pico	p
10^{15}	peta	P	10^{-15}	femto	f
10^{18}	exa	E	10^{-18}	atto	a

*Pronounced as *jeega*.

Note: (i) The prefixes for factors greater than unity have Greek origin; those for factors less than unity have Latin origin (except femto and atto, recently added, which have Danish origin).

(ii) Almost all abbreviations of prefixes for magnitudes < 1, are English lower-case letters. An exception is *micro* (Greek letter μ).

(iii) Abbreviations of prefixes for magnitudes > 1 are English upper-case letters. Exceptions are kilo, hecto, and deca.

(iv) The prefixes hecto, deca, deci and centi should not be used unless there is a strongly-felt need.

1. Multiples of the fundamental unit should be chosen in powers of $\pm 3n$ where n is an integer. Centimetre, owing to its established usage and its convenient size, cannot be given up lightly.

2. Double or compound prefixes should be avoided, e.g. instead of micromicrofarad ($\mu\mu$F) or millinanofarad (mnF), use picofarad (pF).

3. To simplify calculations, attach the prefix to the numerator and not to the denominator. Example: use MN/m^2 instead of N/mm^2; even though mathematically, both forms are equivalent.

4. The rules for *binding-in indices* are not those of ordinary algebra, e.g. cm^2 means $(cm)^2 = (0.01)^2\, m^2 = 0.0001\, m^2$, and not $c \times (m)^2 = 0.01\, m^2$.

1.5.2 Other Accepted Units

It has been recognized at the international level, that some departures from strict purity and coherence are acceptable for practical reasons. For instance, pure SI would acknowledge only decimal multiples and submultiples of the second for time measurement; whereas minute, hour, day, month and year are in everyday use internationally, and will clearly continue to be used. Similarly, the division of the circle into 360 degrees is an internationally recognized practice. Some symbols, other than SI. that are commonly used to express physical quantities are given in Table 1.9.

Table 1.9 Symbols Other than SI that are Commonly Used

Name	Abbreviation	Name	Abbreviation
angström	Å	inch	in
British thermal unit	Btu	kilowatt-hour	kW h
calorie	cal	mile	mi
day	d	minute (of arc)	′
degree	°	minute (of time)	min
dyne	dyn	pound	lb
electron volt	eV	revolution	rev
foot	ft	second (of arc)	″
gauss	G	standard atmosphere	atm
horse power	hp	atomic mass unit	amu
hour	h	year	y

1.5.3 Guidelines for Using SI Units

Following are the rules and conventions regarding the use of SI units:

1. Full names of units. even when they are named after a person, are not written with a capital (or upper-case) initial letter, e.g. kelvin, newton, joule, watt, volt, ampere, etc.

2. The symbols for a unit, named after a person, has a capital initial letter, e.g. W for watt (after James Watt) and J for joule (after James Prescott Joule).

3. Symbols for other units are not written with capital letter, e.g. m for metre.

4. Units may be written out in full or using the agreed symbols, but no other abbreviation may be used. They are printed in full or abbreviated, in roman (upright) type, e.g. amp. is not a valid abbreviation for ampere.

5. Symbols for units do not take a plural form with added 's'; the symbol merely names the unit in which the preceding magnitude is measured, e.g. 50 kg, and not 50 kgs.

6. No full stops or hyphens or other punctuation marks should be used within or at the end of the symbols for units. However, when a unit symbol prefix is identical to a unit symbol, a raised dot may be used between the two symbols to avoid confusion. For example, while writing, say, metre second it should be abbriviated as m·s to avoid confusion with ms, the symbol for millisecond.

7. There is a mixture of capital and lower-case letters in the symbols for the prefixes as shown in Table 1.8, but the full names of the prefixes commence with lower-case letters only, e.g. 5 MW (5 megawatt), 2 ns (2 nanosecond).

8. A space is left between a numeral and the symbol except in case of the permitted non-SI units for angular measurements, e.g. 57° 16′ 44″.

9. A space is left between the symbols for compound units, e.g. N m for newtons × metres and kW h for killowatt hour. This reduces the risk of confusion when an index notation instead of the solidus (/) is used. In the former notation, a velocity in metres per second is written as m s^{-1} instead of m/s, but ms^{-1} may mean 'per millisecond'. This type of confusion will not occur if we follow the rule that the denominators of compound units are always expressed in the base units and not in their multiples or submultiples. Thus a heat flow rate will not be given as J/ms but only as kJ/s = kW.

10. When a compound unit is formed by dividing one unit by another, this may be indicated in one of the two forms as m/s or m s^{-1}. In no case, should more than one solidus sign (/) on the same line be included in such a combination unless a parenthesis be inserted to avoid all ambiguity. In complicated cases, negative powers or parenthesis should be used.

11. Algebraic symbols representing "quantities" are written in *italics*, while symbols for "units" are written as upright characters, e.g.

    a current $\quad\quad I = 3$ A
    an energy $\quad\quad E = 2.75$ J
    a terminal voltage $V = 1.5$ V

12. When expressing a quantity by a numerical value and a certain unit, it has been found suitable in most applications to use units resulting in numerical values between 1.0 and 1000. To facilitate the reading of numerals, the digits may be separated into groups of three—counting from the decimal sign towards the left and the right. The groups should be.separated by a small space, but not by a comma or a point. In numerals of four digits, the space is usually not necessary. (It is recognized, however, that to drop the comma from commercial accounting will involve difficulties, particularly with the adding machines in use at present). A few examples are given below:

Incorrect	*Correct*
(i) 40,000 or 40000	(i) 40 000
(ii) 81234.765	(ii) 81 234.765
(iii) 764213.876	(iii) 764 213.876
(iv) 6 543.21	(iv) 6543.21

Note: (a) The recommended decimal sign is a full stop (.). The sign of multiplication of numbers is a cross (×).

(b) If the magnitude of a number is less than unity, the decimal sign should be preceded by a zero

REVIEW QUESTIONS

1.1 What is electronics ?

1.2 How has electronics affected our daily life ?

1.3 Write at least two important applications of electronics in the field of (a) communications and entertainment, (b) industry and (c) medical sciences.

1.4 State what is meant by radar ? Mention some of its important applications.

1.5 What are the modern trends in electronics ?

1.6 Before understanding electronic circuits, one must first have an understanding of the components that make up those circuits. Justify the statement (in about 7-8 lines).

1.7 Write the unit of resistance ? If a resistor is rated at 1000 Ω and 10 W, what is the maximum current it can carry ?

1.8 Explain constructional features of a wire-wound resistor. What is the range of wattage for wire-wound resistors ?

1.9 Explain in brief: What is (a) a capacitor, (b) a dielectric ?

1.10 Name three primary uses of capacitors.

1.11 Explain briefly the basic construction of a ceramic capacitor. What is the range of capacitance values available in ceramic capacitors ?

1.12 Why are paper capacitors not used in the filters of rectifier power supplies ?

1.13 What forms the dielectric of an electrolytic capacitor ? Why is the electrolytic capacitor polarized ?

1.14 While tuning your radio receiver to a desired station, which component inside the set are you varying ?

1.15 When you adjust the volume control knob of your radio receiver, which component is varied inside the set ?

1.16 What is a trimmer capacitor ? Describe the basic construction of a mica trimmer capacitor.

1.17 What is an inductor ? What is the unit of inductance ?

1.18 Give some important applications of inductors.

1.19 For what purpose can a transformer be used in an electronic circuit ?

1.20 Name a few active components (devices) used in electronic circuits.

1.21 Write down the seven base units in SI units.

OBJECTIVE-TYPE QUESTIONS

Here are some incomplete statements. Four alternatives are provided below each. Tick the alternative that completes the statements correctly:

1. Electronics is that branch of engineering which deals with the application of

   (a) high-current machines
   (b) production of electronic components
   (c) electronic devices
   (d) fission of uranium nuclei

2. One of the examples of an active device is a/an

   (a) electric bulb
   (b) transformer
   (c) loudspeaker
   (d) silicon controlled rectifier (SCR)

3. Which one of the following is used as a passive component in electronic circuits ?

   (a) Resistor
   (b) Transistor
   (c) Vacuum triode
   (d) Tunnel diode

4. The term IC, as used in electronics, denotes
   (a) internal combustion   (c) industrial control
   (b) integrated circuits   (d) Indian culture

5. An example of a solid-state divice is a
   (a) thyratron   (c) field-effect transistor
   (b) pentode   (d) triode

6. A 100-μF capacitor is required in fabricating an electronic circuit. Such a large value of capacitance is possible if the capacitor is a/an
   (a) mica capacitor   (c) air-gang capacitor
   (b) ceramic capacitor   (d) electrolytic capacitor

7. A resistor has a colour band sequence: brown, black, green, and gold. Its value is
   (a) 1 kΩ $\pm$ 10 %   (c) 1000 kΩ $\pm$ 5 %
   (b) 10 kΩ $\pm$ 5 %   (d) 1 MΩ $\pm$ 10 %

8. We need a resistor of value 47 kΩ with $\pm$ 5 % tolerance. The sequence of the colour band on this resistor should be
   (a) yellow, violet, yellow, and gold   (c) yellow, violet, orange, and silver
   (b) yellow, violet, orange, and gold   (d) yellow, violet, brown, and silver

9. By rotating the volume control in a radio receiver, you can change the volume (level) of sound. When you rotate this control, a resistance is varied inside the receiver. Similarly, you can tune in any desired station by rotating the tuning control. When we rotate the tuning control, we vary
   (a) a resistance   (c) an inductance
   (b) a capacitance   (d) only the position of the indicating needle

10. With the help of radar, we can
    (a) listen to more melodious music
    (b) perform mathematical calculations very fast
    (c) cure the damaged tissues in the human body
    (d) detect the presence of an aircraft as well as locate its position

11. With the help of a computer, we can
    (a) perform mathematical calculations very fast
    (b) transmit messages to a distant place
    (c) amplify very weak signals
    (d) see the details of a photograph by magnifying it more than million times

12. Ratings on a capacitor are given as: 25 μF, 12 V. Also, a plus sign is written near one of its terminals. From this information, we can definitely say that the capacitor is a
    (a) mica capacitor   (c) electrolytic capacitor
    (b) ceramic capacitor   (d) any of the above

13. The colour bands on a fixed carbon resistor are: brown, red, and black (given sequentially). Its value is
    (a) 12 Ω (c) 21 Ω
    (b) 120 Ω (d) 210 Ω

14. An ac voltage can be converted into a unidirectional voltage by using
    (a) a power amplifier circuit   (c) a multivibrator circuit
    (b) an oscillator circuit   (d) a rectifier circuit

Ans. 1. c; 2. d; 3. a; 4. b; 5. c; 6. d; 7. c; 8. b; 9. b; 10. d; 11. a; 12. c; 13. a; 14. d.

Current and Voltage Sources

OBJECTIVES: After completing this unit, you will be able to: ○ Name a few sources of electrical energy. ○ Draw the symbols of ideal voltage source, practical voltage source, ideal current source and practical current source. ○ Name examples of some voltage sources. ○ Name examples of some current sources. ○ Draw the graphical representation for the voltage and current sources (both ideal as well as practical). ○ Explain the difference between ideal and practical voltage source. ○ Explain the difference between ideal and practical current source. ○ Convert a given voltage source into an equivalent current source and vice-versa. ○ Solve simple problems involving voltage and current sources.

2.1 SOURCES OF ELECTRICAL POWER

The basic purpose of a source is to supply power to a load. A source is, therefore, connected to the load as shown in Fig. 2.1. The source may supply either dc (direct current) or ac (alternating current). The terminology dc as employed here stands for any quantity that is steady, unchanging and unidirectional in nature. Similarly, the terminology ac stands to specify any quantity which is alternating in nature, i.e. its magnitude is changing in both the positive and negative directions with time. Unless stated otherwise, the term ac represents *sinusoidal* variations. Some dc sources are battery, dc generator and rectification-type dc supply. Similarly, examples of ac sources are alternators, and oscillators or signal generators.

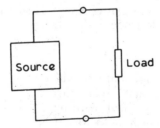

Fig. 2.1 Transfer of energy from
source to load.

2.1.1 Batteries

The battery is the most common dc voltage source. The term *battery* is derived from the expression "battery of cells". A battery consists of a series or parallel combination of two or more similar cells. A *cell* is the fundamental source of electrical energy. Cells can be divided into *primary* and *secondary* types. The secondary cell is rechargeable, whereas the primary is

Fig. 2.2 A battery and some cells.

not. The battery used in a car is of secondary type, since it can be recharged. But the cells used in a torch are of primary type, as they cannot be recharged. Figure 2.2 shows a battery and some typical cells.

Cells* and batteries produce electrical power at the expense of chemical energy, and all have the same basic construction. Each has two electrodes (one positive and the other negative) which are immersed in an electrolyte. Electrolytes are chemical compounds. When dissolved in a solution, they decompose into positive and negative ions. These ions carry the charge inside the cell from one electrode to the other.

2.1.2 Generators

The dc generator is quite different from the battery. It has a rotating shaft. When this shaft is rotated at the specified speed by some external agency (such as a steam turbine or water turbine), a voltage of rated value appears across its terminals (see Fig. 2.3). Generally speaking, a generator is capable of giving higher voltage and power than a battery.

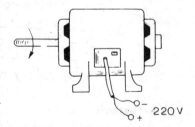

Fig. 2.3 DC generator

---

*An exception is the *solar cell*, which converts light energy into electrical energy. Solar cells are in the developmental stage; and very soon, inexpensive solar cells will be available.

2.1.3 Rectification-Type Supply

The dc supply most frequently used in an electronics laboratory is of this type. It contains a rectifier which converts time-varying voltage, i.e. ac (such as that available from the domestic power-mains) into a voltage of fixed value. This process will be discussed in detail in Unit 4. Two dc laboratory supplies of this type are shown in Fig. 2.4. The battery eliminator used with a transistor radio or calculator is also of this type.

Fig. 2.4   DC laboratory supplies

2.1.4 Alternators

The alternator is quite similar, both in construction and mode of operation, to the dc generator. When its shaft is rotated, an alternating (sinusoidal) voltage is generated across its terminals. These type of alternators are used in most electric power stations. Electronics has hardly anything to do with such alternators, except that in most of the cases it is these alternators, in the power generating station, which give power for the operation of the electronic equipment.

2.1.5 Oscillators or Signal Generators

An oscillator is the equipment which supplies ac voltages. This voltage is used as a signal to test the working of different electronic circuits (such as an amplifier). The frequency of the ac signal supplied by this instrument can be varied. Some signal generators are capable of giving other type of waveforms, such as triangular, square, etc. in addition to the sinusoidal wave. Figure 2.5 shows a laboratory signal generator.

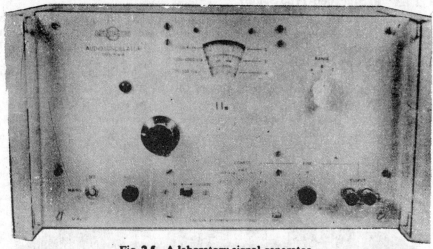

Fig. 2.5 A laboratory signal generator

2.2 INTERNAL IMPEDANCE OF A SOURCE

All electrical energy sources nave some internal impedance (or resistance*). It is due to this internal impedance that the source does not behave ideally. When a voltage source supplies power to a load, its terminal voltage (voltage available at its terminals) drops. A cell used in a torch has a voltage of 1.5 V across its two electrodes when nothing is connected to it. However, when connected to a bulb, its voltage becomes less than 1.5 V. Such a reduction in the terminal voltage of the cell may be explained as follows.

Figure 2.6a shows a cell of 1.5 V connected to a bulb. When we say "cell of 1.5 V", we mean a cell whose open-circuit voltage is 1.5 V. In the equivalent circuit of Fig. 2.6b, the bulb is replaced by a load resistor R_L (of, say, 0.9 Ω), and the cell is replaced by a constant voltage source of 1.5 V in series with the internal resistance R_S (of, say, 0.1 Ω). The total resistance in the circuit is now $0.1 + 0.9 = 1.0$ Ω. Since the net voltage that sends current into the circuit is 1.5 V, the current in the circuit is

$$I = \frac{V}{R} = \frac{1.5}{1.0} = 1.5 \text{ A}$$

The terminal voltage (the voltage across the terminals AB) of the cell is same as the voltage across the load resistor R_L. Therefore,

$$V_{AB} = I \times R_L = 1.5 \times 0.9 = 1.35 \text{ V}$$

The voltage that drops because of the internal resistance is

$$= 1.5 - 1.35 = 0.15 \text{ V}$$

Note that, if the internal resistance of the cell were smaller (compared to the load resistance), the voltage drop would also have been smaller than 0.15 V. The internal resistance (or impedance in case of ac source) of a source may be due to one or more of the following reasons:

---

*In case of dc circuits, the impedance simply reduces to resistance.

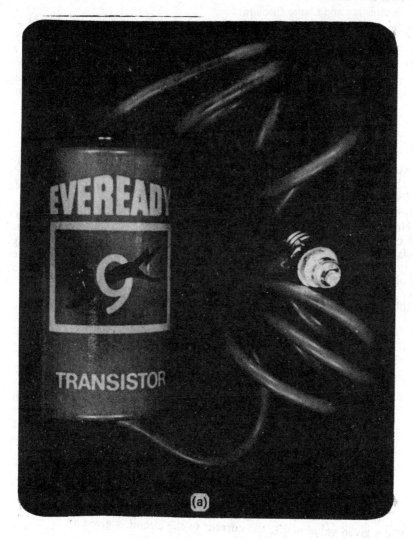

(a)

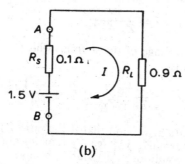

(b)

Fig. 2.6 A cell connected to a bulb

(i) The resistance of the electrolyte between the electrodes, in case of a cell.

(ii) The resistance of the armature winding in case of an alternator or a dc generator.

(iii) The output impedance of the active device like a transistor or vacuum tube in case of an oscillator (or signal generator), and rectification-type dc supply.

2.3 CONCEPT OF VOLTAGE SOURCE

Consider an ac source. Let V_S be its open-circuit voltage (i.e. the voltage which exists across its terminals when nothing is connected to it), and Z_S be its internal impedance. Let it be connected to a load impedance Z_L whose value can be varied, as shown in Fig. 2.7.

Now, suppose Z_L is infinite. It means that the terminals AB of the source are open-circuited. Under this condition, no current can flow. The terminal voltage V_T is obviously the same as the emf V_S, since there is no voltage drop across Z_S. Let us now connect a finite load impedance Z_L, and then go on reducing its value. As we do this, the current in the circuit goes on increasing. The voltage drop across Z_S also goes on increasing. As a result, the terminal voltage V_T goes on decreasing.

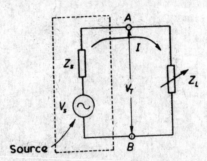

Fig. 2.7 A variable load connected to an ac source

For a given value of Z_L, the current in the circuit is given as

$$I = \frac{V_S}{Z_S + Z_L}$$

Therefore, the terminal voltage of the source, which is the same as the voltage across the load, is

$$V_T = I \times Z_L = \frac{V_S}{Z_S + Z_L} \times Z_L = \frac{V_S}{1 + Z_S/Z_L} \qquad (2.1)$$

From the above equation, we find that if the ratio Z_S/Z_L is small compared to unity, the terminal voltage V_T remains almost the same as the voltage V_S. Under this condition, the source behaves as a good voltage source. Even if the load impedance changes, the terminal voltage of the source remains practically constant (provided the ratio Z_S/Z_L is quite small). Such a source can then be said to be a *"good (but not ideal) voltage source"*.

2.3.1 Ideal Voltage Source

It would have been *ideal,* if the terminal voltage of a source remains fixed whatever be the load connected to it. In other words, a voltage source should ideally provide a fixed terminal voltage even though the current drain (or load resistance) may vary. In Eq. 2.1, to make the terminal voltage V_T fixed for any value of Z_L, the only way is to make the internal impedance Z_S zero. Thus, we infer that *an ideal voltage source must have zero internal impedance.* The symbolic representation of dc and ac ideal voltage sources are given in Fig. 2.8. Figure 2.9 gives the characteristics of an ideal voltage source. The terminal voltage V_T is seen to be constant at V_S for all values of load current*.

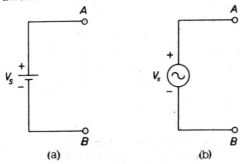

Fig. 2.8 Symbolic representation of an ideal
voltage source:
(a) DC voltage source
(b) AC voltage source

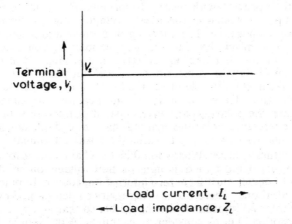

Fig. 2.9 *V-I* characteristics of an ideal voltage source

2.3.2 Practical Voltage Source

An ideal voltage source is not practically possible. There is no source which can maintain its terminal voltage constant when its terminals are short-circuited. If it could do so, it would mean that it can supply an infinite

---

*Load current varies as the load impedance is changed. When we reduce the value of load impedance, the current increases.

amount of power to a short-circuit. This is not possible. Hence, an ideal voltage source does not exist in practice. However, the concept of an ideal voltage source is very helpful in understanding the circuits containing a practical voltage source.

A practical voltage source can be considered to consist of an ideal voltage source in series with an impedance. This impedance is called the *internal impedance* of the source. The symbolic representation of practical voltage sources are shown in Fig. 2.10.

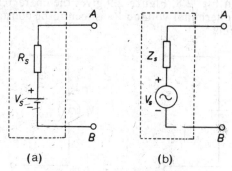

Fig. 2.10 Practical voltage source:
(a) DC voltage source
(b) AC voltage source

It is not possible to reach any other terminal except A and B. These are the terminals available for making external connections. In the dc source, since the upper terminal of the ideal voltage source is marked positive, the terminal A will be positive with respect to terminal B. In the ac source in Fig. 2.10b, the upper terminal of the ideal voltage source is marked as positive and lower as negative. The marking of positive and negative on an ac source does not mean the same thing as the markings on a dc source. Here (in ac), it means that the upper terminal (terminal A) of the ideal voltage source is positive with respect to the lower terminal *at that particular instant*. In the next half-cycle of ac, the lower terminal will be positive and the upper negative. Thus, the positive and negative markings on an ac source indicate the polarities at a given instant of time. In some books you will find the reference polarities marked by—instead of positive and negative signs—an arrow pointing towards the positive terminal.

The question naturally arises: What should be the characteristics of a source so that it may be considered a good enough constant voltage source? An ideal voltage source, of course, must have zero internal impedance. In practice, no soure can be an ideal one. Therefore it is necessary to determine how much the value of the internal impedance Z_S should be, so that it can be called a practical voltage source. Let us consider an example. A dc source has an open-circuit voltage of 2 V, and internal resistance of only 1 Ω. It is connected to a load resistance R_L as shown in Fig. 2.11a. The load resistance can assume any value ranging from 1 Ω to 10 Ω. Let us now find the variation in the terminal voltage of the source.

When the load resistance R_L is 1 Ω, the total resistance in the circuit is $1 + 1 = 2$ Ω. The current in the circuit is

$$I_1 = \frac{V_S}{R_S + R_{Li}} = \frac{2}{1 + 1} = 1 \text{ A}$$

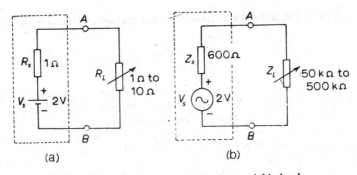

Fig. 2.11 Voltage sources connected to variable loads

The terminal voltage is then

$$V_{T1} = I_1 \times R_{L1} = \frac{V_S}{R_S + R_{L1}} \times R_{L1}$$

$$= \frac{2}{1+1} \times 1 = 1.0 \text{ V}$$

When the load resistance becomes 10 Ω, the total resistance in the circuit becomes $10 + 1 = 11$ Ω. We can again find the terminal voltage as

$$V_{T2} = \frac{V_S}{R_S + R_{L2}} \times R_{L2}$$

$$= \frac{2}{1+10} \times 10 = \frac{20}{11} = 1.818 \text{ V}$$

Thus, we find that the maximum voltage available across the terminals of the source is 1.818 V. When the load resistance varies between its extreme limits—from 1 Ω to 10 Ω—the terminal valtage varies from 1 V to 1.818 V. This is certainly a large variation. The variation in the terminal voltage is more than 40 % of the maximum voltage.

Let us consider another example. A 600-Ω, 2-V ac source is connected to a variable load, as shown in Fig. 2.11*b*. The load impedance Z_L can vary from 50 kΩ to 500 kΩ—again a variation having the same ratio of 1 : 10, as in the case of the first example. We can find the variation in the terminal voltage of the source. When the load impedance is 50 kΩ, the terminal voltage is

$$V_{T1} = \frac{V_S}{Z_S + Z_{L1}} \times Z_{L1}$$

$$= \frac{2}{600 + 50\,000} \times 50\,000 = 1.976 \text{ V}$$

When the load impedance is 500 kΩ, the terminal voltage is

$$V_{T2} = \frac{V_S}{Z_S + Z_{L2}} \times Z_{L2}$$

$$= \frac{2}{600 + 500\,000} \times 500\,000 = 1.997 \text{ V}$$

With respect to the maximum value, the percentage variation in terminal voltage

$$= \frac{1.997 - 1.976}{1.997} \times 100 = 1.05 \%$$

We can now compare the two examples. In the first case, although the internal resistance of the dc source is only 1 Ω, yet it is not justified to call it a constant voltage source. Its terminal voltage varies by more than 40 %. In the second case, although the internal impedance of the ac source is 600 Ω, it may still be called a practical constant voltage source, since the variation in its terminal voltage is quite small (only 1.05 %). Thus, we conclude that it is not the absolute value of the internal impedance that decides whether a source is a good constant voltage source or not. It is the value of the internal impedance relative to the load impedance that is important. The lesser the ratio Z_S/Z_L (in the first example, this ratio varies from 1 to 0.1, whereas in the second example it varies from 0.012 to 0.0012), the better is the source as a constant voltage source.

No practical voltage source can be an ideal voltage source. Thus, no practical voltage source can have the V-I characteristic as shown in Fig. 2.9. When the load current increases, the terminal voltage of a practical voltage source decreases. The characteristic is then modified to that shown in Fig. 2.12a. It is sometimes preferred to take voltage on the x-axis and current on the y-axis. The V-I characteristic of a practical voltage source then looks like the one shown in Fig. 2.12b.

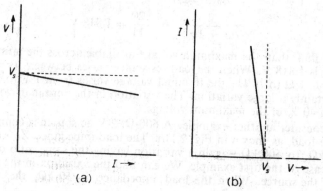

Fig. 2.12 Two ways of drawing V-I characteristics of a practical voltage source

2.4 CONCEPT OF CURRENT SOURCE

Like a constant voltage source, there may be a constant current source—a source that supplies a constant current to a load even if its impedance varies. Ideally, the current supplied by it should remain constant, no matter what the load impedance is. A symbolic representation of such an *ideal current source* is shown in Fig. 2.13a. The arrow inside the circle indicates the direction in which current will flow in the circuit when a load is connected to the source. Figure 2.13b shows the V-I characteristic of an ideal current source.

Let us connect a variable load impedance Z_L to a constant current source, as shown in Fig. 2.13c. As stated above, the current supplied by the source

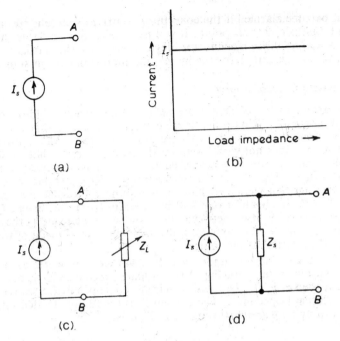

Fig. 2.13 (a) Symbol for an ideal current source
  (b) V-I characteristic of an ideal current source
  (c) A variable load connected to an ideal current source
  (d) Symbol for a practical current source

should remain constant at I_S for all values of load impedance. It means even if Z_L is made infinity, the current through this should remain I_S. Now, we must see if any practical current source could satisfy this condition. The load impedance $Z_L = \infty$ means no conducting path, external to the source, exists between the terminals A and B. Hence, it is a physical impossibility for current to flow between terminals A and B. If the source could maintain a current I_S through an infinitely large load impedance, there would have been an infinitely large voltage drop across the load. It would then have consumed infinite power from the source. Of course, no practical source could ever supply infinite power.

A practical current source supplies current I_S to a short-circuit (i.e. when $Z_L = 0$). That is why the current I_S is called *short-circuit current*. But, when we increase the load impedance, the current falls below I_S. When the load impedance Z_L is made infinite (i.e. the terminals A and B are open-circuited), the load current reduces to zero. It means there should be some path (inside the source itself) through which the current I_S can flow. When some finite load impedance is connected, only a part of this current I_S flows through the load. The remaining current goes through the path inside the source. This inside path has an impedance Z_S, and is called the *internal impedance*. The symbolic representation of such a practical current source is shown in Fig. 2.13d.

Now, if terminals AB are open-circuited ($Z_L = \infty$) in Fig. 2.13d, the terminal voltage does not have to be infinite. It is now a finite value, $V_T = I_S Z_S$. It means that the source does not have to supply infinite power!

Do not become alarmed if the concept of a current source is strange and somewhat confusing at this point. It will become clearer in later chapters. The introduction of semiconductor devices such as the transistor is responsible, to a large extent, for the increasing interest in current sources.

2.4.1 Practical Current Source

An ideal current source is merely an *idea*. In practice, an ideal current source cannot exist. Obviously, there cannot be a source that can supply constant current even if its terminals are open-circuited. The reason why an actual source does not work as an ideal current source is that its internal impedance is not infinite. A practical current source is represented by the symbol shown in Fig. 2.13d. The source impedance Z_S is put in parallel with the ideal current source I_S. Now, if we connect a load across the terminals A and B, the load current will be different from the current I_S. The current I_S now divides itself between two branches—one made of the source impedance Z_S inside the source itself, and the other made of the load impedance Z_L external to the source.

Let us find the conditions under which a source can work as a good (practical) current source. In Fig. 2.14a, a load impedance Z_L is connected to a current source. Let I_S be the short-circuit current of the source, and Z_S be its internal impedance. The current I_S is seen to be divided into two parts—I_1 through Z_S and I_L through Z_L. That is,

$$I_S = I_1 + I_L$$

or

$$I_1 = I_S - I_L$$

Since the impedance Z_S and Z_L are in parallel, the voltage drop across each should be equal, i.e.,

$$I_1 Z_S = I_L Z_L$$

or

$$(I_S - I_L) Z_S = I_L Z_L$$

or

$$I_L = \frac{I_S Z_S}{Z_S + Z_L}$$

or

$$I_L = \frac{I_S}{1 + (Z_L/Z_S)} \qquad (2.2)$$

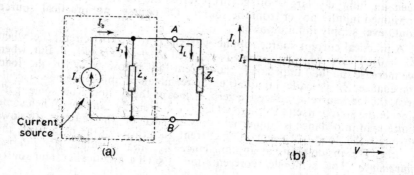

Fig. 2.14 (a) Practical current source feeding current to a load impedance
(b) V-I characteristic of a practical current source

This equation tells us that the load current I_L will remain almost the same as the current I_S, provided the ratio Z_L/Z_S is small compared to unity. The source then behaves as a good current source. In other words, the larger the value of internal impedance Z_S (compared to the load impedance Z_L), the smaller is the ratio Z_L/Z_S, and the better it works as a constant current source.

From Eq. 2.2, we see that the current $I_L = I_S$, when $Z_L = 0$. But, as the value of load impedance is increased, the current I_L is reduced. For a given increase in load impedance Z_L, the corresponding reduction in load current I_L is much smaller. Thus, with the increase in load impedance, the terminal voltage ($V = I_L Z_L$) also increases. The *V-I* characteristic of a practical current source is shown in Fig. 2.14*b*.

2.5 EQUIVALENCE BETWEEN VOLTAGE SOURCE AND CURRENT SOURCE

Practically, a voltage source is not different from a current source. In fact, a source can either work as a current source or as a voltage source. It merely depends upon its working conditions. If the value of the load impedance is very large compared to the internal impedance of the source, it proves advantageous to treat the source as a voltage source. On the other hand, if the value of the load impedance is very small compared to the internal impedance, it is better to represent the source as a current source. From the circuit point of view, it does not matter at all whether the source is treated as a current source or a voltage source. In fact, it is possible to convert a voltage source into a current source and vice-versa.

2.5.1 Conversion of Voltage Source into Current Source and vice versa

Consider an ac source connected to a load impedance Z_L. The source can either be treated as a voltage source or a current source, as shown in Fig. 2.15. The voltage-source representation consists of an ideal voltage source V_S in series with a source impedance Z_{S1}. And the current-source representation consists of an ideal current source I_S in parallel with source impedance Z_{S2}. These are the two representations of the same source. Both types of representations must appear the same to the externally connected load impedance Z_L. They must give the same results.

In Fig. 2.15*b*, if the load impedance Z_L is reduced to zero (i.e. the terminals A and B are short-circuited), the current through this *short* is given as

$$I_L \text{ (short-circuit)} = \frac{V_S}{Z_{S1}} \qquad (2.3)$$

We want both the representations (voltage-source and current-source) to give the same results. This means that current source in Fig. 2.15*c* must also give the same current (as given by Eq. 2.3) when terminals A and B are shorted. But the current obtained by shorting the terminals A and B of Fig. 2.15*c* is simply the source current I_S (the source impedance Z_{S2} connected in parallel with a short-circuit is as good as not being present). Therefore, we conclude that the current I_S of the equivalent current source must be the same as that given by Eq. 2.3. Thus

$$I_L \text{ (short-circuit)} = I_S = \frac{V_S}{Z_{S1}} \qquad (2.4)$$

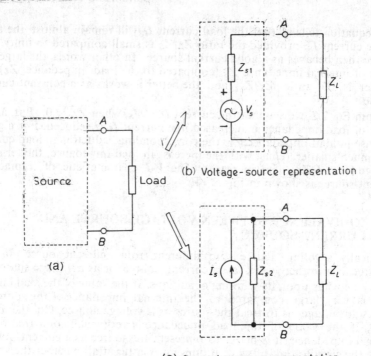

(b) Voltage-source representation

(a)

(c) Current-source representation

Fig. 2.15 A source connected to a load

Again, the two representations of the source must give the same terminal voltage when the load impedance Z_L is disconnected from the source (i.e. when the terminals A and B are open-circuited. In Fig. 2.15b, the open-circuit terminal voltage is simply V_S. There is no voltage drop across the internal impedance Z_{S1}. Let us find out the open-circuit voltage in the current-source representation of Fig. 2.15c. When the terminals A and B are open-circuited, the whole of the current I_S flows through the impedance Z_{S2}. The terminal voltage is then the voltage drop across this impedance. That is

$$V_T \text{ (open-circuit)} = I_S Z_{S2} \qquad (2.5)$$

Therefore, if the two representations of the source are to be equivalent, we must have

$$V_T = V_S$$

Using Eqs. 2.4 and 2.5, we get

$$I_S Z_{S1} = I_S Z_{S2}$$
$$Z_{S1} = Z_{S2} = Z_S \text{ (say)}$$

or

Then both Eqs. 2.4 and 2.5 reduce to

$$V_S = I_S Z_S \qquad (2.6)$$

It may be noted (see Eq. 2.5) that in both the representations of the source, the source impedance as faced by the load impedance at the terminals AB, is the same (impedance Z_S). Thus we have established the equivalence between the voltage-source representation and current-source representation of Fig.2. 15, for short-circuits and for open-circuits. But, we are not sure that the equivalence is valid for any other value of load impedance. To test this, let us check whether a given impedance Z_L draws the same amount of current when connected either to the voltage-source representation or to the current-source representation.

In Fig. 2.15b, the current through the load impedance is

$$I_{L1} = \frac{V_S}{Z_S + Z_L} \tag{2.7}$$

In Fig. 2.15c, the current I_S divides into two branches. Since the current divides itself into two branches in inverse proportion of the impedances, the current through the load impedance Z_L is

$$I_{L2} = I_S \times \frac{Z_S}{Z_S + Z_L} = \frac{I_S Z_S}{Z_S + Z_L}$$

By making use of Eq. 2.6, the above equation can be written as

$$I_{L2} = \frac{V_S}{Z_S + Z_L} \tag{2.8}$$

We now see that the two currents I_{L1} and I_{L2} as given by Eqs. 2.7 and 2.8 are exactly the same. Thus, the equivalence between the voltage-source and current-source representations of Fig. 2.15 is completely established. *We may convert a given voltage source into its equivalent current source by using Eq. 2.6. Similarly, any current source may be converted into its equivalent voltage source by using the same equation.*

Example 2.1 Figure 2.16 shows a dc voltage source having an open circuit voltage of 2 V and an internal impedance of 1 Ω. Obtain its equivalent current-source representation.

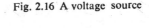

Fig. 2.16 A voltage source

Solution: If we short-circuit the terminals A and B of the voltage source, the current supplied by the source is

$$I \text{ (short circuit)} = \frac{V_S}{R_S} = \frac{2}{1} = 2 \text{ A}$$

In the equivalent current-source representation, the current source is of 2 A. The source impedance of 1 Ω is connected in parallel with this current source. The equivalent current source obtained is shown in Fig. 2.17.

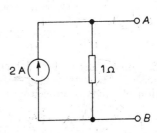

Fig. 2.17 Equivalent current source

Example 2.2 Obtain an equivalent voltage source of the ac current source shown in Fig. 2.18.

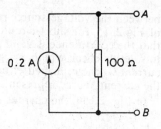

Solution: The open-circuit voltage across terminals A and B is given as

$$V \text{ (open-circuit)} = I_S Z_S$$
$$= 0.2 \times 100 = 20 \text{ V}$$

Fig. 2.18 An ac current source

This will be the value of the "ideal voltage source" in the equivalent voltage-source representation. The source impedance Z_S is put in series with the ideal voltage source. Thus, the equivalent voltage-source representation of the given current source is as given in Fig. 2.19.

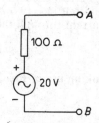

Fig. 2.19 Equivalent voltage source

Example 2.3 In the circuit in Fig. 2.20, an ac current source of 1.5 mA and 2 kΩ is connected to a load consisting of two parallel branches; one of 10 kΩ and other of 40 kΩ. Determine the current I_4 flowing in the 40 kΩ impedance. Now convert the given current source into its equivalent voltage source and then again calculate the current I_4 in the 40 kΩ impedance. Check whether you get the same results in the two cases.

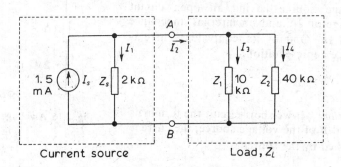

Fig. 2.20 A current source connected to a load

Solution: Let us first determine the net load impedance that is connected across the source terminals A and B. This would be the parallel combination of the two impedances. Thus

$$Z_L = \frac{Z_1 Z_2}{Z_1 + Z_2} = \frac{10 \times 40}{10 + 40} \text{ k}\Omega = 8 \text{ k}\Omega$$

Now the circuit of Fig. 2.20 can be re-drawn as given in Fig. 2.21. A net impedance of 8 kΩ is shown to be connected across the source terminals A and B.

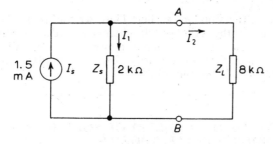

Fig. 2.21

It is clear that the current I_S divides itself into two branches—one consisting of Z_S and other consisting of Z_L. Therefore, the current I_2 is given by

$$I_2 = I_S \times \frac{Z_S}{Z_S + Z_L} = 1.5 \times 10^{-3} \times \frac{2 \times 10^3}{(2+8) \times 10^3} = 0.3 \times 10^{-3} \text{ A}$$

Again, look at Fig. 2.20. The current I_2 divides into two parallel branches. The current in the 40 kΩ impedance can be determined as follows:

$$I_4 = I_2 \times \frac{Z_1}{Z_1 + Z_2} = 0.3 \times 10^{-3} \times \frac{10 \times 10^3}{(10+40) \times 10^3} = 0.06 \text{ mA}$$

We shall again solve this problem following another approach. Here we convert the given current source into its equivalent voltage source. The open-circuit voltage of the source is given as

$$V_S = I_S Z_S = 1.5 \times 10^{-3} \times 2 \times 10^3 = 3.0 \text{ V}$$

Therefore, the equivalent voltage-source representation will be an ideal voltage source of 3.0 V in series with an impedance of 2 kΩ. We can connect the net load impedance Z_L (of 8 kΩ, as calculated above) to this voltage source, as in Fig. 2.22.

The circuit in Fig. 2.22 is a single loop circuit. The loop current can be calculated by applying Kirchhoff's voltage law.

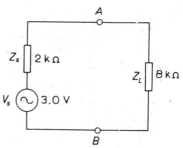

$$3 = I (2 \times 10^3 + 8 \times 10^3)$$

$$\therefore \quad I = \frac{3}{10 \times 10^3} = 0.3 \text{ mA}$$

Fig. 2.22

(Note that the current I turns out to be the same as current I_2 of Fig. 2.21) This current gets divided into two parallel branches of the load impedance Z_L (see Fig. 2.20). The current through the 40 kΩ impedance is

$$I_4 = I \times \frac{Z_1}{Z_1 + Z_2} = 0.3 \times 10^{-3} \times \frac{10 \times 10^3}{(10+40) \times 10^3} = 0.06 \text{ mA}$$

This is the same result as obtained earlier. Thus, we find that solving of an electrical circuit gives the same result whether we treat the source in the circuit as voltage source or current source. However, as we shall see later, the solving of a particular circuit sometimes becomes simpler if we treat the source as one type rather than the other.

2.6 USEFULNESS OF THE CONCEPT OF VOLTAGE AND CURRENT SOURCE IN ELECTRONICS

As we proceed with the study of electronics, we find that the concept of voltage and current sources are of great help. For example, in determining the performance of any electronic circuit (such as an amplifier, which is used for amplifying electrical signals), we convert the original circuit into its equivalent ac circuit. In this equivalent circuit, the active device (such as a transistor or vacuum triode) is replaced by its current-source equivalent or voltage-source equivalent. We can now apply the basics of circuit theory to determine the characteristic behaviour of the electronic circuit.

Figure 2.23a shows the *V-I* characteristics of a semiconductor device called the *zener diode*. Its symbol is shown in Fig. 2.23b. For the time being we will ignore the details of this device. But, if you compare its characteristics with that of an ideal dc voltage source (as shown in Fig. 2.9), you will find a marked similarity. The only difference is that the characteristic curve of the zener diode is inverted. It is *shown* inverted to emphasize that the zener diode is operated with reverse bias (the term *reverse bias* is explained in detail in Unit 4). This means that the current through the zener diode flows in a direction opposite to that of the arrow (in its symbol).

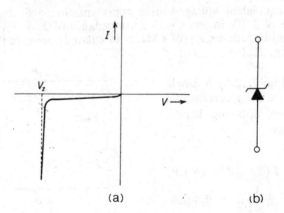

(a)   (b)

Fig. 2.23 Zener diode: (*a*) Characteristics; (*b*) Symbol

Let us see what happens if we connect a zener diode across a practical voltage source, as in Fig. 2.24. If the load impedance R_L varies, the current I_L through it also varies. If the zener diode were not there, the terminal voltage V_T would also vary, because the voltage drop across the source impedance varies. But now, since a zener diode is connected across the terminals A and B, and it has characteristics quite similar to that of *an ideal voltage source*, the situation is different. The terminal voltage V_T remains constant at V_Z whatever be the current flowing through the zener diode.

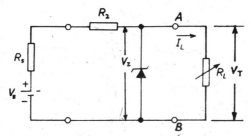

Fig. 2.24 Zener diode connected across a
practical voltage source

When the load current varies, the zener diode current adjusts itself so that its terminal voltage remains constant. This is an example of simple *voltage regulator circuit*. The resistance R_2 is put in the circuit so as to *limit* the current through the zener diode. It ensures safe operation of the zener diode.

A zener diode when connected in the circuit of Fig. 2.24 works as a voltage source. Strictly speaking, it is not a source, because it cannot supply any power of its own. We need another voltage source for its operation. Once it is connected in an electrical circuit, it has *V-I* characteristics similar to that of a constant voltage source. Loosely speaking, we can say that a zener diode is a constant dc-voltage source.

Another device that has similar characteristics to that of a constant dc-voltage source is a *VR tube* (voltage regulator tube). It is a simple gas diode. Its characteristics and symbol are shown in Fig. 2.25. The characteristics of the VR tube and that of a constant dc-voltage source (as shown in Fig. 2.9) are similar. We, therefore, say that a VR tube (or gas diode) acts as a constant dc voltage source. A gas diode conducts when an external voltage is supplied to it. The plate *P* of the diode is made positive with respect to cathode *K*. This device will work as a constant voltage source only when the external voltage is greater than the voltage V_F (see the characteristics in Fig. 2.25).

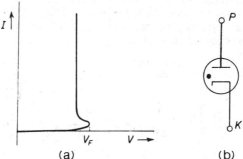

(a)   (b)

Fig. 2.25 VR tube or a gas diode:
(a) Characteristics; (b) Symbol

Another important device is the *transistor*. It is a three-terminal device. These terminals are called *emitter, base* and *collector*. Its symbol is shown in Fig. 2.26a. The transistor is extensively used as an amplifying device. When connected in the amplifier circuit, any of its terminals can be made common between input and output. Figure 2.26b shows the output characteristics of a transistor connected in common-base mode.

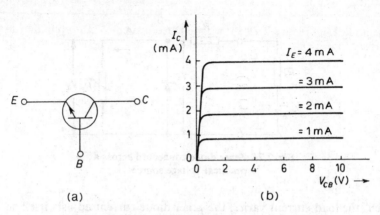

Fig. 2.26 Transistor: (a) Symbol of a transistor (*NPN*)
(*b*) Output characteristics

The characteristics of the transistor are almost horizontal lines. For a given value of the input current (emitter current I_E), the collector current I_C remains constant when the collector voltage V_{CB} is varied. Such characteristics are very similar to the characteristics of a constant current source as shown in Fig. 2.12*b*. Thus, we say that a transistor behaves as a constant current source. Between its output terminals (collector and base), it can be represented by a current source as shown in Fig. 2.27. Here, the current source $I_C (= \alpha I_E)$ is dependent upon the input current I_E. The resistance R_O represents static (dc) output resistance.

The equivalent representation of the transistor as shown in Fig. 2.27 is not of much significance to us. It is meant for the dc operation of the transistor. Since, in an amplifier circuit, the voltages and currents are changing all the time (because of the input signal), we are interested in the transistor's ac behaviour. The ac behaviour of the transistor can be represented by the circuit shown in Fig. 2.28. Here, r_o is the ac resistance of the transistor between its collector and base. This resistance has high value (typically, 1 MΩ). This resistance represents the source resistance when we look upon a transistor as a current source. The value of the ac current source i_c depends upon the input ac current i_e.

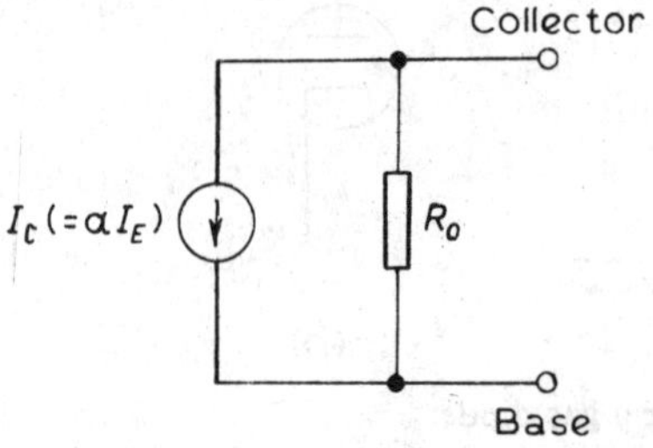

Fig. 2.27 DC representation of a transistor by a current source between its output terminals

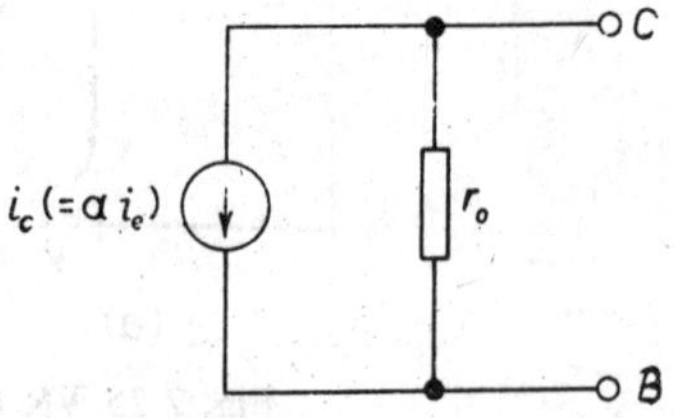

Fig. 2.28 AC representation of a transistor between its output terminals

The other electronic devices which behave like current sources are FET and pentode tube. Since the internal resistance (ac output resistance) of a triode is not very large, it is better represented as a voltage source rather than a current source.

Example 2.4   Figure 2.29 shows the ac equivalent of an amplifier using a transistor. Calculate the output voltage v_o.

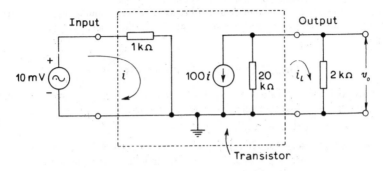

Fig. 2.29 AC equivalent of a transistor amplifier

Solution: The value of the current source in ac equivalent of the transistor is 100 i. The current i can be calculated from the input circuit.

$$i = \frac{10 \text{ mV}}{1 \text{ k}\Omega} = \frac{10 \times 10^{-3}}{1 \times 10^3} = 10 \times 10^{-6} \text{ A}$$

Therefore, the current source in the output circuit is

$$100\ i = 100 \times 10 \times 10^{-6} = 10^{-3} \text{ A}$$

This current divides itself into two branches. The current through the 2 kΩ resistance is

$$i_L = 10^{-3} \times \frac{20 \times 10^3}{(20 + 2) \times 10^3} = 0.909 \times 10^{-3} \text{ A}$$

Therefore, the output voltage v_o is given as

$$v_o = i_L \times 2 \times 10^3 = 0.909 \times 10^{-3} \times 2 \times 10^3$$
$$= \mathbf{1.818 \text{ V}}$$

REVIEW QUESTIONS

2.1 Name two sources of electrical power. Are they voltage sources or current sources ?

2.2 Draw the symbol of an ideal dc voltage source.

2.3 Draw the symbolic representation of a practical ac voltage source. Explain the necessity of including an impedance in this representation.

2.4 Explain the condition under which a practical voltage source is considered to be a good voltage source.

2.5 Name at least one electronic device whose characteristics are very close to that of an ideal voltage source. Explain in one or two lines its characteristics.

2.6 State an application of an electronic device whose characteristics are similar to that of an ideal voltage source.

2.7 Draw the symbol of an ideal current source.

2.8 Draw the symbolic representation of a practical current source. Explain the reason for putting an impedance in this symbolic representation.

2.9 Name two electronic devices which have characteristics similar to that of an ideal current source. Justify your answer in about three lines.

2.10 A practical source can be represented either as a voltage source or as a current source. How can you convert one representation to the other ?

OBJECTIVE-TYPE QUESTIONS

I. Here are some incomplete statements. Four alternatives are provided below each. Tick the alternative that completes the statement correctly:

1. An ideal voltage source is one which has

   (a) very high internal resistance   (c) zero internal resistance
   (b) very low internal resistance   (d) infinite internal resistance

2. An ideal current source is one whose internal resistance is

   (a) very high   (c) zero
   (b) very low   (d) infinite

3. In a practical voltage source, the source resistance is

   (a) very low compared to load resistance
   (b) very high compared to load resistance
   (c) equal to the load resistance
   (d) zero

4. A device whose characteristics are very close to that of an ideal voltage source is a

   (a) vacuum diode   (c) field-effect transistor
   (b) transistor in common-base mode   (d) zener diode

5. A device whose characteristics are very close to that of an ideal current source is a

   (a) transistor in common-base mode   (c) gas diode
   (b) crystal diode   (d) vacuum triode

6. Fig. O.2.1 shows the circuit of a simple constant-voltage supply, using a zener diode. The constant voltage available across the zener diode is 5 V. The current flowing through the 1-kΩ load is

   (a) 20 mA   (c) 25 mA
   (b) 15 mA   (d) 5 mA

7. The current flowing through the zener diode in Fig. O.2.1 is

   (a) 20 mA   (c) 25 mA
   (b) 15 mA   (d) 5 mA

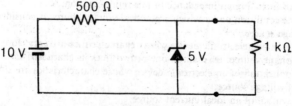

Fig. O.2.1

8. In Fig. O.2.1, the current drain from the battery is

(a) 20 mA   (c) 10 mA
(b) 15 mA   (d) 5 mA

9. A constant current source supplies a current of 300 mA to a load of 1 kΩ. When the load is changed to 100 Ω, the load current will be

(a) 3 A   (c) 300 mA
(b) 30 mA   (d) 600 mA

10. An ideal voltage source of 12 V provides a current of 120 mA to a load. If the load impedance is doubled, the new load current becomes

(a) 60 mA   (c) 240 mA
(b) 120 mA   (d) none of the above

II. Here are some statements. Indicate against each, whether it is TRUE(T) or FALSE(F).

1. A practical current source has low internal resistance.   _____
2. An ideal current source has low internal resistance.   _____
3. An ideal voltage source has low internal resistance.   _____
4. An ideal voltage source has zero internal resistance.   _____
5. An ideal current source has infinite internal resistance.   _____
6. A practical voltage source has very high internal resistance.   _____
7. Solving of an electrical circuit will give the same results whether the source is treated as voltage source or as current source.   _____
8. A resistance is connected to a practical source. For finding the current through this resistance, the only way is to represent the source as a current source. _____
9. The output side of a transistor connected in common-base mode should be treated as a constant voltage source, since its output impedance is very high. _____
10. A zener diode has characteristics similar to that of an ideal current source. _____

Ans. I. 1. c; 2. d; 3. a; 4. d; 5. a; 6. d; 7. d; 8. c; 9. c; 10. a
II. 1. F; 2. F; 3. F; 4. T; 5. T; 6. F; 7. T; 8. F; 9. F; 10. F

TUTORIAL SHEET 2.1

1. Figure T.2.1.1a represents a voltage source. Convert it into an equivalent current source.

[**Ans.** I_S = 1.25 mA; R_S = 8 kΩ]

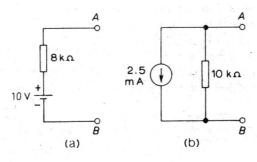

(a)   (b)

Fig. T.2.1.1

2. Figure T.2.1.1*b* represents an ac current source. Convert it into an equivalent voltage source.

[**Ans.** $V_S = 25$ V, $Z_S = 10$ kΩ]

3. Calculate the voltage available between the points A and B in the two situations represented in Fig. T.2.1.2*a* and *b*. State which situation represents a better voltage source condition.

[**Ans.** (*a*) 0.645 V; (*b*) 13.3 V; second case represents a better voltage-source condition]

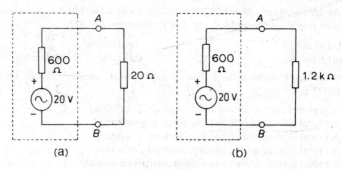

Fig. T.2.1.2

4. An electronic amplifier is used to amplify electrical signals. The voltage gain of an amplifier is defined as the ratio of its output voltage to its input voltage. Fig. T.2.1.3 represents two amplifiers having different values of input impedances Z_{in}. Calculate the output voltage in the two cases. Assume a voltage gain $A = 50$ in both the cases.

[**Ans.** $v_{o1} = 384.6$ mV, $v_{o2} = 497.02$ mV]

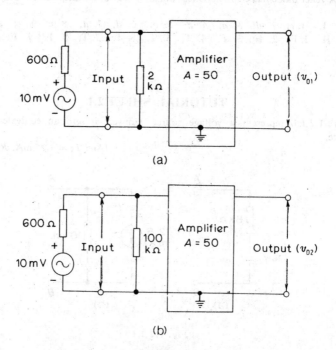

Fig. T.2.1.3

TUTORIAL SHEET 2.2

1. Calculate the current through the 10-kΩ resistor shown in Fig. T.2.2.1. Assume that the Zener diodes have ideal voltage-source characteristics.

[Ans: 1 mA]

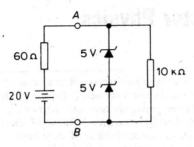

Fig. T.2.2.1

2. Calculate the voltage v_0 in Fig. T.2.2.2. Identify the source impedance Z_S.

[Ans. 7.5 V, $Z_S = 5$ kΩ]

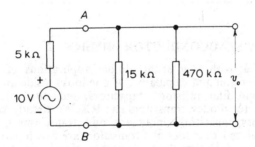

Fig. T.2.2.2

3. Calculate the voltage v_0 in Fig. T.2.2.3.

[Ans. 161.29 mV]

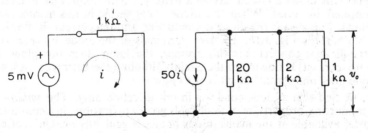

Fig. T.2.2.3

Semiconductor Physics

OBJECTIVES : After completing this unit, you will be able to : ○ State the names of a few conductors, insulators and semiconductors. ○ Explain the differences in conductors, insulators and semiconductors using energy-band diagrams. ○ Explain in brief the meaning of covalent bond, thermal generation, lifetime of charge carriers, recombination, forbidden-energy gap, valence band and conduction band, doping, donor impurity, acceptor impurity, majority and minority carriers, drif current and diffusion current. ○ Explain the mechanism of flow of current in an intrinsic semiconductor on the basis of movement of electrons and holes. ○ Explain how extrinsic (*P*- and *N*-type) semiconductor material is obtained from intrinsic semiconductor material. ○ Explain the mechanism of flow of current in an extrinsic (*P*- and *N*-type) semiconductor. ○ Explain the effect of temperature on the conductivity of an extrinsic semiconductor.

3.1 WHY STUDY SEMICONDUCTOR PHYSICS

All of us are familiar with some of the simple applications of electronics like the radio, television and calculator. If one looks inside any electronic equipment, one will find resistors, capacitors, inductors, transformers, valves, semiconductor diodes, transistors and ICs. We already know something about resistors, capacitors, inductors, and transformers, but small semiconductor devices like diodes and transistors are new to most of us. In modern electronic systems, the whole electronic circuit, containing many diodes, transistors, resistors, etc , is fabricated on a single chip. This is known as an *integrated circuit* (IC).

Let us take a simple example of a semiconductor diode. It is a two terminal device. It has a very important property of conducting in one direction only. Fig. 3.1a shows a circuit having a battery, small lamp and a diode, all cnnnected in series. When the diode is connected in this manner, the lamp glows. This means the diode is conducting and the current is flowing in the circuit. In Fig. 3.1b, the two terminals of the diode are reversed. When the diode is connected in this manner, the lamp does not glow. It means no current is flowing in the circuit, although the battery is present. The diode does not permit the current to flow.

Thus, we find that a diode conducts in one direction only. The *unidirectional* conducting property of a diode finds great applications in electronics. The power available at the mains in our homes is generally ac. Quite often we require dc power to operate some appliances. The diode makes it posssible to convert ac into dc.

The diode is one of the many components used in electronic circuits. Another important component is the transistor. It is used for *amplifying* weak electrical signals. Relatively newer devices, like junction field-effect transistor (JFET), metal-oxide semiconductor field-effect transistor

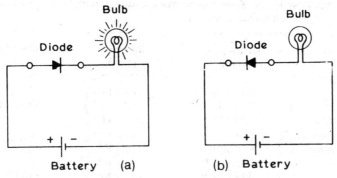

Fig. 3.1 Unidirectional conducting property of a diode:
(a) Diode conducts; (b) Diode does not conduct

(MOSFET), silicon controlled rectifier (SCR), unijunction transistor (UJT), etc. are finding wide applications in electronics. *All these devices are made of semiconductor materials.* To understand the operation of these devices (and many more that are likely to come in future), it is necessary to study the semiconductor materials in some detail.

3.2 SEMICONDUCTOR MATERIALS

We are familiar with conducting and insulating materials. Conducting materials are good conductors of electricity. Examples of good conductors are copper, silver, aluminium, etc. Insulating materials are bad conductors of electricity. Examples of insulators are porcelain, glass, quartz, rubber, bakelite, etc. The electrical wire used in houses consists of a core made of conducting material like copper or aluminium. The core provides an easy path for the flow of electric current. This core is covered with some insulating material such as rubber, cotton, PVC, plastic, etc. These coverings provide protection against short-circuits and also against electrical-shock hazards.

There is another group of materials, such as germanium and silicon. These are neither good conductors nor good insulators. At room temperature these materials have conductivities considerably lower than that of conductors, but much higher than that of insulators. It is for this reason that these materials are classified as *semiconductors*. Table 3.1 gives the resistivities of some commonly used conductors, semiconductors and insulators at room temperature.

When we increase the temperature of a metal conductor such as copper, its resistivity increases. In this respect, the semiconductors behave in an opposite way. When we raise the temperature of a semiconductor, its resistivity decreases (at higher temperature, a semiconductor conducts better), i.e. the semiconductors have *negative temperature coefficient of resistance.*

The semiconductors have another very important property. The conductivity (or resistivity) of a semiconductor can be changed, to a very large extent, by adding a very small amount of some specific materials (called *impurities*). The conductivity of the semiconductor can also be controlled by controlling the amount of impurity added to it. To understand the important properties of semiconductors, it is necessary to study the structure of atom.

Table 3.1 Resistivities of Some Conductors, Semiconductors and Insulators (at Room Temperature)

Materials	Conductivity (S/m)	Resistivity (Ωm)	Classification
Silver	6.25×10^7	1.6×10^{-8}	
Copper	5.88×10^7	1.7×10^{-8}	Conductors
Aluminium	3.85×10^7	2.6×13^{-8}	
Germanium (pure)	1.54	6.5×10^{-1}	
Silicon (pure)	5.0×10^{-4}	2.0×10^3	Semiconductors
Porcelain	3.33×10^{-10}	3.0×10^9	
Glass	5.88×10^{-12}	1.7×10^{11}	Insulators
Hard rubber	1.0×10^{-16}	1.0×10^{16}	

3.3 STRUCTURE OF AN ATOM

We know that the most fundamental unit of matter which is capable of independent existence is the *atom*. According to a simplified picture, an atom consists of a central body, called the *nucleus*, about which a number of smaller particles (called *electrons*) move in approximately elliptical orbits. The nuclei of all elements (except that of hydrogen, which has only one proton in its nucleus) contains two types of particles, called *protons* and *neutrons*. A proton and a neutron have almost same mass. But the proton is a positively charged particle whereas the neutron is electrically neutral. Almost all the mass of the atom is concentrated in its nucleus. The electrons revolving round the nucleus are very light in weight. An electron is about 1850 times lighter than a proton (or neutron). An electron has the same amount of charge as a proton. But the charge on an electron is *negative*. Since matter in its normal state is electrically neutral, the atom (which is the basic building block of matter) should also be neutral. It means, that in an atom (in its normal state), the number of orbiting electrons must be the same as the number of protons in its nucleus.

There is no difference between an electron in an atom of copper and an electron in an atom of aluminium, or any other element. Similarly, there is no difference between a proton in one atom and a proton in another atom of a different element. Likewise, the neutrons in the atoms of various elements are identical. Thus, it follows that electrons, protons and neutrons are the fundamental particles of the universe. If it is so, then why do various elements behave differently? This is because of the difference in the *number* and *arrangement* of the electrons, protons, and neutrons of which each atom is composed.

The number of protons in an atom (or of electrons in a neutral atom) is called its *atomic number*. All the electrons of an atom do not move in the same orbit. The electrons are arranged in different *orbits* or *shells*. The maximum number of electrons that can exist in the first shell (the one nearest to the nucleus) is two. The second shell can accommodate not more than eight electrons, the third not more than eighteen, the fourth not more than thirty-two, and so on. In general, a shell can contain a maximum of $2n^2$ electrons, where n is the number of the shell. But to this rule, there is an exception. The outermost orbit in an atom cannot accommodate more than eight electrons. The electrons present in the outermost orbit are called *valence electrons*.

All the elements have been arranged in a *periodic table* according to the electronic arrangements in their atoms. The elements placed in one vertical column (called *group*) have very similar properties. A part of this periodic table is shown in Table 3.2. The atomic number of each element is shown within brackets along with its symbol. We shall study in the next section, the atomic structure of some elements useful for semiconductor devices.

Table 3.2 A PART OF THE PERIODIC TABLE OF ELEMENTS

Group No.→	III	IV	V
	B (5)	C (6)	
	Al (13)	Si (14)	P (15)
	Ga (31)	Ge (32)	As (33)
	In (49)		Sb (51)

3.3.1 Atomic Structure of Some Elements

Figure 3.2 shows the representations of the atomic structures of different elements. Fig. 3.2*a* is the simplest of all. It represents the hydrogen atom. It contains one electron revolving around one proton which is the nucleus. Note that the nucleus contains no neutrons.

Figure 3.2*b* shows the structure of an aluminium atom. The nucleus of the aluminium atom contains 13 protons (13 *P*) and 14 neutrons (14 *N*). The positive charge of 13 protons is just balanced by the negative charge of the 13 electrons (13 *e*) revolving around the nucleus in different shells. The atom as a whole is electrically neutral. Note that there are two electrons in the first shell, eight electrons in the second, and only *three* electrons in the third (the outermost). The importance of the outermost shell having only three electrons is explained in Sec. 3.6.2.

The electrons in the inner shells do not normally leave the atom. But the electrons which revolve at a great speed in the outermost shell (near the edge of the atoms) do not always remain confined to the same atom. Some of them move in a random manner and may travel from atom to atom. Figure 3.3 shows how the electrons may move from one atom to another in a random manner. Electrons that are able to move in this fashion are known as *free electrons*. It is due to the presence of these free electrons in a material, that it is able to conduct electric current. The electrons in the inner orbits remain bound to the nucleus and are therefore, called *bound electrons*.

Figure 3.2*c* represents a silicon atom. Its nucleus contains 14 protons and 14 neutrons. There are 14 electrons revolving around the nucleus—two in the first shell, eight in the second, and four in the third. Thus, there are four valence electrons. The importance of this arrangement is explained in Sec. 3.5.1.

Figure 3.2*d* represents a phosphorus atom. There are 15 protons and 16 neutrons in its nucleus. It has five valence electrons. The importance of this arrangement is explained in Sec. 3.6.1.

Figure 3.2*e* shows a more complex structure. It represents a germanium atom (atomic number 32). Its nucleus contains 32 protons and 41 neutrons. There are two electrons in the first shell, eight in the second, eighteen in the third and *four* in the fourth (outermost) shell. Thus, germanium, like

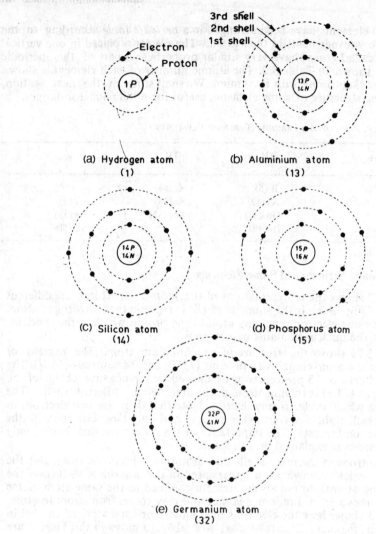

(a) Hydrogen atom
(1)

(b) Aluminium atom
(13)

(C) Silicon atom
(14)

(d) Phosphorus atom
(15)

(e) Germanium atom
(32)

Fig. 3.2 Diagrammatic representation of a few atoms (figures inside brackets represent atomic numbers)

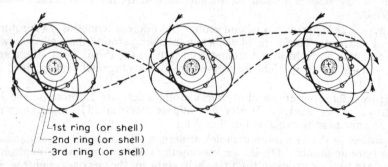

1st ring (or shell)
2nd ring (or shell)
3rd ring (or shell)

Fig. 3.3 Random movement of outermost electrons in aluminium atoms

silicon, has four valence electrons. Because of this similarity in atomic structure, many properties of the two materials are similar.

In Table 3.2 the elements B(5), Al(13), Ga(31) and In(49) are placed in one group (group III) because all these have *three* valence electrons. Similarly, the elements P(15), As(33) and Sb(51) are all in group V, as they have *five* valence electrons. The elements Si(14) and Ge(32) are in group IV, since they have *four* valence electrons.

3.3.2 Electron Energies

Each isolated atom has only a certain number of orbits available. These available orbits represent energy levels for the electrons. Modern physics tells us that only discrete values of electron energies are possible. An electron cannot have *any* value of energy (usually expressed in eV*), but only certain permissible values. No electron can exist at an energy level other than a permissible one. For a single atom, a diagram can be drawn showing the different energy levels available for its electrons. Fig. 3.4 is the energy-level diagram for the hydrogen atom. The permissible energy levels are numbered $n = 1, 2, \ldots$ in increasing order of energy.

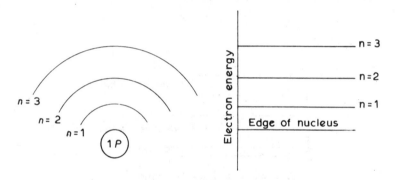

Fig. 3.4 Permissible energy levels in an isolated hydrogen atom

In any atom, the greater the distance of an electron from the nucleus, the greater is its total energy (the total energy includes kinetic and potential energies). An electron orbiting very close to the nucleus in the first shell is tightly bound to the nucleus and possesses only a small amount of energy. It would be difficult to knock out this electron. On the other hand, an electron orbiting far from the nucleus would have a greater energy, and hence it could easily be knocked out of its orbit. This is why it is the valence electrons (i.e. the electrons in the outermost orbit, having maximum energy) that take part in chemical reactions and in bonding the atoms together to form solids.

When energy like heat, light, or other radiations impinge on an atom, the energy of the electrons increases. As a result, they are lifted to higher energy levels (larger orbits). The atom is then said to be *excited*. This state

---

*eV is the abbreviation of *electron volt*; a unit of energy. It is defined as that energy which an electron acquires in moving through a potential difference of 1 V. This unit is commonly used in electronics and particle physics. In terms of *joules*, a more common unit of energy, an electron volt is equivalent to 1.6×10^{-19} J.

does not last long. Very soon, the electrons fall back to the original energy level. In this process, the electrons gives out energy in the form of heat, light, or other radiations.

3.3.3 Energy Bands in Solids

When atoms bond together to form a solid, the simple diagram of Fig. 3.4, for the electron energies, is no longer applicable. In a solid, the orbit of an electron is influenced not only by the charges in its own atom but by nuclei and electrons of every atom in the solid. Since each electron occupies a different position inside the solid, no two electrons can see exactly the same pattern of surrounding charges. As a result, the orbits of the electrons are different.

The simple energy-level diagram in Fig. 3.4 now modifies to that shown in Fig. 3.5. There are millions of electrons, belonging to the first orbits of atoms in the solid. Each of them has different energy. Since there are millions of first-orbit electrons, the closely spaced energy levels differing very slightly in energy, form a cluster or *band*. Similarly the second-orbit and higher-orbit electrons also form bands. We now have first energy band, second energy band, third energy band, etc.

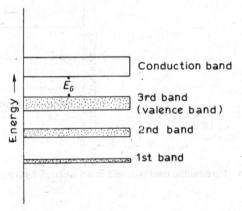

Fig. 3.5 Energy-band diagram of a solid
(silicon)

Silicon is a material commonly used in making transistors. Since, the atomic number of silicon is 14, and each of its atoms has only four electrons in the third (outermost) orbit, the third band becomes the *valence band*. Figure 3.5 then represents the energy-band diagram of silicon. An additional band, called *conduction band*, is also shown above the valence band. All the three lower bands, including the valence band, are shown completely filled. Although the third shell of an isolated atom of silicon is not completely filled (it has only four electrons whereas it could accommodate a maximum of eight electrons), the third energy band (valence band) of solid silicon is completely filled. It is so, because in solid silicon each atom positions itself between four other silicon atoms, and each of these neighbours share an electron with the central atom. In this way, each atom now has eight electrons filling the valence band completely (for details see Sec. 3.5.1). When we say that a band is filled, it means that all the permissible energy levels in the band are occupied by electrons. No electron in a filled band can move, because there is no place to move. Thus, *an electron in a completely filled band cannot contribute to electric current.*

The conduction band represents the next larger group of permissible energy levels. There is an energy gap, E_G, between the valence band and the conduction band. An electron can be lifted from the valence band to the conduction band by adding to silicon, some energy. This energy must be more than the energy gap E_G. If we add energy less than E_G, silicon will not accept it because no permissible energy level exists between the conduction band and valence band to which an electron can be lifted. For this reason, the gap between the valence band and the conduction band is called the *forbidden energy gap.* (For silicon, $E_G = 1.12$ eV, and for germanium it is 0.72 eV).

The orbits in the conduction band are very large. An electron in the conduction band experiences almost negligible nuclear attraction. In fact, an electron in the conduction band does not belong to any particular atom. But, it moves randomly throughout the solid. This is why the electrons in the conduction band are called *free electrons.*

3.4 METALS, INSULATORS AND SEMICONDUCTORS

A material is able to conduct electricity, if it contains movable charges in it. The free electrons (that is, the electrons that exist in the conduction band) move randomly inside a solid, and can carry charge from one point to another, when an external field is applied. The free electrons thus work as *charge carriers.*

A metal such as copper or silver contains a large number of free electrons at room temperature. In fact, there is no forbidden-energy gap between the valence and conduction bands. The two bands actually overlap as shown in Fig. 3.6a. The valence-band energies are the same as the conduction-band energies in the metal. It is very easy for a valence electron to become a conduction (free) electron. Therefore, without supplying any additional energy such as heat or light, a metal already contains a large number of free electrons and that is why it works as a good conductor.

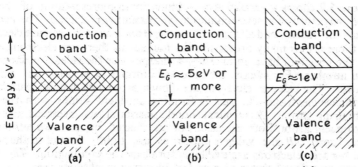

Fig. 3.6 Energy-band diagram for the three types of materials:
(a) Metals (conductors); (b) Insulators; (c) Semiconductors

An insulating material has an energy-band diagram as shown in Fig. 3.6b. It has a very wide forbidden-energy gap (5 eV or more). Because of this, it is practically impossible for an electron in the valence band to jump the gap, to reach the conduction band. Only at very high temperatures or under very stressed (electrically) conditions, can an electron jump the gap. At room temperature, an insulator does not conduct because there are no conduction electrons in it. However, it may conduct if its temperature is very high or if a high voltage is applied across it. This is termed as the *breakdown* of the insulator.

The energy-band diagram for a semiconductor is shown in Fig. 3.6c. In this case, the forbidden energy gap is not wide. It is of the order of 1 eV (for germanium, $E_G = 0.72$ eV; and for silicon $E_G = 1.12$ eV). The energy provided by the heat at room temperature is sufficient to lift electrons from the valence band to the conduction band. Some electrons do jump the gap and go into the conduction band. Therefore, at room temperature, semiconductors are capable of conducting some electric current.

3.5 INTRINSIC SEMICONDUCTORS

Semiconductor devices, such as diodes and transistors, are made from a single crystal of semiconductor material (germanium or silicon). To make a semiconductor device, the very first step is to obtain a sample of semiconductor in its purest form. Such a semiconductor (in pure form) is called an *intrinsic semiconductor*. A semiconductor is not truly intrinsic unless its impurity content is less than one part impurity in 100 million parts of semiconductor. To understand the phenomenon of conduction of current in a semiconductor, it is necessary to study its crystal structure.

3.5.1 Crystal Structure of Semiconductors

When atoms bond together to form molecules of matter, each atom attempts to acquire eight electrons in its outermost shell. If the outermost shell of an atom has eight electrons, it is said to be filled. It then becomes a stable structure. An intrinsic semiconductor (such as pure Ge or Si), has only four electrons in the outermost orbit of its atoms. To fill the valence shell, each atom requires four more electrons. This is done by *sharing* one electron from each of the four neighbouring atoms. The atoms align themselves in a uniform three-dimensional pattern so that each atom is surrounded by four atoms. Such a pattern is called a *crystal*.

Figure 3.7 shows a simplified two-dimensional representation of the crystalline structure of a semiconductor (germanium or silicon). The *core* represents the nucleus and all the orbiting electrons except the valence electrons. Since there are as many protons in the nucleus as there are electrons orbiting it, the core will have an excess +4 charge since the valence electrons are four in number. (For silicon, the core will contain 14 protons but only 10 electrons). The valence electrons are shown around each core. Each of the four valence electrons take part in forming covalent bonds with the four neighbouring atoms. A covalent bond consists of two electrons, one from each adjacent atom. Both the electrons are shared by the two atoms. At absolute zero, *all* the valence electrons are tightly bound to the parent atoms. No free electrons are available for electrical conduction. *The semiconductor therefore behaves as a perfect insulator at absolute zero.*

3.5.2 Charge Carriers in Intrinsic Semiconductors

We have seen that an intrinsic semiconductor behaves as an insulator at absolute zero, because all the electrons are bound to the atoms. Let us now see what happens at room temperature. Room temperature (say, 300 K) may be sufficient to make a valence electron of a semiconductor atom to

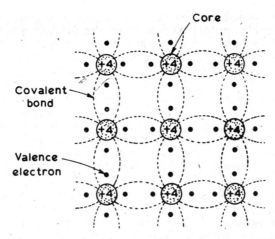

Fig. 3.7 Simplified representation of the crystalline structure of a semiconductor at absolute zero

move away from the influence of its nucleus. Thus, a covalent bond is broken. When this happens, the electron becomes free to move in the crystal. This is shown in Fig. 3.8a.

When an electron breaks a covalent bond and moves away, a vacancy is created in the broken covalent bond. This vacancy is called a *hole*. Whenever a free electron is generated, a hole is created simultaneously. That is, free electrons and holes are always generated in *pairs*. Therefore, the concentration of free electrons and holes will always be equal in an intrinsic semiconductor. This type of generation of free electron-hole pairs is referred to as *thermal generation*.

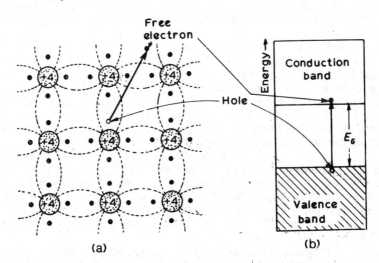

Fig. 3.8 Generation of electron-hole pair in an intrinsic semiconductor:
(a) Crystal structure; (b) Energy-band diagram

Let us examine whether a hole has any charge associated with it. The crystal is electrically neutral. As soon as an electron-hole pair is generated, the electron leaves the covalent bond and moves away from it. Since, an electron is negatively charged, the site of a hole will be left with a net positive charge (equal in magnitude to the charge of an electron). Thus, we say that a positive charge is associated with a hole, or *a hole is positively charged*. We shall see in the next section how a hole moves randomly in the same way as does a free electron. The hole too carries charge from one point to another. Although, strictly speaking, a hole is not a particle; for all practical purposes we can view it as a positively charged particle capable of conducting current. This concept of a hole as a positively charged particle merely helps in simplifying the explanation of current flow in semiconductors.

Figure 3.8a shows the generation of an electron-hole pair in a crystal. The amount of energy required to break a covalent bond is 0.72 eV in case of germanium and 1.12 eV in case of silicon. Equivalently, we say that the energy needed for lifting an electron from the valence band to the conduction band is 0.72 eV for germanium and 1.12 eV for silicon. When an electron jumps the forbidden gap, it leaves a hole in the valence band as shown in Fig. 3.8b.

Note that the value of E_G is more in case of silicon ($E_G = 1.12$ eV) than in case of germanium ($E_G = 0.72$ eV). Therefore, less number of electron-hole pairs will be generated in silicon than in germanium at room temperature. *The conductivity of silicon will be less than that of germanium at room temperature.*

3.5.3 Random Movement of Carriers

Both types of charge carriers move randomly or haphazardly in the crystal. The random movement of a free electron is easy to understand. A free electron moves in the crystal because of the thermal energy. Its path deviates whenever it collides with a nucleus (or other free electrons). This gives rise to a zig-zag or random motion similar to gas molecules moving in a gas container.

Let us see how the hole-movement takes place. A hole is generated whenever an electron breaks a covalent bond and becomes free. Consider that an electron-hole pair is generated at point A in Fig. 3.9a. The free electron goes elsewhere in the crystal leaving behind a hole at point A. The broken bond now has only one electron. This unpaired electron has a tendency to acquire an electron (whenever it can) and to complete the pair, forming the covalent bond. Due to thermal energy, an electron from the neighbouring bond may get sufficiently excited to break its own bond. It may then jump into the hole. In Fig. 3.9b, the valence electron at B breaks its bond and jumps into the hole at A. When this happens, the original hole at A vanishes and a new hole appears at B. The original hole has apparently moved from A to B, as shown in Fig. 3.9c. An instant later, the hole at B attracts and captures the valence electron from the neighbouring bond at C (see Fig. 3.9d). Apparently the hole has moved from B to C, as shown in Fig. 3.9e.

Figure 3.9f shows, by means of solid arrows, how the valence electrons successively jump from B to A and then from C to B. The net effect is as though the hole at A has moved through the crystal from A to C. This movement of holes is shown by dotted arrows in Fig. 3.9f. Thus, we find that the movement of a hole in a particular direction actually consists of a series of discontinuous electron movements in the opposite direction. It is for this reason that the holes appear to travel more slowly than the free electrons.

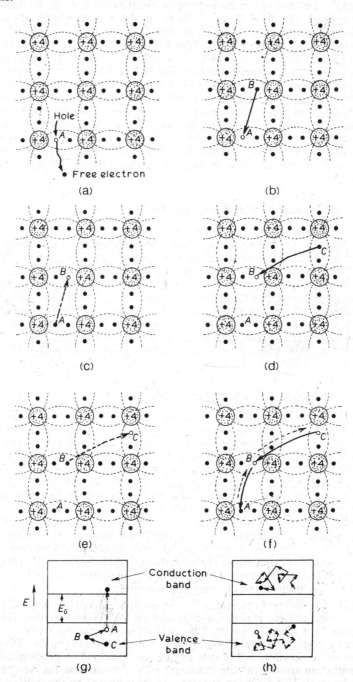

Fig. 3.9 Movement of a hole through a semiconductor crystal

Although the movement of holes actually consists of the movement of electrons, this movement of electrons is different from the movement of free electrons. The free electrons move in the conduction band, but the holes move because of the movement of electrons in the valence band. The movement of the hole from A to C in the crystal can be shown in the energy-band diagram as in Fig. 3.9g. An electron jumps from the valence band to the conduction band leaving behind a hole at A. The electron at B moves to hole at A. An instant later, another electron at C moves to point B. The effect is as though the hole has moved from A to C. Actually the holes move because of the jumpy movement of valence electrons from one position to the other. This jumpy movement of valence electrons need not be considered at all, since we are concerned about the net effect (i.e. the movement of holes). Therefore, in future discussions, whenever we talk of electron movement, it would imply the movement of free electrons and not of the valence electrons. Free electrons move randomly in the conduction band, whereas holes move randomly in the valence band, as shown in Fig. 3.9h.

3.5.4 Recombination of Electrons and Holes

In an intrinsic semiconductor, electrons and holes are produced continuously on account of thermal agitation. Since the electrons and holes move in the crystal in a random manner, there is a possibility of an electron meeting a hole. When it happens, both the electron and hole disappear because the electron occupies the position of a hole in a broken covalent bond. The covalent bond is again established. At any temperature, the rate of this recombination is equal to the rate of generation of electrons and holes. However, an electron (or a hole) travels some distance before it recombines. The average time an electron (or hole) remains free is called its *lifetime*. At any instant, both types of charge carriers are present in equal numbers at a given temperature.

3.5.5 Conduction in Intrinsic Semiconductors

Let us see what happens when we connect a battery across a semiconductor, as in Fig. 3.10. The electrons experience a force towards the positive terminal of the battery; and holes towards the negative terminal. The random motion of electrons and holes gets modified. Over and above the random motion, there also occurs a net movement, called *drift*. Since the random motion (of electrons or holes) does not contribute to any electric current, we need not consider it. The free electrons drift towards the positive terminal of the battery, and the holes towards the negative terminal. The electric current flows through the semiconductor in the same direction as in which the holes are moving (the holes have positive charge). Since the electrons are negatively charged, the direction of electric current (conventional) is opposite to the direction of their motion. Although, the two types of charge carriers move in opposite directions, the two currents are in the same direction, i.e. they add together.

When the flow of carriers is due to an applied voltage (as in Fig. 3.10), the resultant current is called a *drift current*. A second type of current may also exist in a semiconductor. This current is called *diffusion current*, and it flows as a result of a gradient of carrier concentration (i.e. the difference of carrier concentration from one region to another). A gradient of carrier concentration arises near the boundary of a *PN*-junction (as we shall see in next chapter). The diffusion current is also be due to the motion of both holes and electrons.

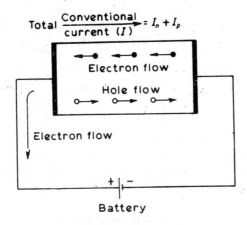

Fig. 3.10 Conduction of current in an intrinsic
semiconductor

3.5.6   Effect of Temperature on Conductivity of Intrinsic Semiconductors

A semiconductor (germanium or silicon) at absolute zero, behaves as a perfect insulator. At room temperature, because of thermal energy, some electron-hole pairs are generated. For example, in a sample of germanium at room temperature (300 K) the intrinsic carrier concentration (i.e. the concentration of free electrons or of holes) is 2.5×10^{19} per m^3. The semiconductor has a small conductivity. Now, if we raise the temperature further, more electron-hole pairs are generated. The higher the temperature, the higher is the concentration of charge carriers. As more charge carriers are made available, the conductivity of an intrinsic semiconductor increases with temperature. In other words, the resistivity (inverse of conductivity) decreases as the temperature increases. That is, *the semiconductors have negative temperature-coefficient of resistance.*

3.6   EXTRINSIC SEMICONDUCTORS

Intrinsic (pure) semiconductors are of little use (it may only be used as a heat or light-sensitive resistance). Practically all the semiconductor devices are made of a semiconductor material to which certain specified types of impurities have been added. The process of deliberately adding impurities to a semiconductor material is called *doping*. Doping is done after the semiconductor material has been refined to a high degree of purity. A doped semiconductor is called an *extrinsic semiconductor*.

3.6.1   *N*-Type Semiconductors

Let us consider what happens if a small amount of pentavalent impurity, for example, phosphorus is added to a sample of intrinsic silicon. The size of the impurity atoms is roughly the same as that of silicon. An impurity atom replaces a silicon atom in its crystalline structure. If the amount of impurity is very small (say, one part in one million), we can safely assume that each impurity atom is surrounded, all around, by silicon atoms. This is shown in Fig. 3.11, which represents a part of the crystal.

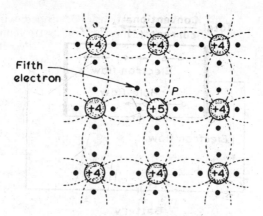

Fig. 3.11 N-type semiconductor

Let us now focus our attention on an impurity atom in the crystalline structure. Unlike a silicon atom, the phosphorus atom has five valence electrons. Four of these form covalent bonds with four neighbouring silicon atoms. The fifth electron has no chance of forming a covalent bond. It is this electron that is important to us. Since it is not associated with any covalent bond and is quite far from the nucleus, it is very loosely bound. It requires very little energy to free itself from the attractive force of its nucleus (this energy is only 0.01 eV in the case of germanium and 0.05 eV in the case of silicon). This energy is so small that at room temperature practically all such electrons become free. In other words, at room temperature each impurity atom *donates* one electron to the conduction band. That is the reason why this type of impurity is called *donor* type. These donated electrons are called *excess electrons*, since they are in excess to the electrons which are thermally generated (by breaking covalent bonds).

All the electrons which have been donated by the impurity atoms can take part in the conduction of electric current. Besides, there will also be some electron-hole pairs generated because of the breaking of covalent bonds. The number of thermally generated electron-hole pairs will be very small compared to the number of free electrons due to the impurity atoms. Further, as the number of electrons is very large, the chances of their recombination with holes also increases. Consequently, the net concentration of holes is much less than its intrinsic value. Thus, the number of free electrons becomes far greater than the number of holes. That is why we say that an N-type semiconductor has electrons (negatively charged) as *majority* carriers, and holes as *minority* carriers.

Now, let us see what happens to the core of the impurity atom, when the fifth electron leaves it. The core represents the atom without the valence electrons. Since there are five valence electrons in the impurity atom, a charge of +5 is shown in its core. When the fifth electron leaves the impurity atom, it then has +1 excess charge. It then becomes a positively charged *immobile ion*. It is immobile because it is held tightly in the crystal by the four covalent bonds.

Representation of N-Type Semiconductor In the designation "N-type semiconductor", the letter N stands for negative charges (electrons), because the electrons are the *major* charge carriers. But it does not mean that a sample of N-type semiconductor is negatively charged. It is important to note that whether a semiconductor is intrinsic or doped with impurity, **it remains electrically neutral**. Free electrons and holes are generated in **pairs** due to thermal energy. The negative charge of free electrons thus generated is exactly balanced by the positive charge of the holes. In an N-type semiconductor, there are additional free electrons created because of the addition of donor atoms. *The negative charge of these electrons is again balanced by the positive charge of the immobile ions.* (The total number of holes and immobile ions is exactly same as the total number of free electrons created.)

As we shall see in the chapters that follow, the N-type semiconductor (and also P-type semiconductor, which is explained in the next section) is used in the fabrication of diodes and transistors. To understand the mechanism of current flow through these devices, we should consider all type of charged particles in the semiconductor. In an N-type semiconductor, there are a large number of free electrons, a few holes, and a sufficiently large number of immobile positive ions. In this book, we shall be representing an electron by a black circle, a hole by a white circle, and an immobile positive donor ion by an encircled plus sign. Thus, we can represent an N-type semiconductor as shown in Fig. 3.12.

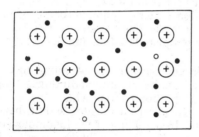

Legends :
- Free electron (negative charge)
- Hole (positive charge)
- Immobile ion (positive charge)

Fig. 3.12 Representation of an N-type semiconductor

Note that no silicon (or germanium) atoms are shown in this figure. They should be assumed as a continuous structure over the whole background. The fixed ions are regularly distributed in the crystal structure. But the holes and electrons, being free to move, are randomly distributed at moment.

3.6.2 P-Type Semiconductor

For making an N-type semiconductor, we add a pentavalent impurity to an intrinsic semiconductor. Instead, if we add a trivalent impurity (such as boron, aluminium, gallium and indium) to the intrinsic semiconductor, the result is a P-type semiconductor. As an example, let us consider a sample of intrinsic (pure) silicon to which a very small amount of boron is added. Since the impurity ratio is of the order of one part in one million, each impurity atom is surrounded by silicon atoms. The boron atom in the crystal has only three valence electrons. These electrons form covalent bonds with the three neighbouring silicon atoms (Fig. 3.13). The fourth neighbouring silicon atom is unable to form a covalent bond with the boron atom because the boron atom does not have the fourth electron in its valence orbit.

There is a deficiency of an electron around the boron atom. The single electron in the incomplete bond has a great tendency to snatch an electron

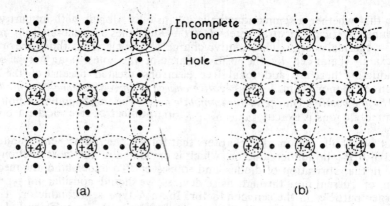

Fig. 3.13 *P*-type semiconductor: (*a*) Boron added to silicon
(*b*) Creation of a hole

from the neighbouring atom. This tendency is so great that an electron in an adjacent covalent bond, having very small additional energy, can jump to occupy the vacant position. This electron then completes the covalent bond around boron atom. The additional energy required for this is of the order of 0.01 eV. At room temperature, the thermal energy is sufficient to provide this energy so as to fill the incomplete bonds around all the boron atoms.

When an electron from the adjacent covalent bond jumps to fill the vacancy in the incomplete bond around the boron atom, two things happen. First, a vacancy is created in the adjacent bond from where the electron had jumped. This vacancy has a positive charge associated with it, hence it is a *hole* (see Fig. 3.13*b*). Second, due to the filling of the incomplete bond around boron, it now becomes a *negative ion*. It is immobile, since it is held tightly in the crystal structure by covalent bonds. The boron atom becomes negative ion by *accepting* one electron from the crystal. That is why this type of impurity is called *acceptor* type.

Besides the excess holes created due to the addition of acceptor-type impurity, there are some holes (and also equal number of free electrons) generated by breaking covalent bonds. Summarizing; a *P*-type material has holes (positively charged carriers) in *majority*, and free electrons in *minority*. In addition, there are also negative immobile ions.

Representation of P-Type Semiconductor Following the same convention as explained earlier, we can represent a sample of *P*-type semiconductor by a diagram (Fig. 3.14). The white circles represent holes, black circles the electrons, and encircled minus signs the immobile negative ions. The majority charge carriers in a *P*-type semiconductor are holes which are positively charged.

3.6.3 Effect of Temperature on Extrinsic Semiconductors

We have seen that addition of a small amount of donor or acceptor impurity produces a large number of charge carriers in an extrinsic semiconductor. In fact, this number is so large that the conductivity of an extrinsic semiconductor is many times that of an intrinsic semiconductor at room temperature.

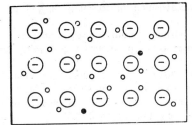

Legends:
o   Hole (positive charge)

•   Electron (negative charge)

⊖   Immobile ion
    (negative charge)

Fig. 3.14 Representation of a *P*-type semiconductor

Let us see what happens if we raise the temperature of an *N*-type semiconductor. Since all the donors have already donated their free electrons (at room temperature), the additional thermal energy only serves to increase the thermally generated carriers. As a result, the concentration of minority carriers increase. Eventually, a temperature is reached when the number of covalent bonds that are broken is very large, so that the number of holes is approximately the same as the number of electrons. The extrinsic semiconductor now behaves essentially like an intrinsic semiconductor (of course, with higher conductivity). This *critical temperature* is 85 °C for germanium and 200 °C for silicon.

This concludes our study of semiconductor physics. In the chapters that follow we will study some important semiconductor devices. Practically, all of these contain extrinsic semiconductors of both types in one crystal. The simplest combination, called a *PN*-junction, is the subject for the next chapter.

REVIEW QUESTIONS

3.1 Name at least two conductors. Give the order of their conductivities.

3.2 Name any two insulators and give the order of their conductivities.

3.3 Name two commonly used semiconductors. Give the order of conductivities of these materials.

3.4 When atoms share electrons, what type of bonding is it called ?

3.5 Explain why the discrete energy levels of an isolated atom split into a band of energy when atoms combine together to form a crystal.

3.6 Explain the difference in conductors, insulators and semiconductors using the energy-band diagrams.

3.7 What will happen to the number of electrons in the conduction band of a semiconductor as the temperature of the material is increased ?

3.8 Sketch the two dimensional crystal structure of intrinsic silicon at the absolute zero of temperature. Also, sketch its energy-band diagram. Sketch the same crystal structure at room temperature. Also sketch its energy-band diagram.

3.9 Explain the reason why the conductivity of germanium is more than that of silicon at room temperature.

3.10 Why is a valence electron at the top of the valence band more apt to thermal excitation than the one at the lower level in that band ?

3.11 Explain what a hole is ? How do they move in an intrinsic semiconductor ?

3.12 Explain why the temperature coefficient of resistance of a semiconductor is negative ?

3.13 What process is the opposite of thermal generation of electron-hole pairs ?

3.14 Explain, say within five lines, why the concentration of free electrons and holes is equal in an intrinsic semiconductor.

3.15 Explain what is the need of adding an impurity to an intrinsic semiconductor.

3.16 Which of the following atoms could be used as N-type impurities and which P-type impurities ?

(a) Phosphorous; (b) Antimony; (c) Boron; (d) Arsenic; (e) Aluminium; (f) Indium

3.17 Explain why a pentavalent impurity atom is known as donor-type impurity.

3.18 In an N-type semiconductor, does it take more energy to excite a valence electron thermally, or to liberate an electron from the impurity atom ? (5-7 lines).

3.19 What are the majority current carries in an N-type semiconductor ? Why should there be any holes in this material ?

3.20 Explain how holes are created in a P-type semiconductor.

3.21 Of what polarity are the impurity ions in N-type and P-type semiconductors ? Justify your answer in brief (within 8 lines).

3.22 Explain what happens to the concentration of the majority carriers when the temperature of an extrinsic semiconductor is increased.

3.23 Explain why at high temperatures, an extrinsic semiconductor behaves like an intrinsic semiconductor.

3.24 In a P-type semiconductor, can the electrons ever outnumber the holes ? Explain within 5-6 lines.

3.25 Explain, within 8 lines, why electrons are the majority carriers in an N-type semiconductor.

OBJECTIVE-TYPE QUESTIONS

1. Here are some incomplete statements. Four alternatives are provided below each. Tick the alternative that completes the statement correctly:

1. The conductivity of materials found in nature varies between extreme limits of, say, 10^{-18} S/m to 10^9 S/m. The probable value of conductivity of silicon is

    (a) 0.5×10^{-3} S/m
    (b) 1.0×10^2 S/m
    (c) 0.7×10^5 S/m
    (d) 1.8×10^{-12} S/m

2. A germanium atom contains

    (a) four protons
    (b) four valence electrons
    (c) only two electron orbits
    (d) five valence electrons

3. When atoms are held together by the sharing of valence electrons

    (a) they from a covalent bond
    (b) the valence electrons are free to move away from the atom
    (c) each atom becomes free to move
    (d) each shared electron leaves a hole

4. An electron in the conduction band

    (a) is bound to its parent atom
    (b) is located near the top of the crystal
    (c) has no charge
    (d) has a higher energy than an electron in the valence band

5. An intrinsic semiconductor at the absolute zero of temperature

    (a) behaves like an insulator
    (b) has a large number of holes
    (c) has a few holes and same number of electrons
    (d) behaves like a metallic conductor

6. When a voltage is applied to an intrinsic semiconductor which is at room temperature

    (a) electrons move to the positive terminal and holes move to the negative terminal
    (b) holes move to the positive terminal and electrons move to the negative terminal
    (c) both holes and electrons move to the positive terminal
    (d) both holes and electrons move to the negative terminal

7. When the temperature of an intrinsic semiconductor is increased

    (a) resistance of the semiconductor increases
    (b) heat energy decreases the atomic radius
    (c) holes are created in the conduction band
    (d) energy of the atoms is increased

8. The movement of a hole is brought about by

    (a) the vacancy being filled by a free electron
    (b) the vacancy being filled by a valence electron from a neighbouring atom
    (c) the movement of an atomic core
    (d) the atomic core changing from a $+4$ to a $+5$ charge

9. If a small amount of antimony is added to germanium

    (a) the resistance in increased
    (b) the germanium will be a P-type semiconductor
    (c) the antimony becomes an acceptor impurity
    (d) there will be more free electrons than holes in the semiconductor

10. Donor-type impurities

    (a) create excess holes
    (b) can be added to germanium, but not to silicon
    (c) must have only five valence electrons
    (d) must have only three valence electrons

11. The conduction band

    (a) is always located at the top of the crystal
    (b) is also called the forbidden energy gap
    (c) is a range of energies corresponding to the energies of the free electrons
    (d) is not an allowed energy band

12. The forbidden energy gap in semiconductors

    (a) lies just below the valence band
    (b) lies just above the conduction band
    (c) lies between the valence band and the conduction band
    (d) is the same as the valence band

13. In an N-type semiconductor, the concentration of minority carriers mainly depends upon

    (a) the doping technique
    (b) the number of donor atoms
    (c) the temperature of the material
    (d) the quality of the intrinsic material, Ge or Si

14. If the amount of impurity, either P-type or N-type, added to the intrinsic is controlled to 1 part in one million, the conductivity of the sample

    (a) increases by a factor of 10^6
    (b) increases by a factor of 10^3
    (c) decreases by a factor of 10^{-3}
    (d) is not affected at all

15. A semiconductor that is electrically neutral

    (a) has no majority carriers
    (b) has no free charges
    (c) has no minority carriers
    (d) has equal amounts of positive and negative charges

16. When a normal atom loses an electron, the atom

    (a) becomes a positive ion
    (b) becomes a negative ion
    (c) becomes electrically neutral
    (d) is then free to move about

17. Excess minority carriers are the minority carriers that

    (a) are thermally generated
    (b) are impurity generated
    (c) are in excess of the equilibrium number
    (d) are in excess of the number of majority carriers

18. Resistivity is a property of a semiconductor that depends on

    (a) the shape of the semiconductor
    (b) the atomic nature of the semiconductor
    (c) the shape and the atomic nature of the semiconductor
    (d) the length of the semiconductor

II. Read each of the following statements, and complete them by filling in the blanks with appropriate words:

1. The electrons in the outermost orbit are called _____ electrons.
2. The larger the orbit, the _____ is the energy of the electron.
3. The forces holding the silicon atoms together in a crystal are called _____ bonds.
4. The merging of a free electron and a hole is called _____.
5. A pure germanium crystal is an _____ semiconductor and a doped crystal is an _____ semiconductor.
6. To get excess electrons in an intrinsic semiconductor, we can add _____ atoms. These atoms have _____ valence electrons.
7. Free electrons are the _____ carriers in N-type semiconductors, and holes are the _____ carriers.

Ans. I. 1. a ; 2. b ; 3. a ; 4. d ; 5. a ; 6. a ; 7. d ; 8. b ; 9. d ; 10. c ; 11. c ; 12. c ; 13. c ; 14. b ; 15. d ; 16. a ; 17. c ; 18. b.

II. 1. valence ; 2. greater ; 3. covalent ; 4. recombination ; 5. intrinsic, extrinsic ;
6. pentavalent, five ; 7. majority, minority

Semiconductor Diode

OBJECTIVES: After completing this unit, you will be able to: ○ Explain how barrier potential is set up in a *PN*-junction diode. ○ State the approximate value of barrier potential in a germanium and a silicon diode. ○ Explain the meaning of space-charge region (depletion region), zener breakdown, avalanche breakdown, static resistance and dynamic resistance. ○ Explain the conduction property of a forward-biased and reverse-biased diode. ○ Draw the forward and reverse characteristics of germanium and silicon diodes. ○ Explain the difference between germanium diodes and silicon diodes on the basis of forward voltage drop and reverse saturation current. ○ Explain the effect of temperature on the reverse saturation current. ○ Calculate the static and the dynamic resistance of a diode from its *V-I* characteristics. ○ Draw the characteristics of an ideal diode. ○ Explain the need of rectifiers in electronics. ○ Draw the circuit diagram and explain the working of half-wave rectifier, centre-tapped full-wave rectifier, and bridge rectifier. ○ Compare the performance of half-wave rectifier and full-wave rectifier (both centre-tapped and bridge type). ○ Derive in case of half-wave and full-wave rectifiers, the expressions for output dc voltage, average or dc current, rms current, ripple factor, and rectification efficiency. ○ State the important ratings of a rectifying diode. ○ Explain the need of filters in dc power supply. ○ Explain with the help of suitable waveforms, the working of half-wave and full-wave rectifiers using shunt-capacitor, series-inductor and π-filter. ○ State typical applications of light-emitting diode (LED), varactor diode, and zener diode.

4.1 *PN*-JUNCTION

By themselves, *P*-type and *N*-type materials taken separately are of very limited use. If we join a piece of *P*-type material to a piece of *N*-type material *such that the crystal structure remains continuous at the boundary*, a *PN*-junction is formed. Such a *PN*-junction makes a very useful device. It is called a *semiconductor* (or *crystal*) *diode*.

A *PN*-junction cannot be made by simply pushing the two pieces together; this would not lead to a single crystal structure. Special fabrication techniques are needed to form a *PN*-junction. For the time being let us not bother how a *PN*-junction is formed.

A *PN*-junction itself is an important device. Furthermore, practically all semiconductor devices contain at least one *PN*-junction. For this reason it is very necessary to understand how a *PN*-junction behaves when connected in an electrical circuit. In this chapter we shall discuss the properties of a *PN*-junction. It will help us to understand even those devices which have more than one *PN*-junction.

4.2 JUNCTION THEORY

The most important characteristic of a *PN*-junction is its ability to conduct current *in one direction only*. In the other (reverse) direction it offers very high resistance. How this happens is explained in the sections that follow.

4.2.1 PN- Junction with no External Voltage

Figure 4.1 shows a PN-junction just immediately after it is formed. Note that it is a single crystal. Its left half is P-type and right half is N-type. The P region has holes and negatively charged impurity ions. The N region has free electrons and positively charged impurity ions. (For simplicity, minority charge carriers are not shown in the figure.) Holes and electrons are the mobile charges, but the ions are immobile. The sample as a whole is electrically neutral and so are the P region and N region considered separately. Therefore in the P region, the charge of moving holes equal the total charges on its free electrons and immobile ions. Similarly, in the N region, the negative charge of its majority carriers is compensated by the charge of its minority carriers and immobile ions.

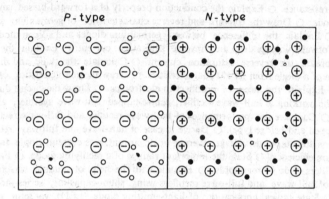

Fig. 4.1 . A PN-junction when just formed

Note that no external voltage has been connected to the PN-junction of Fig. 4.1. As soon as the PN-junction is formed, the following processes are initiated:

(i) Holes from the P region diffuse into the N region. They then combine with the free electrons in the N region.

(ii) Free electrons from the N region diffuse into the P region. These electrons combine with the holes.

(iii) The diffusion of holes (from P region to N region) and electrons (from N region to P region) takes place because they move haphazardly due to thermal energy, and also because there is a difference in their concentrations in the two regions. The P region has more holes and the N region has more free electrons.

(iv) One would normally expect the holes in the P region and free electrons in the N region to flow towards each other and combine. Thus, all the holes and free electrons would have been eliminated. But in practice this does not occur. The diffusion of holes and free electrons across the junction occurs for a very short time. After a few recombinations of holes and electrons in the immediate neighbourhood of the junction, a restraining force is set up automatically. This force is called a *barrier*. Further diffusion of holes and electrons from one side to the other is stopped by this barrier. How this barrier force is developed is explained in the paragraphs that follow.

(v) Some of the holes in the *P* region and some of the free electrons in *N* region diffuse towards each other and recombine. Each recombination eliminates a hole and a free electron. In this process, the negative acceptor ions in the *P* region and positive donor ions in the *N* region in the immediate neighbourhood of the junction are left uncompensated. This situation is shown in Fig. 4.2. Additional holes trying to diffuse into the *N* region are repelled by the uncompensated positive charge of the donor ions. The electrons trying to diffuse into the *P* region are repelled by the uncompensated negative charges on the acceptor ions. As a result, *total* recombination of holes and electrons cannot occur.

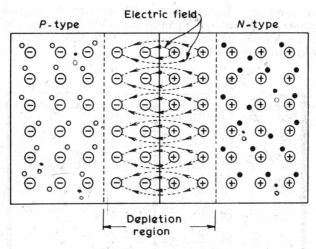

Fig. 4.2 Space-charge region or depletion region is formed in the vicinity of the junction

(vi) The region containing the uncompensated acceptor and donor ions is called *depletion region*. That is, there is a depletion of mobile charges (holes and free electrons) in this region. Since this region has immobile (fixed) ions which are electrically charged it is also referred to as the *space-charge region*. The electric field between the acceptor and the donor ions is called a *barrier*. The physical distance from one side of the barrier to the other is referred to as the *width* of the barrier. The difference of potential from one side of the barrier to the other side is referred to as the *height* of the barrier. With no *external* batteries connected, the barrier height is of the order of tenths of a volt. For a silicon *PN*-junction, the barrier potential is about 0.7 V, whereas for a germanium *PN*-junction it is approximately 0.3 V.

(vii) The barrier discourages the diffusion of majority carriers across the junction. But what happens to the minority carriers? There are a few free electrons in the *P* region and a few holes in the *N* region. The barrier helps these minority carriers to drift across the junction. The minority carriers are constantly generated due to thermal energy. Does it mean there would be a current due to the movement of these minority carriers? Certainly not. Electric current cannot flow since no circuit has been connected to the *PN*-junction. The drift of minority carriers across the junction is counterbalanced by the diffusion of the same number of majority carriers across the junction. These few majority carriers have sufficiently high kinetic energy* to overcome the barrier and cross the junction. In fact, the barrier height adjusts itself so that the flow of minority carriers is exactly balanced by the flow of majority carriers across the junction.

Thus, we conclude that a barrier voltage is developed across the *PN*-junction even if no external battery is connected.

4.2.2 *PN*-Junction with Forward Bias

Suppose we connect a battery to the *PN*-junction diode such that the positive terminal of the battery is connected to the *P*-side and the negative terminal to the *N*-side, as shown in Fig. 4.3. In this condition the *PN*-junction is said to be *forward-biased*.

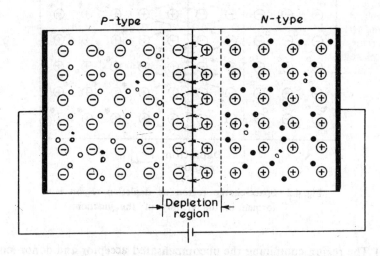

Fig. 4.3 *PN*-junction showing forward bias

When the *PN*-junction is forward-biased, the holes are repelled from the positive terminal of the battery and are compelled to move towards the junction. The electrons are repelled from the negative terminal of the battery and drift towards the junction. Because of their acquired energy, some of the holes and the free electrons penetrate the depletion region. This reduces the potential barrier. The width of the depletion region reduces and so does the barrier height. As a result of this, more majority carriers diffuse across the junction. These carriers recombine and cause movement of charge carriers in the space-charge region.

For each recombination of free electron and hole that occurs, an electron from the negative terminal of the battery enters the *N*-type material. It then drifts towards the junction. Similarly, in the *P*-type material near the

---

*In a semiconductor, all the charge carriers do not have same kinetic energy. Some have very high energy, whereas some have very low energy. The average energy depends upon the temperature of the sample.

positive terminal of the battery, an electron breaks a bond in the crystal and enters the positive terminal of the battery. For each electron that breaks its bond, a hole is created. This hole drifts towards the junction. Note that there is a continuous electron current in the external circuit. The current in the *P*-type material is due to the movement of holes. The current in the *N*-type material is due to the movement of electrons. The current continues as long as the battery is in the circuit. If the battery voltage is increased, the barrier potential is further reduced. More majority carriers diffuse across the junction. This results in an increased current through the *PN*-junction.

4.2.3 *PN*-Junction with Reverse Bias

Figure 4.4 shows what happens when a battery with the indicated polarity is connected to a *PN*-junction. Note that the negative terminal of the battery is connected to the *P*-type material and the positive terminal of the battery to the *N*-type material. The holes in the *P* region are attracted towards the negative terminal of the battery. The electrons in the *N* region are attracted to the positive terminal of the battery. Thus the majority carriers are drawn away from the junction. This action widens the depletion region and increases the barrier potential (compare this with the unbiased *PN*-junction of Fig. 4.1).

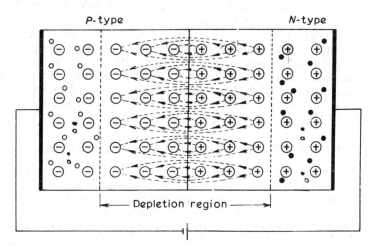

Fig. 4.4 *PN*-junction showing reverse bias

The increased barrier potential makes it more difficult for the majority carriers to diffuse across the junction. However, this barrier potential is helpful to the minority carriers in crossing the junction. In fact, as soon as a minority carrier is generated, it is swept (or drifted) across the junction because of the barrier potential. The rate of generation of minority carriers depends upon temperature. If the temperature is fixed, the rate of generation of minority carriers remains constant. Therefore, the current due to the flow of minority carriers remains the same whether the battery voltage is low or high. For this reason, this current is called *reverse saturation current*. This current is very small as the number of minority carriers is small. It is of the order of nanoamperes in silicon diodes and microamperes in germanium diodes.

There is another point to note. The reverse-biased PN-junction diode has a region of high resistivity (space-charge or depletion region) sandwiched in between two regions (P and N regions away from the junction) of relatively low resistivity. The P and N regions act as the plates of a capacitor, and the space-charge region acts as the dielectric. Thus, the PN-junction in reverse-bias has an effective capacitance, called *transition* or *depletion capacitance*.

Reverse Breakdown We have seen that a PN-junction allows a very small current to flow when it is reverse-biased. This current is due to the movement of minority carriers. It is almost independent of the voltage applied. However, if the reverse bias is made too high, the current through the PN-junction increases abruptly (see Fig. 4.5). The voltage at which this phenomenon occurs is called *breakdown voltage*. At this voltage, the crystal structure breaks down. In normal applications, this condition is avoided. The crystal structure will return to normal when the excess reverse bias is removed, provided that overheating has not permanently damaged the crystal.

There are two processes which can cause junction breakdown. One is called *zener breakdown* and the other is called *avalanche breakdown*. When reverse bias is increased, the electric field at the junction also increases. High electric field causes covalent bonds to break. Thus a large number of carriers are generated. This causes a large current to flow. This mechanism of breakdown is called *zener breakdown*.

In case of avalanche breakdown, the increased electric field causes increase in the velocities of minority carriers. These high energy carriers break covalent bonds, thereby generating more carriers. Again, these generated carriers are accelerated by the electric field. They break more covalent bonds during their travel. A chain reaction is thus established, creating a large number of carriers. This gives rise to a high reverse current. This mechanism of breakdown is called *avalanche breakdown*.

4.3 *V-I* CHARACTERISTICS OF A *PN*-JUNCTION DIODE

We would like to know how a device responds when it is connected in an electrical circuit. This information is obtained by means of a graph, known as its *V-I* characteristics, or simply characteristics. It is a graph between the voltage applied across its terminals and the current that flows through it. For a typical PN-junction diode, the characteristic is shown in Fig. 4.5. It tells us how much diode current flows for a particular value of diode voltage.

To obtain this graph, we set up a circuit in the laboratory. This circuit is shown in Fig. 4.6a. Note that in this circuit, the PN-junction is represented by its schematic *symbol*. The details of the diode symbol appear in Fig. 4.6b. The P region of the diode is called the *anode*, and the N region the *cathode*. The symbol looks like an arrow pointing from the P region to the N region. It serves as a reminder to us that the *conventional* current flows easily from the P region to the N region of the diode.

In the circuit (Fig. 4.6a), the dc battery V_{AA} is connected to the diode through the potentiometer P. Note that the dc battery is pushing the conventional current in the same direction as the diode arrow. Hence, the diode is *forward-biased*. Since current flows easily through a forward-biased diode, a resistance R is included in the circuit so as to limit the current. If excessive current is permitted to flow through the diode, it may

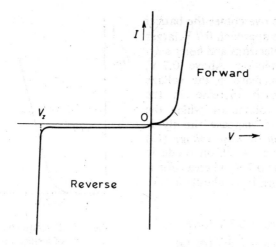

Fig. 4.5   *V-I* characteristic of a *PN*-junction diode

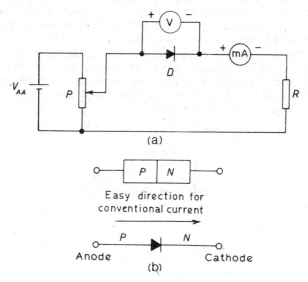

Fig. 4.6   (*a*) Circuit used to obtain the *V-I* characteristic of
a diode for forward bias;   (*b*) Symbol of the diode

get permanently damaged. The potentiometer helps in varying the voltage applied to the diode. The milliammeter measures the current in the circuit. The voltmeter measures the voltage across the diode.

Figure 4.7 shows the magnified view of a silicon-diode characteristic when the diode is forward-biased. Note that the voltage is plotted along the horizontal axis, as voltage is the independent variable. Each value of the diode voltage produces a particular current. The current, being the dependent variable, is plotted along the vertical axis.

From the curve of Fig. 4.7, we find that the diode current is very small for the first few tenths of a volt. The diode does not conduct well until the external voltage overcomes the barrier potential. As we approach 0.7 V, larger number of free electrons and holes start crossing the junction. Above 0.7 V, even a small increase in, the voltage produces a sharp increase in the current. The voltage at which the current starts to increase rapidly is called the *cut-in* or *knee voltage* (V_0) of the diode. For a silicon diode it is approximately 0.7 V, whereas for a germanium diode it is about 0.3 V. That is

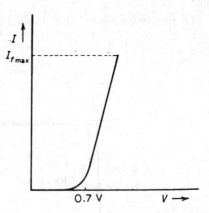

$$V_0 = 0.7 \text{ V for Si}$$
$$V_0 = 0.3 \text{ V for Ge}$$

Fig. 4.7 Forward characteristics of a silicon diode

If too large a current passes through the diode, excessive heat will destroy it. For this reason, the manufacturer's data sheet specifies the maximum current $I_{F\,max}$ that a diode can safely handle. For instance, the silicon junction-diode BY126 has a maximum current rating of 1 A.

To obtain the reverse-bias characteristics, we use the same circuit as in Fig. 4.6a, except for a few changes. First, we reverse the terminals of the diode. Second, the milliammeter is replaced by a microammeter. The resulting circuit is as shown in Fig. 4.8a. The magnified view of the reverse characterstics of the diode is shown in Fig. 4.8b.

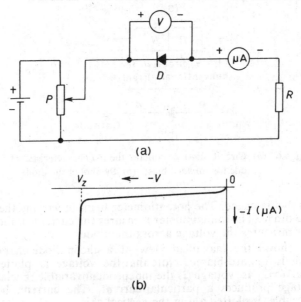

(a)

(b)

Fig. 4.8 (a) Circuit to plot reverse-bias characteristics of a diode; (b) Reverse-bias characteristics

In the reverse-bias, the diode current is very small—only few μA for germanium diodes and only a few nA for silicon diodes. It remains small and almost constant for all voltages less than the breakdown voltage V_z. At breakdown, the current increases rapidly for small increase in voltage. (See Sec. 4.2.3.)

4.4 THE IDEAL DIODE

We have seen that a diode has a very important property. It permits only *unidirectional conduction*. It conducts well in the forward direction and poorly in the reverse direction. It would have been ideal if a diode acted as a perfect conductor (with zero voltage across it) when forward-biased, and as a perfect insulator (with no current through it) when reverse-biased. The *V-I* characteristics of such an *ideal diode* would be as shown in Fig. 4.9a. An ideal diode acts like an *automatic switch*. When the current tries to flow in the forward direction, the switch is *closed*. On the other hand, when the current tries to flow the other way (against the direction of the diode arrow) the switch is *open*.

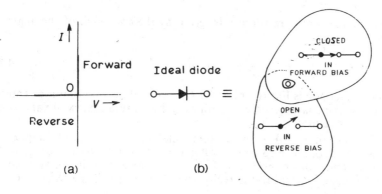

Fig. 4.9   (*a*) Ideal-diode characteristics; (*b*) Switch analogy

4.5 STATIC AND DYNAMIC RESISTANCE OF A DIODE

No diode can act as an ideal diode. An actual diode does not behave as a perfect conductor when forward-biased, and as a perfect insulator when reverse-biased. It does not offer zero resistance when forward-biased. Also its reverse-resistance, through very large, is not infinite.

Figure 4.10 shows the forward characteristics of a typical silicon diode. This diode may be connected in a dc circuit. When forward biased, it offers a definite resistance in the circuit. This resistance is known as the dc or *static resistance* (R_F) of the diode. It is simply the ratio of the dc voltage across the diode to the dc current flowing through it. For instance, if the dc voltage across the diode is 0.7 V, the current through it can be found from Fig. 4.10. The operating point of the diode is at point P, and the corresponding current can be read as 14 mA. The static resistance of the diode at this operating point will be given as

$$R_F = \frac{OA}{AP} = \frac{0.7\,\text{V}}{14\,\text{mA}} = 50\,\Omega$$

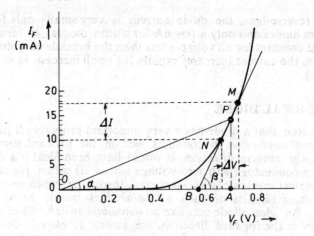

Fig. 4.10 Calculation of static and dynamic resistance
of a diode

In general, the static resistance is given by the cotangent of the angle α. That is

$$R_F = \frac{OA}{AP} = \cot \alpha \qquad (4.1)$$

If the characteristic is linear, this ratio OA/AP will be a constant quantity. But, in case the characteristic is nonlinear, the dc resistance will vary with the point of measurement.

In addition to 14 mA of dc current, small ac current may be superimposed in the circuit. The resistance offered by the diode to this ac signal is called its *dynamic* or *ac resistance*. The ac resistance of a diode, at a particular dc voltage, is equal to the reciprocal of the slope of the characteristic at that point, i.e.

$$r_f = \frac{\text{change in voltage}}{\text{resulting change in current}} = \frac{\Delta V}{\Delta I}$$

[*Note*: The Greek letter Δ (delta) means "a change of", wherever it appears in formulae. So, ΔI is a change in current. Generally, it indicates a small-scale (or incremental) change.]

We can calculate the ac resistance of a diode as follows:

Around the operating point P, take two points M and N very near to it, as shown in Fig. 4.10. These two points will then indicate incremental changes in voltage and current. The dynamic resistance is related to the slope of the line MN and is calculated as follows.

$$r_f = \frac{\Delta V}{\Delta I} = \frac{(0.73 - 0.66) \text{ V}}{(17.5 - 10) \text{ mA}} = \frac{0.07 \text{ V}}{7.5 \text{ mA}} = 9.46 \ \Omega$$

The smaller the incremental changes ΔV and ΔI, the closer is the above result to the exact value of the dynamic resistance. For making these incremental values smaller, the points M and N have to be closer. It then becomes difficult to read the voltage and current values accurately from the graph. We can circumvent this difficulty if we remember that as ΔV becomes smaller and smaller, the slope of the line MN becomes the same as that of the tangent to the curve at point *P*. In this alternative approach, we first draw a tangent to the curve at point *P*. This tangent meets the *x*-axis at point *B* (see Fig. 4.10). The dynamic resistance of the diode is then given as

$$r_f = \frac{BA}{AP} = \cot \beta \qquad (4.2)$$

From the graph, we can calculate the dynamic resistance as

$$r_f = \frac{BA}{AP} = \frac{(0.7 - 0.57)\ \text{V}}{14\ \text{mA}} = \frac{0.13\ \text{V}}{14\ \text{mA}} = 9.3\ \Omega$$

This may be seen to be almost the same as the value obtained earlier.

Now look at the reverse characteristic of the *PN*-junction diode (Fig. 4.8*b*). We find that even for a large reverse voltage (but below breakdown) the current is very small. The reverse current may be 1 μA at a voltage of 5 V. Then the static resistance of the diode is

$$R_R = \frac{5\ \text{V}}{1\ \mu\text{A}} = 5\ \text{M}\Omega$$

This is sufficiently high. It is much higher than the forward resistance R_F. Since the diode curve in the reverse bias is almost horizontal, its dynamic resistance r_r will be extremely high in this region of operation.

4.6 USE OF DIODES IN RECTIFIERS

Electric energy is available in homes and industries in India, in the form of alternating voltage. The supply has a voltage of 220 V (rms) at a frequency of 50 Hz. In the USA, it is 110 V at 60 Hz. For the operation of most of the devices in electronic equipments, a dc voltage is needed. For instance, a transistor radio requires a dc supply for its operation. Usually, this supply is provided by dry cells. But sometime we use a battery eliminator in place of dry cells. The battery eliminator converts the ac voltage into dc voltage and thus *eliminates* the need for dry cells. Nowadays, almost all electronic equipments include a circuit that converts ac voltage of mains supply into dc voltage. This part of the equipment is called *power supply*. In general, at the input of the power supply, there is a power transformer. It is followed by a diode circuit called *rectifier*. The output of the rectifier goes to a *smoothing filter*, and then to a *voltage* regulator circuit. A block diagram of such a power supply is shown in Fig. 4.11. The rectifier circuit is the heart of a power supply.

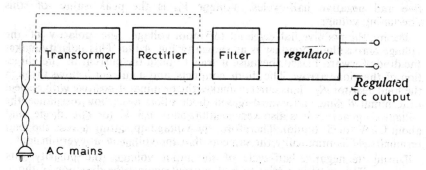

Fig. 4.11 Block diagram of a power supply

4.6.1   Half-Wave Rectifier

The unidirectional conducting property of a diode finds great application in rectifiers. These are the circuits which convert an ac voltage into dc voltage.

Figure 4.12 shows the circuit of a half-wave rectifier. Most electronic equipments have a transformer at the input. The transformer serves two purposes. Firstly, it allows us to step the voltage up or down. This way we can get the desired level of dc voltage. For example, the battery eliminator used with a transistor radio gives a dc voltage of about 6 V. We can use a step down transformer to get such a low ac voltage at the input of the rectifier. On the other hand, the cathode-ray tube used in an oscilloscope needs a very high dc voltage—of the order of a few kV. Here, we may use a step-up transformer. The second advantage of the transformer is the isolation it provides from the power line. It reduces the risk of electrical shock.

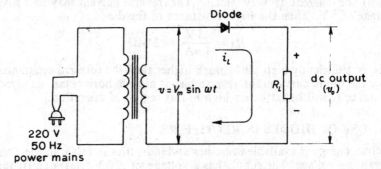

Fig. 4.12 Half-wave rectifier circuit

In Fig. 4.12, the diode forms a series circuit with the secondary of the transformer and the load resistor R_L. Let us see how this circuit rectifies ac into dc.

The primary of the transformer is connected to the power mains. An ac voltage is induced across the secondary of the transformer. This voltage may be less than, or equal to, or greater than the primary voltage depending upon the turns ratio of the transformer. We can represent the voltage across the secondary by the equation

$$v = V_m \sin \omega t \tag{4.3}$$

Figure 4.13a shows how this voltage varies with time. It has alternate positive and negative half-cycles. Voltage V_m is the peak value of this alternating voltage.

During the positive half-cycle of the input voltage, the polarity of the voltage across the secondary is as shown in Fig. 4.14a. This polarity makes the diode forward biased, because it tries to push the current in the direction of the diode arrow. The diode conducts, and a current i_L flows through the load resistor R_L. This current makes the terminal A positive with respect to terminal B. Since a forward-biased diode offers a very low resistance, the voltage drop across it is also very small (about 0.3 V for Ge diode and about 0.7 V for Si diode). Therefore, the voltage appearing across the load terminals AB is practically the same as that the voltage v_i at every instant.

During the negative half-cycle of the input voltage, the polarity gets reversed. The voltage v tries to send current against the direction of diode

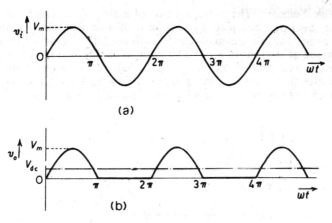

Fig. 4.13 Half-wave rectifier: (*a*) Input voltage waveform
(*b*) Output voltage waveform

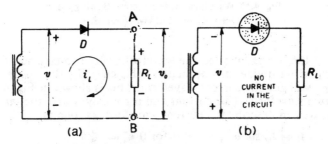

Fig. 4.14 Half-wave rectifier circuit: (*a*) During positive
half-cycle; (*b*) During negative half-cycle

arrow. See Fig. 4.14*b*. The diode is now reverse biased. It is shown shaded
in the figure to indicate that it is non-conducting. Practically no current
flows through the circuit. Therefore, almost no voltage is developed across
the load resistance. All the input voltage appears across the diode itself.
This explains how we obtain the output waveshape as shown in Fig. 4.13*b*.

To sum up, when the input voltage is going through its positive half-
cycle, the voltage of the output is almost the same as the input voltage.
During the negative half-cycle, no voltage is available across the load. The
complete waveform of the voltage v_o across the load is shown in Fig. 4.13*b*.
This voltage, though not a perfect dc, is at least *unidirectional*.

Peak Inverse Voltage Let us again focus our attention on the diode in
Fig. 4.14*b*. During the negative half-cycle of the input, the diode is
reverse biased. The whole of the input voltage appears across the diode (as
there is no voltage across the load resistance). When the input reaches its
peak value V_m, in the negative half-cycle, the voltage across the diode is
also maximum. This maximum voltage is known as the *peak inverse voltage*
(PIV). It represents the maximum voltage the doide must withstand
during the negative half-cycle of the input. Thus, for a half-wave rectifier,

$$\text{PIV} = V_m \qquad (4.4)$$

Output dc Voltage The average value of a sine wave (such as that in Fig. 4.13a) over one complete cycle is zero. If a dc ammeter (moving coil type) is connected in an ac circuit, it will read zero. (The dc meter reads average value of current in a circuit.) Now, if the dc ammeter is connected in the half-wave rectifier circuit (Fig. 4.12), it will show some reading. This indicates that there is some dc current flowing through the load R_L. We can find out the value of this current in a half-wave rectifier circuit.

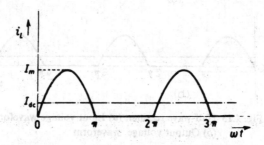

Fig. 4.15 Waveform of the current flowing through
load R_L in a half-wave rectifier

In Fig. 4.13b, we had plotted the waveform of the voltage across the load resistor R_L. If we divide each ordinate of this curve by the value of resistance R_L, we get the current waveform. This is shown in Fig. 4.15. Note that the two waveforms (for current and for voltage) are similar. Mathematically, we can describe the current waveform as follows:

$$i_L = I_m \sin \omega t \; ; \qquad \text{for } 0 \leqslant \omega t \leqslant \pi \qquad (4.5)$$

and $\qquad i_L = 0 \qquad ; \qquad \text{for } \pi \leqslant \omega t \leqslant 2\pi \qquad (4.6)$

Here, I_m is the peak value of the current i_L. It is obviously related to the peak value of voltage V_m as

$$I_m = \frac{V_m}{R_L} \qquad (4.7)$$

since the diode resistance in the conducting state is assumed to be zero. To find the dc or average value of current, we find the net area under the curve in Fig. 4.15 over one complete cycle, i.e. from 0 to 2π (curve repeats itself after the first cycle), and then divide this area by the base, i.e. 2π. We first integrate and then use Eqs. 4.5 and 4.6 to find the area.

$$\text{Area} = \int_0^{2\pi} i_L \, d(\omega t)$$

$$= \int_0^{\pi} I_m \sin \omega t \, d(\omega t) + \int_{\pi}^{2\pi} 0 \, d(\omega t)$$

$$= I_m \left[- \cos \omega t \right]_0^{\pi} + 0$$

$$= I_m \left[- \cos \pi - (- \cos 0) \right]$$

$$= 2I_m$$

Average value of the load current is then

$$I_{avg} = I_{dc} = \frac{area}{base} = \frac{2I_m}{2\pi}$$

or

$$I_{dc} = \frac{I_m}{\pi} \tag{4.8}$$

The dc voltage developed across the load R_L is

$$V_{dc} = I_{dc} \times R_L = \frac{I_m}{\pi} \times R_L \tag{4.9}$$

While writing Eq. 4.7, we had assumed that

(i) the diode resistance in forward bias is zero, and
(ii) the secondary winding of transformer has zero resistance.

The second assumption is often very near the truth. The winding resistance is almost zero. But, the forward diode-resistance r_d is sometimes not so small. If it is comparable to the load resistance R_L, we must take it into consideration. Equation 4.7 for peak current then gets modified to

$$I_m = \frac{V_m}{(R_L + r_d)} \tag{4.10}$$

The dc voltage across the load resistor R_L can now be written with the help of Eq. 4.9 as

$$V_{dc} = \frac{V_m R_L}{\pi(R_L + r_d)} = \frac{V_m}{\pi(1 + r_d/R_L)}$$

$$\cong \frac{V_m}{\pi} \ (if \ r_d \ll R_L) \tag{4.11}$$

Example 4.1 The turns ratio of a transformer used in a half-wave rectifier (such as in Fig. 4.12) is $n_1 : n_2 = 12 : 1$. The primary is connected to the power mains : 220 V, 50 Hz. Assuming the diode resistance in forward bias to be zero, calculate the dc voltage across the load. What is the PIV of the diode ?

Solution : The maximum (peak value) primary voltage is

$$V_p = \sqrt{2} \ V_{rms} = \sqrt{2} \times 220 = 311 \ V$$

Therefore, the maximum secondary voltage is

$$V_m = \frac{n_2}{n_1} V_p = \frac{1}{12} \times 311 = 25.9 \ V$$

The dc load voltage is

$$V_{dc} = \frac{V_m}{\pi} = \frac{25.9}{\pi} = 8.24 \ V$$

The peak inverse voltage is

$$PIV = V_m = 25.9 \ V$$

4.6.2 Full-Wave Rectifier

In a half-wave rectifier, discussed above, we utilize only one half-cycle of the input wave. In a full-wave rectifier we utilize both the half cycles. Alternate half cycles are inverted to give a unidirectional load current. There are two types of rectifier circuits that are in use. One is called centre-tap rectifier and uses two diodes. The other is called bridge rectifier and uses four diodes.

Centre-Tap Rectifier The circuit of a centre-tap rectifier is shown in Fig. 4.16a. It uses two diodes $D1$ and $D2$. During the positive half-cycles of secondary voltage, the diode $D1$ is forward biased and $D2$ is reverse biased. The current flows through the diode $D1$, load resistor R_L and the upper half of the winding, as shown in Fig. 4.16b. During negative half-cycles diode $D2$ becomes forward biased and $D1$ reverse biased. Now $D2$ conducts and $D1$ becomes open. The current flows through diode $D2$, load resistor R_L and the lower half of the winding, as shown in Fig. 4.16c. Note that the load current in both Figs. 4.16b and c is in *the same direction*. The waveform of the current i_L, and hence of the load voltage v_o, is shown in Fig. 4.16d.

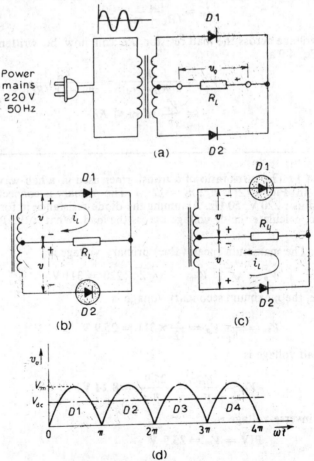

Fig. 4.16 Centre-tap full-wave rectifier

Peak Inverse Voltage Figure 4.17 shows the centre-tap rectifier circuit at the instant the secondary voltage reaches its positive maximum value. The voltage V_m is the maximum (peak) voltage across half of the secondary winding. At this instant, the diode $D1$ is conducting and it offers almost zero resistance. The whole of the voltage V_m across the upper half winding appears across the load resistor R_L. Therefore, the reverse voltage that appears across the nonconducting diode is the summation of the voltage across the lower half winding and the voltage across the load resistor R_L. From the figure, this voltage is $V_m + V_m = 2V_m$. Thus,

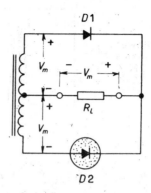

Fig. 4.17 The PIV across the non-conducting diode $D2$ in a centre-tap rectifier is $2V_m$

$$PIV = 2V_m \qquad (4.12)$$

Bridge Rectifier A more widely used full-wave rectifier circuit is the bridge rectifier, shown in Fig. 4.18. It requires four diodes instead of two, but avoids the need for a centre-tapped transformer. During the positive half-cycle of the secondary voltage, diodes $D2$ and $D4$ are conducting and diodes $D1$ and $D3$ are nonconducting. Therefore, current flows through the secondary winding, diode $D2$, load resistor R_L and diode $D4$, as shown in Fig. 4.18b. During negative half-cycles of the secondary voltage, diodes $D1$ and $D3$ conduct, and the diodes $D2$ and $D4$ do not conduct. The current therefore flows through the secondary winding, diode $D1$, load resistor R_L and diode $D3$, as shown in Fig. 4.18c. In both cases, the current passes through the load resistor *in the same direction*. Therefore, a fluctuating, unidirectional voltage is developed across the load. The load voltage waveform is shown in Fig. 4.18d.

Peak Inverse Voltage Let us now find the peak inverse voltage that appears across a nonconducting diode in a bridge rectifier. Figure 4.19 shows the bridge rectifier circuit at the instant the secondary voltage reaches its positive peak value, V_m. The diodes $D2$ and $D4$ are conducting, whereas diodes $D1$ and $D3$ are reverse biased and are nonconducting. The conducting diodes $D2$ and $D4$ have almost zero resistance (and hence zero voltage drops across them). Point B is at the same potential as the point A. Similarly, point D is at the same potential as the point C. The entire voltage V_m across the secondary winding appears across the load resistor R_L. The reverse voltage across the nonconducting diode $D1$ (or $D3$) is also V_m. Thus,

$$PIV = V_m \qquad (4.13)$$

Output dc Voltage in Various Rectifiers The voltage waveform in Fig. 4.18d is exactly the same as that in Fig. 4.16d. In both the rectifier circuits, the load voltage is the same. However, there is one difference. In the bridge rectifier, V_m is the maximum voltage across the secondary winding. But in the centre-tap rectifier, V_m represents the maximum voltage across half the secondary winding.

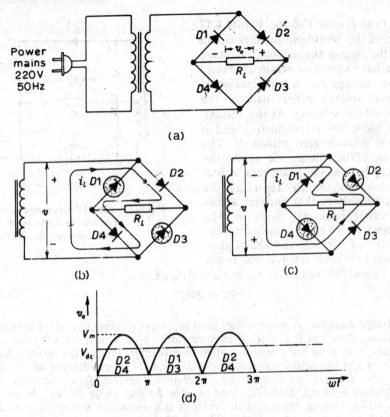

(a)

(b)   (c)

(d)

Fig. 4.18 Bridge rectifier

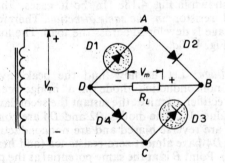

Fig. 4.19 The PIV across the nonconducting
diode D1 or D3 is V_m

Now let us compare the full-wave rectified voltage waveform (of Fig. 4.18d or Fig. 4.16d) with the half-wave rectified voltage waveform (of Fig. 4.13b). In a half-wave rectifier, only positive half-cycles are utilized for the dc output. But a full-wave rectifier utilizes both the half-cycles. Therefore, the dc or average voltage available in a full-wave rectifier will be double the dc voltage available in a half-wave rectifier. If the resistance of a forward biased diode is assumed zero, the dc voltage of a full-wave rectifier (refer Eq. 4.11) is

$$V_{dc} = \frac{2V_m}{\pi} \qquad (4.14)$$

We can mathematically derive Eq. 4.14. The output voltage of a full-wave rectifier (see Fig. 4.18*b*) is described as

$$v_0 = V_m \sin \omega t \qquad 0 \leqslant \omega t \leqslant \pi$$
$$= - V_m \sin \omega t \qquad \pi \leqslant \omega t \leqslant 2\pi$$

A minus sign appears in the second equation because during the second half-cycle the wave is still sinusoidal, but inverted. The average or the dc value of voltage is

$$V_{dc} = \frac{1}{2\pi} \int_0^{2\pi} v_0 \, d(\omega t)$$

$$= \frac{1}{2\pi} \left[\int_0^{\pi} (V_m \sin \omega t) \, d(\omega t) + \int_{\pi}^{2\pi} (- V_m \sin \omega t) \, d(\omega t) \right]$$

$$= \frac{1}{2\pi} \left[\left| - V_m \cos \omega t \right|_0^{\pi} + \left| V_m \cos \omega t \right|_{\pi}^{2\pi} \right]$$

$$= \frac{V_m}{2\pi} [- \cos \pi + \cos 0 + \cos 2\pi - \cos \pi]$$

$$= \frac{2V_m}{\pi}.$$

This is same as Eq. 4.14.

Why Bridge Rectifier Circuits are Preferred As mentioned earlier, the bridge rectifier is the most widely used full-wave rectifier. It has many advantages over a centre-tap rectifier. It does not require centre-tapped secondary winding. (If stepping up or stepping down of voltage is not needed, we may even do away with the transformer.) The peak inverse voltage of each diode is equal to the peak secondary voltage V_m, whereas the PIV of the non-conducting diode in a centre-tap rectifier is $2V_m$. This fact is of vital importance when higher dc voltages are required.

Suppose we need a certain dc output voltage (say, $2V_m/\pi$) from a full-wave rectifier. If it is a bridge rectifier, the transformer secondary voltage need have a peak value of only V_m. But if it is a centre-tap rectifier, the secondary must have $2V_m$ as its peak voltage. This is twice the value needed for a bridge rectifier. It means that for a centre-tap rectifier, the transformer secondary must have double the number of turns. Such a transformer is costlier. Furthermore, each of the two diodes in a centre-tap rectifier must have a PIV rating of $2V_m$. But the diode in a bridge rectifier is required to have PIV rating of only V_m. Hence, the diodes meant for use in a centre-tap rectifier are costlier than those meant for a bridge rectifier

The main disadvantage of a bridge rectifier is that it requires four diodes, two of which conduct on alternate half cycles. This creates a problem when low dc voltages are required. The secondary voltage is low and the two diode-voltage drops (1.4 V, in case of Si diodes) become significant. These diode-voltage drops may be compensated by selecting a transformer with slightly higher secondary voltage. But then the voltage regulation becomes poor. For this reason, in low-voltage applications we prefer the centre-tap rectifier which has only one diode drop (= 0.7 V). By using germanium diodes instead of silicon, the diode drop may further be reduced to 0.3 V.

Example 4.2 The turns ratio of the transformer used in a bridge rectifier is $n_1 : n_2 = 12 : 1$. The primary is connected to 220 V, 50 Hz power mains. Assuming that the diode voltage drops to be zero, find the dc voltage accross the load. What is the PIV of each diode? If the same dc voltage is obtained by using a centre-tap rectifier, what is the PIV?

Solution: The maximum primary voltage is

$$V_p = \sqrt{2} \, V_{rms} = \sqrt{2} \times 220 = 311 \text{ V}$$

Therefore, the maximum secondary voltage is

$$V_m = \frac{n_2}{n_1} V_p = \frac{1}{12} \times 311 = 25.9 \text{ V}$$

The dc voltage accross the load is

$$V_{dc} = \frac{2V_m}{\pi} = \frac{2 \times 25.9}{\pi} = \textbf{16.48 V}$$

The PIV (for bridge rectifier) is

$$\text{PIV} = V_m = \textbf{25.9 V}$$

For the centre-tap rectifier, the PIV is

$$\text{PIV} = 2V_m = 2 \times 25.9 = \textbf{51.8 V}$$

4.7 HOW EFFECTIVELY A RECTIFIER CONVERTS AC INTO DC

If we connect a load resistor R_L directly accross an ac power mains, the current flowing through it will be purely ac (sinusoidal having zero average value). This current is shown in Fig. 4:20a. In some applications, we require a dc current to flow through the load. The dc current* is unidirectional and, ideally, has no fluctuations with time. The ideal dc current is shown in Fig. 4.20b. To see how effectively a rectifier converts ac into dc, we compare its output current waveshape with the ideal dc current.

If the load takes current from a half-wave rectifier, the current waveform will be as in Fig. 4.20c. It is unidirectional, *but fluctuates greatly with time*. The waveform of the load current, when the load is connected to a full-wave rectifier, is shown in Fig. 4.20d. This too is unidirectional and *fluctuates with time*. A unidirectional, fluctuating waveform may be considered as consisting of a number of components. It has an average or dc

---

*The terms ac and dc were originally used as the abbreviations of *alternating current* and *direct current* respectively. Therefore, it may seem odd from the language point of view to use terms "an ac current, a dc current, a dc voltage, etc." However, the adjectives ac and dc have now been adopted for referring to any quantity whose variation with time is of "alternating" and "direct" type, respectively.

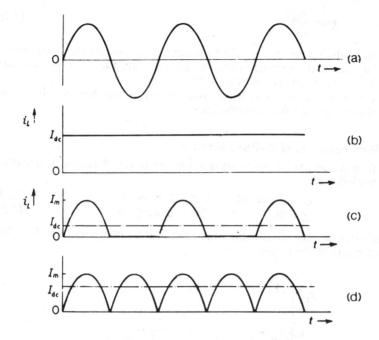

Fig. 4.20 Comparison of half-wave and full-wave rectifiers with an ideal ac-to-dc converter

value over which are superimposed a number of ac (sinusoidal) components of different frequencies. These undesired ac components are called *ripples*. The lowest ripple frequency in case of a half-wave rectifier is the same as the power-mains frequency. But, for full-wave rectifier it is not so. As can be seen from Figs. 4.20d and *a*, the period of the output wave of a full-wave rectifier is half the period of the input wave. The variation in current (or voltage) repeats itself after each angle π of the input wave. Therefore, the lowest frequency of the ripple in the output of a full-wave rectifier is twice the input frequency. That is, the ripple frequency,

$$f_r = f_i = 50 \text{ Hz} \qquad \text{(half-wave rectifier)} \qquad (4.15)$$

and $\qquad f_r = 2f_i = 100 \text{ Hz} \qquad \text{(full-wave rectifier)} \qquad (4.16)$

How effectively a rectifier converts ac power into dc power is described quantitatively by terms such as *ripple factor, rectification efficiency*, etc.

The ripple factor is a measure of purity of the dc output of a rectifier, and is defined as

$$r = \frac{\text{rms value of the components of wave}}{\text{average or dc value}} \qquad (4.17)$$

The rectification efficiency tells us what percentage of total input ac power is converted into useful dc output power. Thus, rectification efficiency is defined as

$$\eta = \frac{\text{dc power delivered to load}}{\text{ac input power from transformer secondary}}$$

or $$\eta = \frac{P_{dc}}{P_{ac}} \qquad (4.18)$$

Here, P_{ac} is the power that would be indicated by a wattmeter connected in the rectifying circuit with its voltage terminals placed accross the secondary winding and P_{dc} is the dc output power.

We shall now analyse half-wave and full-wave rectifiers to find their ripple factor and rectification efficiency.

4.7.1 Performance of Half-Wave Rectifier

The half-wave rectified current wave is plotted in Fig. 4.21 and is described mathematically as

$$i_L = I_m \sin \omega t; \qquad \text{for } 0 \leqslant \omega t \leqslant \pi \qquad (4.19)$$

and $$i_L = 0; \qquad \text{for } \pi \leqslant \omega t \leqslant 2\pi \qquad (4.20)$$

For determining the ripple factor or rectification efficiency, we first find the rms value of the current.

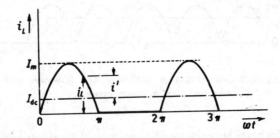

Fig. 4.21 Half-wave rectified current wave. (The instantaneous ac component of current is the difference between instantaneous total current and dc current, i.e., $i' = i_L - I_{dc}$.)

RMS Value of Current The rms or effective value of the current flowing through the load is given as

$$I_{rms} = \sqrt{\frac{1}{2\pi} \int_0^{2\pi} i_L^2 \, d(\omega t)}$$

where current i_L is described by Eqs. 4.19 and 4.20. Therefore,

$$I_{rms} = \sqrt{\frac{1}{2\pi} \left[\int_0^\pi I_m^2 \sin^2 \omega t \, d(\omega t) + \int_\pi^{2\pi} 0 \, d(\omega t) \right]}$$

$$= \sqrt{\frac{I_m^2}{2\pi} \int_0^\pi \frac{(1 - \cos 2\omega t)}{2} \, d(\omega t)}$$

$$= \sqrt{\frac{I_m^2}{2\pi \times 2} \left| \omega t - \frac{\sin 2\omega t}{2} \right|_0^\pi}$$

or $$I_{rms} = \frac{I_m}{2} \qquad (4.21)$$

This is the rms value of the total current (dc value and ac components). As can be seen from Fig. 4.21, the instantaneous value of ac fluctuation is the difference of the instantaneous total value and the dc value. That is, the instantaneous ac value is given as

$$i' = i_L - I_{dc}$$

Therefore, the rms value of ac components is given as

$$I'_{rms} = \sqrt{\frac{1}{2\pi} \int_0^{2\pi} (i_L - I_{dc})^2 \, d(\omega t)}$$

$$= \sqrt{\frac{1}{2\pi} \int_0^{2\pi} (i_L^2 + I_{dc}^2 - 2 i_L I_{dc}) \, d(\omega t)}$$

$$= \sqrt{I_{rms}^2 + I_{dc}^2 - 2 I_{dc}^2}$$

or $\quad I'_{rms} = \sqrt{I_{rms}^2 - I_{dc}^2}$ $\hfill (4.22)$

Ripple Factor From Eq. 4.17, the ripple factor is given as

$$r = \frac{I'_{rms}}{I_{dc}} = \frac{\sqrt{I_{rms}^2 - I_{dc}^2}}{I_{dc}} = \sqrt{\left(\frac{I_{rms}}{I_{dc}}\right)^2 - 1} \qquad (4.23)$$

Using Eqs. 4.8 and 4.21, for half-wave rectifier the ratio

$$\frac{I_{rms}}{I_{dc}} = \frac{I_m/2}{I_m/\pi} = 1.57$$

Therefore, the ripple factor is given as

$$r = \sqrt{(1.57)^2 - 1} = 1.21 \qquad (4.24)$$

Thus, we see that the ripple current (or voltage) exceeds the dc current (or voltage). This shows that the half-wave rectifier is a poor converter of ac into dc.

Rectification Efficiency For a half-wave rectifier, the dc power delivered to the load is

$$P_{dc} = I_{dc}^2 R_L = \left(\frac{I_m}{\pi}\right)^2 R_L$$

and the total input ac power is

$$P_{ac} = I_{rms}^2 (r_d + R_L) = \left(\frac{I_m}{2}\right)^2 (r_d + R_L)$$

Therefore, the rectification efficiency is

$$\eta = \frac{P_{dc}}{P_{ac}} = \frac{(I_m/\pi)^2 R_L}{(I_m/2)^2 (r_d + R_L)} \times 100 \%$$

$$= \frac{40.6}{1 + r_d/R_L} \% \qquad (4.25)$$

If $r_d \ll R_L$, $\eta \to 40.6$ per cent. It means that under the best conditions (i.e. no diode loss), only 40.6 % of the ac input power is converted into dc power. The rest remains as ac power in the load.

4.7.2 Performance of Full-Wave Rectifier

Figure 4.22 shows a full-wave rectified current wave. Its period may be seen to be π. The wave repeats itself after each π. Therefore, while computing the average or rms values, we should take the integration between the limits 0 to π, instead of 0 to 2π. The waveshape between 0 to π is described as

$$i_L = I_m \sin \omega t \qquad (4.26)$$

where ω ($= 2\pi f$) is the angular frequency of the input ac voltage.

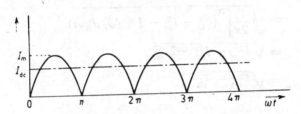

Fig. 4.22 Full-wave rectified current wave

RMS Value of Current Effective or rms value of current is given as

$$I_{rms} = \sqrt{\frac{1}{\pi} \int_0^\pi i_L^2 \, d(\omega t)} = \sqrt{\frac{1}{\pi} \int_0^\pi I_m^2 \sin^2 \omega t \, d(\omega t)}$$

$$= \sqrt{\frac{I_m^2}{\pi} \int_0^\pi \left(\frac{1 - \cos 2\omega t}{2} \right) d(\omega t)} = \sqrt{\frac{I_m^2}{\pi} \left| \frac{\omega t}{2} - \frac{\sin 2\omega t}{4} \right|_0^\pi}$$

$$= \sqrt{\frac{I_m^2}{\pi} \times \frac{\pi}{2}}$$

or $\quad I_{rms} = \dfrac{I_m}{\sqrt{2}}$ $\qquad (4.27)$

Note that this is the same as the rms value of the full sinusoidal ac wave.

The dc or average value of the current is

$$I_{dc} = \frac{1}{\pi} \int_0^\pi i_L \, d(\omega t) = \frac{1}{\pi} \int_0^\pi I_m \sin \omega t \, d(\omega t)$$

$$= \frac{2 I_m}{\pi} \qquad (4.28)$$

This current, as it should be, is double the dc current of a half-wave rectifier.

Ripple Factor Equation 4.22 is valid for a full-wave rectifier too. We can therefore use Eq. 4.23 to calculate the ripple factor of a full-wave rectifier.

$$r = \sqrt{\left(\frac{I_{rms}}{I_{dc}} \right)^2 - 1} = \sqrt{\left(\frac{I_m/\sqrt{2}}{2 I_m/\pi} \right)^2 - 1}$$

$$= 0.482 \qquad (4.29)$$

Rectification Efficiency For a full-wave rectifier, the dc power delivered to the load is

$$P_{dc} = I_{dc}^2 R_L = \left(\frac{2I_m}{\pi}\right)^2 R_L$$

and the total input ac power is

$$P_{ac} = I_{rms}^2 (r_d + R_L) = \left(\frac{I_m}{\sqrt{2}}\right)^2 (r_d + R_L)$$

Therefore, the rectification efficiency is

$$\eta = \frac{P_{dc}}{P_{ac}} = \frac{(2I_m/\pi)^2}{(I_m\sqrt{2})^2 (r_d + R_L)} \times 100\,\%$$

$$= \frac{81.2}{1 + r_d/R_L}\,\% \tag{4.30}$$

This shows that the rectification efficiency of a full-wave rectifier is twice that of a half-wave rectifier under identical conditions. The maximum possible efficiency can be 81.2 % (when $r_d \ll R_L$).

Example 4.3 In a centre-tap full-wave rectifier, the load resistance $R_L = 1\ k\Omega$. Each diode has a forward-bias dynamic resistance r_d of 10 Ω. The voltage across half the secondary winding is 220 sin 314t. Find (a) the peak value of current, (b) the dc or average value of current, (c) the rms value of current, (d) the ripple factor, and (e) the rectification efficiency.

Solution: The voltage across half the secondary winding is given as

$$v = 220 \sin 314t$$

(a)   The peak value of voltage is
$$V_m = 220 \text{ V}$$

Therefore, peak value of current is

$$I_m = \frac{V_m}{r_d + R_L} = \frac{220}{10 + 1000} = 0.2178 \text{ A}$$
$$= 217.8 \text{ mA}$$

(b) The dc or average value of current is

$$I_{dc} = \frac{2I_m}{\pi} = \frac{2 \times 217.8}{\pi} = 138.66 \text{ mA}$$

(c) The rms value of current is

$$I_{rms} = \frac{I_m}{\sqrt{2}} = 154 \text{ mA}$$

(d) The ripple factor is given as

$$r = \sqrt{\left(\frac{I_{rms}}{I_{dc}}\right)^2 - 1} = \sqrt{\left(\frac{154}{138.66}\right)^2 - 1} = 0.482$$

(e) The rectification efficiency is given as

$$\eta = \frac{P_{dc}}{P_{ac}}$$

But, $P_{dc} = I_{dc}^2 \, R_L = (138.66)^2 \times (10^{-3})^2 \times 1000 = 19.2265 \text{ W}$

and $P_{ac} = I_{rms}^2 \, (r_d + R_L) = (154)^2 \times (10^{-3})^2 \times (10 + 1000) = 23.953 \text{ W}$

$\therefore \quad \eta = \frac{P_{dc}}{P_{ac}} = \frac{19.2265}{23.953} = 0.8026 = 0.8026 \times 100 \% = \mathbf{80.26 \%}$

A full-wave rectifier is preferred to a half-wave rectifier, because its rectification efficiency is double and its ripple factor is low. Table 4.1 gives the comparison between different rectifiers discussed so far. Unless otherwise indicated, all rectifiers discussed from now on are full-wave rectifiers (either centre-tap or bridge).

Table 4.1 COMPARISON BETWEEN DIFFERENT AVERAGE RECTIFIERS

	Half-wave	Full-wave	
		Centre-tap	Bridge
Number of diodes	1	2	4
Transformer necessary	No†	Yes	No†
Peak secondary voltage	V_m	V_m^*	V_m
Peak inverse voltage	V_m^{**}	$2\,V_m$	V_m
Peak load current, I_m	$V_m/(r_d+R_L)$	$V_m/(r_d+R_L)$	$V_m/(2r_d+R_L)$
RMS current, I_{rms}	$I_m/2$	$I_m/\sqrt{2}$	$I_m/\sqrt{2}$
DC current, I_{dc}	I_m/π	$2I_m/\pi$	$2I_m/\pi$
Ripple factor, r	1.21	0.482	0.482
Rectification efficiency (max)	40.6 %	81.2 %	81.2 %
Lowest ripple frequency, f_r	f_i	$2f_i$	$2f_i$

*It is the voltage between centre-tap and one of the terminals.
**With a capacitor-input filter, the PIV of a half-wave circuit becomes $2V_m$, as we shall see later.
† Transformer may be used for isolation even if not required for stepping up (or down) the input ac.

4.8 HOW TO GET A BETTER DC

The object of rectification is to provide a steady dc voltage, similar to the voltage from a battery. We have seen that a full-wave rectifier provides a better dc than a half-wave rectifier. But, even a full-wave rectifier does not provide ripple-free dc voltage. The rectifiers provide what we may call "a pulsating dc". We can *filter* or *smooth* out the ac variations from the rectified voltage. For this we use a filter or smoothing circuit (see Fig. 4.11). In this section, we shall discuss different types of filter circuits.

4.8.1 Shunt Capacitor Filter

This is the simplest and cheapest filter. You just have to connect a large value capacitor C in shunt with the load resistor R_L, as shown in Fig. 4.23a. The capacitance offers a low-reactance path to the ac components of current.

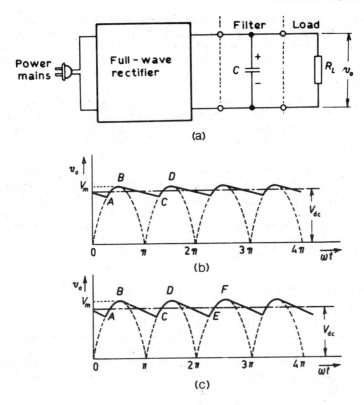

Fig. 4.23 Full-wave rectifier with shunt capacitance filter

To dc (with zero frequency), this is an open circuit. All the dc current passes through the load. Only a small part of the ac component passes through the load, producing a small ripple voltage.

The capacitor changes the conditions under which the diodes (of the rectifier) conduct. When the rectifier output voltage is increasing, the capacitor charges to the peak voltage V_m. Just past the positive peak, the rectifier output voltage tries to fall (see the dotted curve in Fig. 4.23b). But at point B, the capacitor has $+V_m$ V across it. Since the source voltage becomes slightly less than V_m, the capacitor will try to send current back through the diode (of the rectifier). This reverse-biases the diode, i.e., it becomes open-circuited.

The diode (open-circuit) disconnects or separates the source from the load. The capacitor starts to discharge through the load. This prevents the load voltage from falling to zero. The capacitor continues to discharge until the source voltage (the dotted curve) becomes more than the capacitor voltage (at point C). The diode again starts conducting, and the capacitor is again charged to peak value V_m. During the time the capacitor is charging (from point C to point D) the rectifier supplies the charging current i_C through the capacitor branch as well as the load current i_L. When the capacitor discharges (from point B to point C), the rectifier does not supply any current; the capacitor sends current i_L through the load. The current is maintained through the load all the time.

The rate at which the capacitor discharges between points B and C (in Fig. 4.23b) depends upon the time constant CR_L. The longer this time constant is, the steadier is the output voltage. If the load current is fairly small (i.e., R_L is sufficiently large) the capacitor does not discharge very much, and the average load voltage V_{dc} is slightly less than the peak value V_m (see Fig. 4.23b).

An increase in the load current (i.e. decrease in the value of R_L) makes the time constant of the discharge path smaller. The capacitor then discharges more rapidly, and the load voltage is not constant (see Fig. 4.23c). The ripple increases with increase in load current. Also, the dc output voltage, V_{dc} decreases.

A much more steadier load voltage can be obtained if a capacitor of too large a value is used. But, the maximum value of the capacitance that can be employed is limited by another factor. The larger the capacitance value, the greater is the current required to charge the capacitor to a given voltage. The maximum current that can be safely handled by a diode is limited by a figure quoted by the manufacturer. This puts a limit on the maximum value of the capacitance used in the shunt capacitor filter.

4.8.2 Series Inductor Filter

An inductor has the fundamental property of opposing any change in current flowing through it. This property is used in the series inductor filter of Fig. 4.24. Whenever the current through an inductor tends to change, a "back emf" is induced in the inductor. This induced back emf prevents the current from changing its value. Any sudden change in current that might have occurred in the circuit without an inductor is smoothed out by the presence of the inductor.

Since the reactance of the inductor increases with frequency, better filtering of the higher harmonic ripples takes place. The output voltage waveform will therefore consist principally of the second harmonic frequency

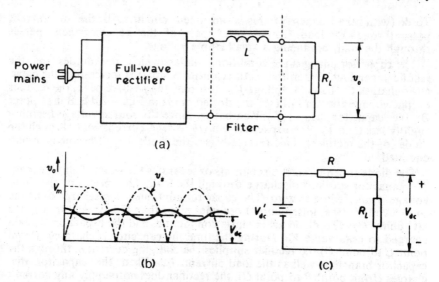

Fig. 4.24 Full-wave rectifier with series inductor filter

(the lowest ripple frequency), as shown in Fig. 4.24b. It shows a large dc component and a small ac component.

For dc (zero frequency), the choke resistance R in series with the load resistance R_L forms a voltage divider as shown in Fig. 4.24c. If V'_{dc} is the dc voltage from a full-wave rectifier, the dc voltage V_{dc} across the load is given as

$$V_{dc} = \frac{R_L}{R + R_L} V'_{dc} \qquad (4.31)$$

Usually, R is much smaller than R_L; therefore, almost all of the dc voltage reaches the load.

The operation of a series inductor filter depends upon the current through it. Therefore, this filter (and also the choke-input LC filter discussed in the next section) can only be used together with a full-wave rectifier (since it requires current to flow at all times). Furthermore, the higher the current flowing through it, the better is its filtering action. Therefore, an increase in load current results in reduced ripple.

4.8.3 Choke-Input LC Filter

Figure 4.25 shows a choke-input filter using an inductor L in series and capacitor C in shunt with load. We have seen that a series inductor filter has the feature of decreasing the ripples when the load current is increased. Reverse is the case with a shunt capacitor filter. In this case, as the load current is increased, the ripples also increase. An LC filter combines the features of both the series inductor filter and shunt capacitor filter. Therefore, the ripples remain fairly the same even when the load current changes.

The choke (iron-core inductor) allows the dc component to pass through easily because its dc resistance R is very small. For dc, the capacitor appears as open circuit and all the dc current passes through the load resistance R_L. Therefore, the circuit acts like a dc voltage divider of Fig. 4.25c, and the output dc voltage is given by Eq. 4.31.

The fundamental frequency of the ac component in the output of the rectifier is 100 Hz (twice the line frequency). For this ac, the reactance X_L ($= 2\pi f L$) is high. The ac current has difficulty in passing through the inductor. Even if some ac current manages to pass through the choke, it flows through the low reactance X_C ($= 1/2\pi f C$) rather than through load resistance R_L. The ripples are reduced very effectively because X_L is much greater than X_C, and X_C is much smaller than R_L. The circuit works like the ac voltage divider of Fig. 4.25d. If V'_r is the rms value of the ripple voltage from the full-wave rectifier, then the rms value of the output ripple is given as

$$V_r \cong \frac{X_C}{X_L} V'_r \qquad (4.32)$$

The reactances X_C and X_L are computed at 100 Hz. Typical values for L are 5 to 30 H and for C, 5 to 40 μF.

In the capacitor-input filter, the current flows through the transformer in a series of pulses. But in the choke-input filter, the current flows continuously. This means, the transformer is utilized more efficiently. A further advantage of the choke-input filter is that the ripple content at the output is not only low but is also less dependent on the load current.

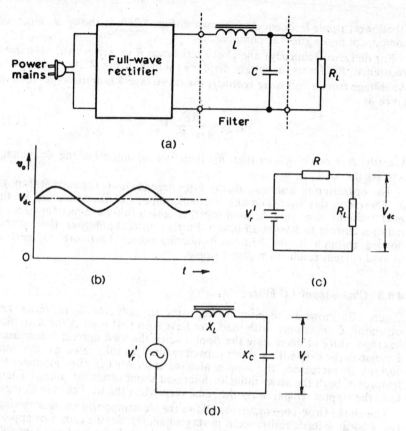

Fig. 4.25 Full-wave rectifier with choke-input filter:
(a) Circuit; (b) Output; (c) DC equivalent
circuit; (d) AC equivalent circuit

Bleeder Resistor Since an inductor depends upon current for its operation, it functions best under large current demands. For optimum functioning, the inductor should have a minimum current flowing at all times. If the
current through the inductor falls below this minimum value, the output
voltage rises sharply. The voltage regulation becomes poor. In order to
provide this minimum current through the choke, a *bleeder resistor* R_b is
usually included in the circuit. Figure 4.26 shows a bridge rectifier with a
choke-input filter using a bleeder resistor.

In Fig. 4.26, even if load resistance R_L becomes open-circuit, the bleeder
resistor R_b maintains the minimum current necessary for optimum inductor-
operation. The bleeder resistor can serve a number of other functions as
well. For example, it can be used as a voltage divider for providing a variable output voltage. It can also serve as a discharge path for the capacitor,
so that voltage does not remain across the output terminals after the load
has been disconnected, and the circuit de-energized. This reduces the hazard
of electrical shock when the load is connected to the output terminals next
time.

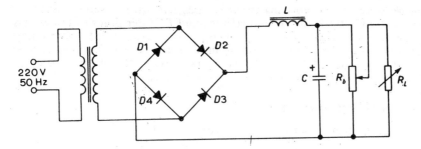

Fig. 4.26 Bridge rectifier with a choke-input filter, using a bleeder resistor.
(The bleeder resistor maintains minimum current in C,
discharges C, and is used for varying the output.)

4.8.4 π Filter

Very often, in addition to the LC filter, we use an additional capacitor C_1 for providing smoother output voltage. This filter is called π filter (its shape is like the Greek letter π). Such a filter is shown in Fig. 4.27. The rectifier now feeds directly into the capacitor C_1. Therefore, the filter is also called *capacitor-input filter.*

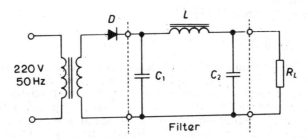

Fig. 4.27 A half-wave rectifier feeding into a π-filter

Since the rectifier feeds into the capacitor C_1, this type of filter can be used together with a half-wave rectifier. (The chock-input filter cannot be used with a half-wave rectifier.) Typical values for C_1 and C_2 for a half-wave rectifier are 32 μF each; and for L, 30 H. The half-wave rectifier ripple frequency being 50 Hz, these components have reactances of $X_L = 100 \ \Omega$ and $X_C = 9492 \ \Omega$ approximately The reactances of L and C_2 act as an ac potential divider. This reduces the ripple voltage to approximately 100/9426 times its original value.

In the full-wave rectifier, the ripple frequency is 100 Hz. It means that a filter using the same component values would be more efficient in reducing the ripple. In other words, for a given amount of ripple smaller components can be used. Typical values are $C_1 = C_2 = 8 \ \mu$F and $L = 15$ H. Electrolytic capacitors have fairly large capacitances values and yet occupy minimum space. Usually both capacitors C_1 and C_2 are made inside one metal container. The metal container serves as the common ground for the two capacitors.

The disadvantages of the capacitor-input LC filter are the cost, weight, size and external field produced by the series inductor. These disadvantages can be overcome by replacing the series inductor with a series resistor of 100 to 200 Ω. It is then called capacitor-input RC filter. But this has the disadvantage of increasing the dc voltage drop in the filter. The voltage regulation becomes poorer. It also requires adequate ventilation to conduct away the heat produced in the resistor. As a result, it is only used to supply dc power to equipments taking only a small current.

4.9 TYPES OF DIODES

The important characteristics of semiconductor diodes are:

1. Maximum forward current
2. PIV rating
3. Forward and reverse ac resistances
4. Junction capacitance
5. Behaviour in breakdown region

One or more of these characteristics may be of prime importance depend ingu pon the intended application of the diode. The main types of diodes used in electronic circuitry are:

1. Signal diodes
2. Power diodes
3. Zener diodes
4. Varactor diodes
5. Light-emitting diodes (LEDs)

4.9.1 Signal Diodes

These diodes are not required to handle large currents and/or voltages. The usual requirements are a large reverse-resistance/forward-resistance ratio and a minimum of junction capacitance. Some of the commercially available signal diodes are listed in the data book as general-purpose diodes. Some are best suited to a particular type of circuit application, such as a radio waves detector, or as an electronic switch in logic circuitry. The maximum reverse voltage, or peak inverse voltage, that the diode may be required to handle is usually not very high, and neither is the maximum forward current. Most types of diodes have a PIV rating in the range 30 V to 150 V. The maximum forward current range may be somewhere between 40 mA and 250 mA.

4.9.2 Power Diodes

Power diodes are mostly used in rectifiers. The important parameters of a power diode are the peak inverse voltage, the maximum forward current, and the reverse-resistance/forward-resistance ratio. The peak inverse voltage rating is likely to be somewhere between 50 V and 1000 V. The maximum forward current may be perhaps 30 A, or even more. As semiconductor technology advances, diodes capable of handling larger and larger power are being made available. Power diodes are usually silicon diodes. A power diode must have a forward resistance as low as possible. This helps in reducing the voltage drop across the diode when a large forward current flows. The forward resistance is usually not very much more than an ohm or two. The reverse resistance of a power diode must be as high as possible. Almost no current should flow through the diode when reverse-biased.

4.9.3 Zener Diodes

Zener diodes are designed to operate in the breakdown region without damage. By varying the doping level, it is possible to produce zener diodes with breakdown voltages from about 2 V to 200 V.

As discussed in Sec. 4.2.3, the large current at breakdown is brought about by two factors, known as the zener and avalanche effects. When a diode is heavily doped, the depletion layer is very narrow. When the voltage across the diode is increased (in reverse bias), the electric field across the depletion layer becomes very intense. When this field is about 3×10^7 V/m, electrons are pulled from the covalent bonds. A large number of electron-hole pairs are thus produced and the reverse current sharply increases. This is known as the *zener effect*.

Avalanche effect occurs because of a cumulative action. The external applied voltage accelerates the minority carriers in the depletion region. They attain sufficient kinetic energy to *ionize* atoms by collision. This creates new electrons which are again accelerated to high-enough velocities to ionize more atoms. This way, an avalanche of free electrons is obtained. The reverse current sharply increases.

The zener effect is predominant for breakdown voltages less than about 4 V. The avalanche breakdown is predominant for voltages greater than 6 V. Between 4 and 6 V, both effects are present. It is the zener effect that was first discovered, and the term "zener diode" is in wide use for a *breakdown diode* whether it uses zener effect or avalanche effect, or both. If the applied reverse voltage exceeds the breakdown voltage, a zener diode acts like a *constant-voltage* source. For this reason, a zener diode is also called *voltage reference diode*.

The circuit symbol of a zener diode is shown in Fig. 4.28. A zener diode is specified by its breakdown voltage and the maximum power dissipation. The most common application of a zener diode is in the voltage stabilizing or regulator circuits.

Anode   Cathode

Fig. 4.28 Circuit symbol of a zener diode

Example 4.4 A zener diode is specified as having a breakdown voltage of 9.1 V, with a maximum power dissipation of 364 mW. What is the maximum current the diode can handle?

Solution: The maximum permissible current is

$$I_{Z \, max} = \frac{P}{V_Z} = \frac{364 \times 10^{-3}}{9.1} = 40 \text{ mA}$$

Zener Diode Voltage Regulator After the ripples have been smoothed or filtered from the rectifier output, we get a sufficiently steady dc output. But for many applications, even this sort of power supply may not serve the purpose. Firstly, this supply does not have a good enough voltage regulation. That is, the output voltage reduces as the load (current) connected to it is increased. Secondly, the dc output voltage varies with the change in the ac input voltage. To improve the constancy of the dc output voltage as the load and/or the ac input voltage vary, a voltage-regulator circuit is used. The stabilizer circuit is connected between the output of the filter and the load (see Fig. 4.11).

The simplest regulator circuit consists merely of a resistor R_S connected in series with the input voltage, and a zener diode connected in parallel with the load (Fig. 4.29). The voltage from an unregulated power supply is used as the input voltage V_I to the *regulator* circuit. As long as the voltage across R_L is less than the zener-breakdown voltage V_Z, the zener diode does not conduct. If the zener diode does not conduct, the resistors R_S and R_L make a potential divider across V_I. At an increased V_I, the voltage across R_L becomes greater than the zener-breakdown voltage. It then operates in its breakdown region. The resistor R_S limits the zener current from exceeding its rated maximum $I_{Z\,max}$.

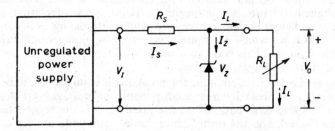

Fig. 4.29 The zener-diode voltage regulator

The current from the unregulated power supply splits at the junction of the zener diode and the load resistor. Therefore,

$$I_S = I_Z + I_L \qquad (4.33)$$

When the zener diode operates in its breakdown region, the voltage V_Z across it remains fairly constant even though the current I_Z flowing through it may vary considerably. If the load current I_L should increase (because of the reduction in load resistance), the current I_Z through the zener diode falls by the same percentage in order to maintain constant current I_S. This keeps the voltage drop across R_S constant. Hence, the output voltage V_O remains constant. If, on the other hand, the load current should decrease, the zener diode passes an extra current I_Z such that the current I_S is kept constant. The output voltage of the circuit is thus stabilized.

Let us examine the other cause of the output voltage variation. If the input voltage V_I should increase, the zener diode passes a larger current so that extra voltage is dropped across R_S. Conversely, if V_I should fall; the current I_Z also falls, and the voltage drop across R_S is reduced. Because of the self-adjusting voltage drop across R_S, the output voltage V_O fluctuates to a much lesser extent than does the input voltage V_I.

4.9.4 Varactor Diodes

A reversed-biased *PN*-junction can be compared to a charged capacitor. The *P* and *N* regions (away from the space charge region) are essentially low resistance areas due to high concentration of majority carriers. The space-charge region, which is depleted of majority carriers, serves as an effective insulation between the *P* and *N* regions. The *P* and *N* regions act as the plates of the capacitor while the space-charge region acts as the insulating dielectric. The reverse-biased *PN*-junction thus has an effective capacitance, whose value is given as

$$C = \frac{\epsilon A}{W} \qquad (4.34)$$

where ϵ (the Greek letter "epsilon") is the permittivity of the semiconductor material, A is the area of the junction, and W is the width of the space-charge region. The width W of the space-charge region is approximately proportional to the square root of the reverse bias voltage V. The area A and permittivity ϵ being constant, we can write Eq. 4.34 as

$$C = \frac{K}{\sqrt{V}} \qquad (4.35)$$

As the reverse bias increases, the space-charge region becomes wider, thus effectively increasing the plate separation and decreasing the capacitance. Silicon diodes optimized for this variable-capacitance effect are called *varactors*. Figure 4.30 shows the two symbols used to represent a varactor diode. It also shows graphically how the capacitance of a varactor diode varies with the reverse-bias voltage. Typically, the capacitance variation might be 2-12 pF, or 20-28 pF, or perhaps 28-76 pF.

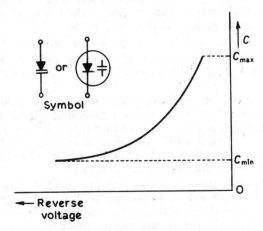

Fig. 4.30 Varactor diode characteristic and
its symbol

Varactor diodes are replacing mechanically tuned capacitors in many applications. A varactor diode in parallel with an inductor gives a resonant-tank circuit. The resonant frequency of this tank circuit can easily be changed by varying the reverse voltage across the diode.

Example 4.5 A varactor diode has a capacitance of 18 pF when the reverse-bias voltage applied across it is 4 V. Determine the capacitance if the diode-bias voltage is increased to 8 V.

Solution: The capacitance of a varactor diode is inversely proportional to the square root of the bias voltage, i.e.

$$C = \frac{K}{\sqrt{V}}$$

Here, $V = 4$ V; $C = 18$ pF $= 18 \times 10^{-12}$ F.

$$\therefore \qquad 18 \times 10^{-12} = \frac{K}{\sqrt{4}}$$

or, $\qquad K = 36 \times 10^{-12}$

Hence, when the voltage has increased to 8 V, the capacitance becomes

$$C = \frac{K}{\sqrt{V}} = \frac{36 \times 10^{-12}}{\sqrt{8}} = 12.728 \times 10^{-12} \text{ F}$$
$$= 12.728 \text{ pF}$$

4.9.5 Light-Emitting Diodes (LEDs)

When a PN-junction diode is forward-biased, the potential barrier is lowered. The majority carriers start crossing the junction. The conduction-band electrons from the N region cross the barrier and enter the P region. Immediately on entering the P region, each electron falls into a hole and recombination takes place. Also, some holes may cross the junction from the P region into the N region. A conduction-band electron in the N region may fall into a hole even before it crosses the junction. In either case, recombinations take place around the junction.

Each recombination radiates energy. In an ordinary diode (power diode or signal diode), the radiated energy is in the form of heat. In the *light-emitting diode* (LED), the radiated energy is in the form of light (or photons).

Germanium and silicon diodes have less probabilities of radiating light. By using materials such as gallium arsenide phosphide (GaAsP) and gallium phosphide (GaP), a manufacturer can produce LEDs that radiate red, green or orange lights. Infrared LEDs use gallium arsenide (GaAs), and they emit invisible (infrared) radiation. These find applications in burglar-alarm systems and other areas requiring invisible radiation.

Figure 4.31a shows the schematic symbol of a LED. The LEDs that emit visible light find applications in instrument displays, panel indicators, digital watches, calculators, multimeters, intercoms, telephone switch boards, etc. A *seven-segment display* unit as shown in Fig. 4.31b, is made by using a number of LEDs. By activating suitable combination of LEDs in this unit, any digit from 0 to 9 can be displayed by it.

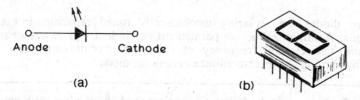

Anode   Cathode

(a)   (b)

Fig. 4.31 (a) Schematic symbol of an LED; (b) A seven-segment display unit using LEDs

LEDs have a number of advantages over ordinary incandescent lamps. They work on low voltages (1 or 2 V) and currents (5 to 10 mA), and thus consume less power. They require no heating, no warm-up time, and hence are very fast in action. They are small in size and light in weight. They are not affected by mechanical vibrations, and have long life (more than 20 years).

REVIEW QUESTIONS

4.1 (a) What causes majority carriers to flow at the moment when a P region and an N region are brought together ? (b) Why does this flow not continue until all the carriers have recombined ?

4.2 Explain the formation of the "depletion region" in an open circuited PN-junction.

4.3 State what you understand by barrier potential across a PN-junction. Also explain its significance.

4.4 The barrier potential developed across an open-circuited PN-junction aids the flow of minority carriers. Explain how this flow of charge carriers is counterbalanced.

4.5 Explain why the peak inverse voltage of a semiconductor diode is an important parameter.

4.6 Sketch, on the same axes, typical static characteristics for germanium and silicon diodes. Label clearly the values of forward voltage drop and reverse saturation current.

4.7 What do you understand by "an ideal diode"? Draw its V-I characteristics.

4.8 What limits the number of reverse current carriers ?

4.9 Why is the reverse current in a silicon diode much smaller than that in a comparable germanium diode ?

4.10 Explain how the process of avalanche breakdown occurs in a PN-junction diode. How is it different from zener breakdown ?

4.11 Explain why a PN-junction possesses capacitance.

4.12 Which carriers conduct forward current in a diode ? Draw the symbol for a PN-junction diode showing the direction of forward current.

4.13 Roughly how much forward voltage is needed to cause current to flow in (a) a silicon diode; (b) a germanium diode ?

4.14 Draw the circuit diagram of a half-wave rectifier. Explain its working. What is the minimum frequency of ripple in its output ?

4.15 If V_m is the peak value of the voltage across the secondary winding in a half-wave rectifier, what is the value of the dc component in its output voltage ? Derive this relationship.

4.16 Draw the circuit of a half-wave rectifier with a capacitor-input filter. Give typical compopent values and describe the operation of the circuit. What is the peak inverse voltage across the diode: (a) without the capacitor connected, and (b) with the capacitor connected ?

4.17 Draw the circuit diagram of a full-wave rectifier using (a) centre-tap connection, and (b) bridge connection. Explain the working of each. What is the PIV in each case ?

4.18 Explain why a bridge rectifier is preferred over a centre-tap rectifier. Is there any application where a centre-tap rectifier is preferred over a bridge rectifier ?

4.19 Prove that the ripple factor of a half-wave rectifier is 1.21 and that of a full-wave rectifier is 0.482.

4.20 Show that the maximum rectification efficiency of a half-wave rectifier is 40.6 %.

4.21 Derive an expression for the rectification efficiency of a full-wave rectifier.

4.22 Draw the output voltage waveform of a half-wave rectifier and then show the effect, on this waveform, of connecting a capacitor across the load resistance.

4.23 Repeat Q. 4.22 for a full-wave rectifier.

4.24 Explain the need of using smoothing circuits in a power supply.

4.25 Explain why a series-inductor filter cannot be used with a half-wave rectifier.

4.26 Explain the working of a choke-input filter.

4.27 What is a bleeder resistor ? What functions can it serve in a power supply ?

4.28 Draw the circuit diagram, including typical component values, for a 12 V, 2 A power supply using bridge rectifier and a π-filter.

4.29 What are the important specifications of semiconductor diodes ?

4.30 State the difference between the specifications of a signal diode and a power diode.

4.31 Name the two types of reverse breakdowns which can occur in a PN-junction diode. Which occurs at lower voltages ?

4.32 Explain why it is necessary to use a voltage regulator circuit in a power supply.

4.33 Draw the block diagram of a regulated power supply. Explain in brief the functioning of each block.

4.34 Draw the circuit diagram of a voltage regulator circuit using a zener diode. Explain its working. Is there any limitation on the value of the series resistor used in this circuit ?

4.35 In what respect is an LED different from an ordinary PN-junction diode ? State applications of LEDs. Why should you prefer LEDs over conventional incandescent lamps ?

OBJECTIVE-TYPE QUESTIONS

Here are some incomplete statements. Four alternatives are provided below each. Tick the alternative that completes the statement correctly:

1. The potential barrier at a PN-junction is due to the charges on either side of the junction. These charges are

 (a) minority carriers
 (b) majority carriers
 (c) both majority and minority carriers
 (d) fixed donor and acceptor ions

2. In an unbiased PN-junction, the junction current at equilibrium is

 (a) due to diffusion of minority carriers only
 (b) due to diffusion of majority carriers only
 (c) zero, because equal but opposite carriers are crossing the junction
 (d) zero, because no charges are crossing the junction

3. In a PN-junction diode, holes diffuse from the P region to the N region because

 (a) the free electrons in the N region attract them
 (b) they are swept across the junction by the potential difference
 (c) there is greater concentration of holes in the P region as compared to N region
 (d) none of the above

4. In a PN-junction diode, if the junction current is zero, this means that

 (a) the potential barrier has disappeared
 (b) there are no carriers crossing the junction
 (c) the number of majority carriers crossing the junction equals the number of minority carriers crossing the junction
 (d) the number of holes diffusing from the P region equals the number of electrons diffusing from the N region

5. In a semiconductor diode, the barrier potential offers opposition to only

    (a) majority carriers in both regions
    (b) minority carries in both regions
    (c) free electrons in the N region
    (d) holes in the P region

6. When we apply reverse bias to a junction diode, it

    (a) lowers the potential barrier
    (b) raises the potential barrier
    (c) greatly increases the minority-carrier current
    (d) greatly increases the majority-carrier current

7. The number of minority carriers crossing the junction of a diode depends primarily on the

    (a) concentration of doping impurities
    (b) magnitude of the potential barrier
    (c) magnitude of the forward-bias voltage
    (d) rate of thermal generation of electron-hole pairs

8. The reverse saturation current in a junction diode is the current that flows when

    (a) only majority carriers are crossing the junction
    (b) only minority carriers are crossing the junction
    (c) the junction is unbiased
    (d) the potential barrier is zero

9. When forward bias is applied to a junction diode, it

    (a) increases the potential barrier
    (b) decreases the potential barrier
    (c) reduces the majority-carrier current to zero
    (d) reduces the minority-carrier current to zero

10. The depletion or space-charge region in a junction diode contains charges that are

    (a) mostly majority carriers
    (b) mostly minority carriers
    (c) mobile donor and acceptor ions
    (d) fixed donor and acceptor ions

11. Avalanche breakdown in a semiconductor diode occurs when

    (a) forward current exceeds a certain value
    (b) reverse bias exceeds a certain value
    (c) forward bias exceeds a certain value
    (d) the potential barrier is reduced to zero

12. When a PN-junction is biased in the forward direction

    (a) only holes in the P region are injected into the N region
    (b) only electrons in the N region are injected into the P region
    (c) majority carriers in each region are injected into the other region
    (d) no carriers move

13. The forward bias applied to a PN-junction diode is increased from zero to higher values. Rapid increase in the current flow for a relatively small increase in voltage occurs

    (a) immediately
    (b) only after the forward bias exceeds the potential barrier
    (c) when the flow of minority carriers is sufficient to cause an avalanche breakdown
    (d) when the depletion area becomes larger than the space-charge area

14. The capacitance of a reverse-biased *PN*-junction

    (a) increases as the reverse bias is decreased
    (b) increases as the reverse bias is increased
    (c) depends mainly on the reverse saturation current
    (d) makes the *PN*-junction more effective at high frequencies

15. In a half-wave rectifier, the load current flows for

    (a) the complete cycle of the input signal
    (b) only for the positive half-cycle of the input signal
    (c) less than half cycle of the input signal
    (d) more than half cycle but less than the complete cycle of the input signal

16. In a full-wave rectifier, the current in each of the diodes flows for

    (a) the complete cycle of the input signal
    (b) half cycle of the input signal
    (c) less than half cycle of the input signal
    (d) zero time

17. In a half-wave rectifier, the peak value of the ac voltage across the secondary of the transformer is $20\sqrt{2}$ V. If no filter circuit is used, the maximum dc voltage across the load will be

    (a) 28.28 V
    (b) 14.14 V
    (c) 20 V
    (d) 9 V

18. If V_m is the peak voltage across the secondary of the transformer in a half-wave rectifier (without any filter circuit), then the maximum voltage on the reverse-biased diode is

    (a) V_m
    (b) $\frac{1}{2} V_m$
    (c) $2 V_m$
    (d) none of the above

19. In the above question, if we use a shunt capacitor filter, the maximum voltage that occurs on the reverse-biased diode is

    (a) V_m
    (b) $\frac{1}{2} V_m$
    (c) $2 V_m$
    (d) none of the above

20. In a centre-tap full-wave rectifier, V_m is the peak voltage between the centre-tap and one end of the secondary. The maximum voltage across the reverse-biased diode is

    (a) V_m
    (b) $\frac{1}{2} V_m$
    (c) $2 V_m$
    (d) none of the above

21. A zener diode

    (a) has a high forward-voltage rating
    (b) has a sharp breakdown at low reverse voltage
    (c) is useful as an amplifier
    (d) has a negative resistance

22. The light-emitting diode (LED)

    (a) is usually made from silicon
    (b) uses a reverse-bised junction
    (c) gives a light output which increases with increase in temperature
    (d) depends on the recombination of holes and electrons

Ans. 1. *d*; 2. *c*; 3. *c*; 4. *c*; 5. *a*; 6. *b*; 7. *d*; 8. *b*; 9. *b*; 10. *d*; 11. *b*; 12. *c*; 13. *b*; 14. *a*; 15. *b*; 16. *b*; 17. *d*; 18. *a*; 19. *c*; 20. *c*; 21. *b*; 22. *d*.

TUTORIAL SHEET 4.1

1. Calculate the maximum dc voltage available from a half-wave rectifier shown in Fig. T.4.1.1. Also find the reading of the milliammeter.   [Ans. 4.95 V; 4.95 mA]

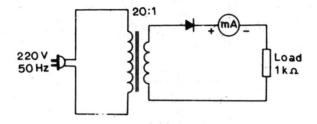

Fig. T.4.1.1

2. Calculate the PIV rating of the diodes used in the full-wave rectifier shown in Fig. T.4.1.2. Also find the maximum dc voltage that can be obtained from this circuit. Mark the polarity on the milliammeter and determine how much dc current it will indicate.   [Ans. 9.9 V; 9.9 mA]

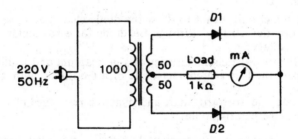

Fig. T.4.1.2

3. It is desired to obtain a maximum of 15 V (dc) from a bridge rectifier circuit. It uses silicon diodes (the voltage drop across each diode is 0.7 V). It is energized from an ac mains supply (220 V, 50 Hz) through a step-down transformer. Find the turns ratio of this transformer.   [Ans. 12 : 1]

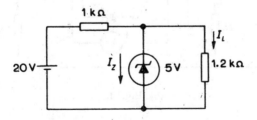

Fig. T.4.1.3

4. In Fig. T.4.1.3 calculate the load current I_L and zener diode current I_Z. Breakdown voltage of the zener diode may be assumed to be 5 V.   [Ans. 4.16 mA; 10.84 mA]

5. The silicon diode shown in Fig. T.4.1.4 is rated for a maximum current of 100 mA. Calculate the minimum value of the resistor R_L. Assume the forward voltage drop across the diode to be 0.7 V.　　　　[Ans. 93 Ω – we can take a safer value of 100 Ω].

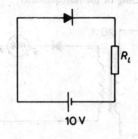

Fig. T.4.1.4

EXPERIMENTAL EXERCISE 4.1

TITLE:　Semiconductor (or crystal) diode characteristics.

OBJECTIVES:　To,

1. trace the circuit meant to draw the diode-characteristics;
2. measure the current through the diode for a particular value of forward voltage;
3. plot the forward characteristics of a germanium and a silicon diode;
4. compare the forward characteristic of a Ge diode with that of a Si diode;
5. calculate the forward static and dynamic resistance of the diode at a particular operating point.

APPARATUS REQUIRED: Experimental board, regulated power supply, milliammeter, electronic multimeter.

CIRCUIT DIAGRAM:　The circuit diagram is given in Fig. E.4.1.1.

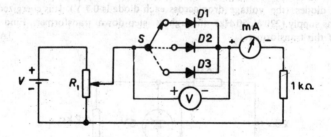

Fig. E.4.1.1

BRIEF THEORY: A diode conducts in forward bias (i.e. when its anode is at higher potential than its cathode). It does not conduct in reverse bias. When the diode is forward-biased, the barrier potential at junction reduces.

The majority carriers then diffuse across the junction. This causes current to flow through the diode. In reverse bias, the barrier potential increases, and almost no current can flow through the diode.

The external battery is connected so that its positive terminal goes to the anode and its negative terminal goes to cathode. The diode is then forward-biased. The amount of forward bias can be varied by changing the externally applied voltage. As shown in Fig. E.4.1.1, the external voltage applied across the diode can be varied by the potentiometer R_1. A series resistor (say, 1 kΩ) is connected in the circuit so that excessive current does not flow through the diode. We can note down different values of the current through the diode for various values of the voltage across it. A plot between this voltage and current gives the *diode forward characteristics*.

At a given operating point we can determine the static resistance (R_d) and dynamic resistance (r_d) of the diode from its characteristic. The static resistance is defined as the ratio of the dc voltage to dc current, i.e.

$$R_d = \frac{V}{I}$$

The dynamic resistance is the ratio of a small change in voltage to a small change in current, i.e.

$$r_d = \frac{\Delta V}{\Delta I}$$

PROCEDURE:

1. Find the type number of the diodes connected in the experimental board.
2. Trace the circuit and identify different components used in the circuit. Read the value of the resistor using the colour code.
3. Connect the milliammeter and voltmeter of suitable ranges, say, 0 to 25 mA for ammeter and 0 to 1.5 V for voltmeter.
4. Switch on the power supply. With the help of the potentiometer R_1, increase the voltage slowly.
5. Note the milliammeter and voltmeter readings for each setting of the potentiometer. Tabulate the observations.
6. Draw the graph between voltage and current.
7. At a suitable operating point, calculate the static and dynamic resistance of the diode, as illustrated in Fig. E.4.1.2.

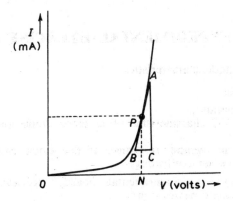

Fig. E.4.1.2

8. Bring the other diode in the circuit and repeat the above.

OBSERVATIONS:

1. *Type number of the diode* = _____
2. *Information from the data book*:
   (a) Maximum forward current rating   = _____ mA
   (b) Maximum peak inverse voltage rating = _____ V
3. *Characteristics*:

Sr. No.	Type No. _____		Type No. _____	
	Voltage (*in* V)	Current (*in* mA)	Voltage (*in* V)	Current (*in* mA)
1.				
2.				
3.				

CALCULATIONS:

i. Static resistance, $R_d = \dfrac{V}{I} = $ ——— = _____ Ω

2. Dynamic resistance, $r_d = \dfrac{\Delta V}{\Delta I} = $ ——— = _____ Ω

RESULTS:

i. The *V-I* characteristics of the diodes are shown in the graph
2. The values of static and dynamic resistance of different diodes is as given below:

	Diode type No. _____	Diode type No. _____
R_d		
r_d		

EXPERIMENTAL EXERCISE 4.2

TITLE: Zener diode characteristics.

OBJECTIVES: To,

1. trace the circuit;
2. plot the *V-I* characteristic of a zener diode under reverse-biased condition;
3. calculate the dynamic resistance of the diode under reverse-biased condition (when conducting).

APPARATUS REQUIRED: Experimental board, milliammeter, electronic multimeter, regulated power supply.

CIRCUIT DIAGRAM: The circuit diagram is shown in Fig. E.4.2.1.

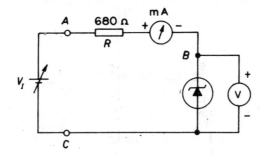

Fig. E.4.2.1

BRIEF THEORY: A *PN*-junction diode normally does not conduct when reverse biased. But if the reverse bias is increased, at a particular voltage it starts conducting heavily. This voltage is called *breakdown voltage*. High current through the diode can permanently damage it. To avoid high current, we connect a resistor in series with it. Once the diode starts conducting, it maintains almost constant voltage across its terminals whatever may be the current through it. That is, it has very low dynamic resistance. A zener diode is a *PN*-junction diode, specially made to work in the breakdown region. It is used in voltage regulators.

PROCEDURE:

1. Note the type number of the zener diode. Find breakdown voltage, wattage and maximum current ratings of the diode from the data book.
2. Trace the circuit. Note the value of the current-limiting resistor.
3. Connect milliammeter and electronic voltmeter of suitable range. (The information obtained from the data book will help in choosing suitable range of the meters).
4. Connect the negative lead of the voltmeter to point C (Fig. E.4.2.1). By connecting positive lead to point A, you can read the input dc voltage V_I. By connecting positive lead to point B, you get the voltage V_Z across the zener diode.
5. Switch on the power supply. Increase slowly the supply voltage in steps. Measure the voltages V_I and V_Z, and current I_Z. Once breakdown occurs, V_Z remains fairly constant even though I_Z increases.
6. Plot graph between V_Z and I_Z. This is the *V-I* characteristic of the zener diode.
7. Calculate the dynamic resistance of zener diode in breakdown region, as illustrated in Fig. E.4.2.2.

OBSERVATIONS:

1. *Type number of the zener diode* = _____
2. *Information from the data book*:
   (*a*) Breakdown voltage   = _____ V
   (*b*) Maximum current rating   = _____ mA
   (*c*) Maximum wattage rating   = _____ W

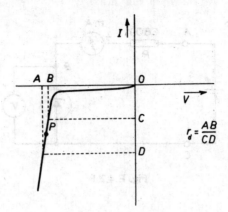

Fig. E.4.2.2

3. V-I characteristics:

S. No	V_I (in V)	V_Z (in V)	I_Z (in mA)
1.			
2.			
3.			
4.			

CALCULATIONS: Dynamic resistance, $r_d = \dfrac{\Delta V_Z}{\Delta I_Z} = \dfrac{AB}{CD} = \underline{\hspace{2cm}} \ \Omega$

RESULTS:

1. The V-I characteristic of the zener diode is shown in the graph.
2. The dynamic resistance of the diode, $r_d = \underline{\hspace{2cm}} \Omega$.

EXPERIMENTAL EXERCISE 4.3

TITLE: Half-wave rectifier.

OBJECTIVES: To,

1. trace the circuit of half-wave rectifier;
2. draw the waveshape of the electrical signal at the input and output points (after observing it in CRO) of the half-wave rectifier;
3. measure the following voltages:
   (a) AC voltage at the input of the rectifier,
   (b) AC voltage at the output points,
   (c) DC voltage at the output points;
4. verify the formula $V_{dc} = \dfrac{V_m}{\pi}$
5. verify that ripple factor for a half-wave rectifier is 1.21.

APPARATUS REQUIRED: Half-wave rectifier circuit, a CRO, an electronic (or ordinary) multimeter.

CIRCUIT DIAGRAM: As shown in Fig. E.4.3.1.

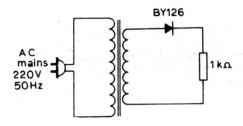

Fig. E.4.3.1

BRIEF THEORY: A diode is a unidirectional conducting device. It conducts only when its anode is at a higher voltage with respect to its cathode. In a half-wave rectifier circuit, during positive half-cycle of the input, the diode gets forward biased and it conducts. Current flows through the load resistor R_L and voltage is developed across it. During negative half-cycle of the input, the diode gets reverse biased. Now no current (except the leakage current which is very small) flows. The voltage across the load resistance during this period of input cycle is zero. Thus a pure ac signal is converted into a unidirectional signal. It can be shown that

(i) $V_{dc} = \dfrac{V_m}{\pi}$

where, V_{dc} is the output dc voltage and V_m is peak ac voltage at the input of rectifier circuit

(ii) Ripple factor $= \dfrac{\text{ac voltage at the output}}{\text{dc voltage at the output}} = 1.21$

PROCEDURE:

1. Look at the given circuit of the half-wave rectifier. Trace the circuit. Note the type number of the diode. Also note the value of the load resistor used in the circuit.
2. Connect the primary side of the transformer to the ac mains. Connect the CRO probe to the output points. Adjust different knobs of the CRO so that a good and stable waveshape is visible on its screen. Plot this waveform in your record book. Take the CRO probes at the input points of the rectifier. Note the waveshape of the signal. Compare them.
3. Now use a multimeter to measure the ac voltage at the secondary terminals of the transformer. This gives the rms value. Also measure the ac and dc voltages at the output points.
4. Multiply this rms value by $\sqrt{2}$ to get the peak value. Calculate the theoretical value of dc voltage using formula

$$V_{dc} = \frac{V_m}{\pi}$$

Compare this value with the practically measured value of output dc voltage.

5. Using the measured values of dc and ac output voltages, calculate ripple factor. This value should be about 1.21.

OBSERVATIONS:

1. *Code number or type of diode =* _____
2. *Information from data book:*
   (a) Maximum forward dc current = _____ mA
   (b) Peak inverse voltage (PIV) = _____ V
3. *Waveforms from CRO:*

4. *Measurement of different voltages:*
   (a) AC voltage at the input = _____ V
   (b) DC voltage at the output = _____ V
   (c) AC voltage at the output = _____ V
5. *Verification of theoretical formula:*

Quantity	Theoretical value	Practical value
1. Output dc voltage		
2. Ripple factor	1.21	

RESULTS:

1. Input and output waveshapes are seen on CRO.
2. Practical value of dc voltage is little less than the theoretical value. Difference is only _____V.
3. The practical value of ripple factor is more than its theoretical value. The difference is _____.

EXPERIMENTAL EXERCISE 4.4

TITLE: Full-wave rectifier (centre-tap type).

OBJECTIVES: To,

1. trace the circuit of a full-wave rectifier;
2. draw the waveshape of the electrical signal at the input and output points (after observing it in CRO) of the full-wave rectifier;
3. measure the following voltages:
   (a) AC voltage at the input points,
   (b) AC voltage at the output points,
   (e) DC voltage at the output points;
4. verify the formula

$$V_{dc} = \frac{2V_m}{\pi}$$

5. verify that the ripple factor for a full-wave rectifier is 0.482.

APPARATUS REQUIRED: Full-wave rectifier circuit, CRO, an electronic (or ordinary) multimeter.

CIRCUIT DIAGRAM: As shown in Fig. E.4.4.1.

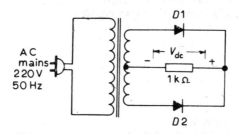

Fig. E.4.4.1

BRIEF THEORY: In a full-wave rectifier circuit there are two diodes, a transformer and a load resistor. The transformer has a centre-tap in its secondary winding. It provides out-of-phase voltages to the two diodes. During the positive half-cycle of the input, the diode $D2$ is reverse biased and it does not conduct. But diode $D1$ is forward biased and it conducts. The current flowing through $D1$ also passes through the load resistor, and a voltage is developed across it. During the negative half-cycle, the diode $D2$ is forward biased and $D1$ is reverse biased. Now, current flows through diode $D2$ and load resistor. The current flowing through load resistor R_L passes in the same direction in both the half-cycles. The dc voltage obtained at the output is given as

$$V_{dc} = \frac{2V_m}{\pi}$$

where V_m is the peak value of the ac voltage between the centre-tap point and one of the diodes. It can be proved that the ripple factor of a full-wave rectifier is 0.482.

PROCEDURE:

1. Trace the circuit. Note the value of the load resistor and the type number of the two diodes.
2. Connect the mains voltage to the primary of the centre-tapped transformer. Connect the output terminals to the vertical plates of the CRO. Adjust different knobs of the CRO and obtain a stationary pattern on its screen. Now touch the CRO probes at the centre-tap and one of the diodes. Observe the waveshape on the CRO. Plot both the waveshapes in your record book. Compare the two voltage waveshapes.
3. Measure ac voltage at the input (centre-tap and one of the diodes) and output points. Also measure the dc voltage across the load resistor.
4. From the measured ac voltage, calculate the dc voltage. Compare it with the measured value of dc output voltage. Now calculate the ripple factor by dividing ac voltage (at the output) by dc voltage at the output. How much does it differ from the theoretical value of 0.482 ?

OBSERVATIONS:

1. *Type numbers of the diodes* = _____
2. *Information from data book.*
   (a) Maximum forward-current rating = _____ mA
   (b) Peak inverse voltage (PIV) rating = _____ V
3. *Waveshape at the input and output points:*

4. *Measurements of voltages:*
   (a) AC voltage at the input points (between centre-tap and one of the diodes) = _____ V
   (b) AC voltage at the output points = _____ V
   (c) DC voltage at the output points = _____ V
5. *Verification of the formula:*

   (a) Output dc voltage

Quantity	Theoretical value	Practical value
1. Output dc voltage	$\dfrac{2V_m}{\pi} =$	
2. Ripple factor	0.482	$\dfrac{V_{ac}}{V_{dc}} =$

RESULTS:

1. The waveshapes at input and output are observed on the CRO and they are plotted.
2. The output dc voltage is a little less than the theoretical value.
3. There is a little difference between the theoretical value and measured value of ripple factor.

EXPERIMENTAL EXERCISE 4.5

TITLE: Bridge rectifier circuit.

OBJECTIVES: To,

1. trace the given circuit of bridge rectifier;
2. draw the electrical waveshape at the input and output points after observing it on CRO;
3. measure the following voltages:
   (a) AC voltage at the input of rectifier,
   (b) AC voltage at the output points,
   (c) DC voltage at the output points;

4. verify the following formula

$$V_{dc} = \frac{2V_m}{\pi}$$

5. verify that the ripple factor for bridge-rectifier circuit is 0.482.

APPARATUS REQUIRED: Bridge-rectifier circuit, a CRO and an electronic multimeter.

CIRCUIT DIAGRAM: As shown in Fig. E.4.5.1.

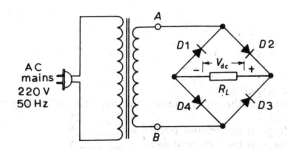

Fig. E.4.5.1

BRIEF THEORY: In a bridge rectifier circuit there are four diodes, a transformer and a load resistor. When the input voltage is positive at point A diodes $D2$ and $D4$ conduct. The current passes through the load resistor R_L. During the other half of the input signal, the point A is negative with respect to the point B. The diodes $D1$ and $D3$ conduct. The current passes through the load resistor in the same direction as during the positive half cycle. DC voltage is developed across the load. It can be proved that the output dc voltage is given by

$$V_{dc} = \frac{2V_m}{\pi}$$

where V_m is the peak ac voltage at the input of the rectifier. Also. we can show that the R.F. = 0.482.

PROCEDURE:

1. Find the type number of the diodes connected in the circuit. Trace the circuit and note down the value of the load resistor.
2. Energize the rectifier with the ac mains. Connect the output of the rectifier to the CRO. Adjust different knobs of CRO till you get a stable pattern on the screen. Similarly observe the voltage waveshape at the input of the rectifier. Compare the two waveshapes.
3. Now, measure the ac voltage at the secondary of the transformer. Also measure ac and dc voltage at the output points.
4. Using the theoretical formula

$$V_{dc} = \frac{2V_m}{\pi}$$

calculate the dc voltage at the output. Compare this value with the measured dc voltage

5. Use the measured values of ac and dc voltage at the output points to calculate the ripple factor. Compare this value with the theoretical value, which is 0.482.

OBSERVATIONS:

1. *Type numbers of the diodes* = _____
2. *Information from data book*:
   (a) Maximum forward current rating = _____ mA
   (b) Peak inverse voltage (PIV) rating = _____ V
3. *Waveshape at the input and output points*:

4. *Measurement of voltages*:
   (a) AC voltage at the input points (across the secondary winding terminals) = _____ V
   (b) AC voltage at the output points = _____ V
   (c) DC voltage at the output points = _____ V

5. *Verification of the formula*:

   (a) Output dc voltage

Quantity	Theoretical value	Practical value
1. Output dc voltage	$\dfrac{2V_m}{\pi} =$	
2. Ripple factor	0.482	$\dfrac{V_{ac}}{V_{dc}} =$

RESULTS:

1. The waveshapes at input and output are observed on CRO and are plotted.
2. The output dc voltage is a little less than the theoretical value. Why? We had not taken the voltage drop across the diodes into consideration while deriving the formula $V_{dc} = 2V_m/\pi$.

EXPERIMENTAL EXERCISE 4.6

TITLE: Different filter circuits.

OBJECTIVES: To,

1. plot the waveshape of the electrical signal at the output point with and without shunt capacitor filter in a half- and full-wave rectifier:

2. plot the waveshape of the electrical signal at the output points, with and without series inductor filter in a half- and full-wave rectifier;

3. plot the waveshape of the electrical signal at the output points, with and without π filter in a half- and full-wave rectifier;

4. measure the output dc voltage, with and without shunt capacitor filter in a half-wave and full-wave rectifier circuit;

5. measure the output dc voltage, with and without series inductor filter in a half- and full-wave rectifier circuit;

6. measure the output dc voltage with and without π filter in a half- and full-wave rectifier circuit;

7. verify that dc voltage at the output is approximately equal to the peak value of the input ac signal when shunt capacitor (and π filter) filter is used in a half- and full-wave rectifier.

APPARATUS REQUIRED: Rectifier circuit with different filters, a CRO, and an electronic (or an ordinary) multimeter.

CIRCUIT DIAGRAM: As shown in Fig. E.4.6.1.

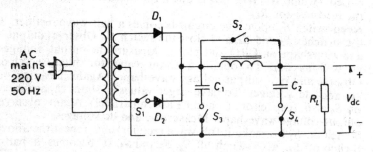

Fig. E.4.6.1

BRIEF THEORY: The output of a half-wave or full-wave rectifier contains an appreciable amount of ac voltage in addition to dc voltage. But, what we desire is pure dc without any ac voltage in it. The ac variations can be *filtered* out or *smoothed* out from the rectified voltage. This is done by *filter circuits*

In a shunt capacitor filter, we put a high-value capacitor in shunt with the load. The capacitor offers a low impedance path to the ac components of current. Most of the ac current passes through the shunt capacitor. All the dc current passes through the load resistor. The capacitor tries to maintain the output voltage constant at V_m. This is shown in Fig. E.4.6.2, for half-wave rectifier.

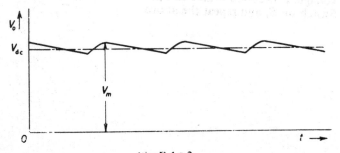

Fig. E.4.6.2

In a series inductor filter, an inductor is used in series with the load. The inductor offers high impedance to ac variations of current and low impedance to dc. As a result, the output across the load has very low ac content. The output becomes a much better dc.

A π filter utilizes the filtering properties of both the inductor and capacitor. It uses two capacitors (in shunt) and one inductor (in series). With this type of filter, the rectified output becomes almost free from ac.

PROCEDURE:

1. Trace the given rectifier circuit with different filter components. Identify every component in the circuit. Note down their values. Identify the switches S_1, S_2, S_3 and S_4.
2. With switch S_1 on, diode D_2 is in the circuit. It behaves as a full-wave rectifier. When switch S_1 is open, it becomes a half-wave rectifier. By closing switches S_3 and S_4, the capacitors C_1 and C_2 respectively can be brought into the circuit. If the switch S_2 is closed, the inductor L becomes out of circuit (the whole of the current passes through the closed switch S_2). When S_2 is open, the inductor comes in series with the load resistor R_L.
3. Keep switch S_1 open. The circuit becomes a half-wave rectifier. Open the switches S_3 and S_4, and close the switch S_2. Observe output voltage waveshape on CRO and plot it. Measure the output voltages (ac as well as dc). To obtain a shunt capacitor filter, switch on S_3. Observe and plot output-voltage waveshape. Again measure output ac and dc voltages. To have larger values of shunt capacitor, switch on S_4 also, (capacitors C_1 and C_2 are in parallel). Again observe the output-voltage waveshape. Measure ac and dc voltages.
4. Switch on S_1 (to make it full-wave rectifier) and repeat the above.
5. Switch off S_1. Also switch off S_2, S_3 and S_4. It becomes a half-wave rectifier with series inductor filter. Observe and plot the output-voltage waveshape. Measure output dc and ac voltages.
6. Switch on S_1 and repeat the above.
7. Switch off S_1 and switch on S_3 and S_4 (switch S_2 is in off position). It becomes a half-wave rectifier with π filter. Observe and plot the output-voltage waveshape. Measure output voltage (ac as well as dc).
8. Switch on S_1 and repeat the above.
9. Measure the ac voltage between the centre-tap and one of the end-terminals of the secondary of the transformer. From this, calculate the peak value V_m of the input voltage. Now, keeping the switch S_1 open, make a shunt capacitor filter by switching on S_3. (Switch S_4 is open and switch S_2 is closed.) Measure the output dc voltage. Compare it with V_m. Now switch on S_4. Again measure the output dc voltage. It becomes nearer to V_m.
10. Switch on S_1 and repeat the above.

OBSERVATIONS:

1. Filters

Rectifier type	Filter type	V_{ac} (volts)	V_{dc} (volts)
Half-wave	1. No filter 2. Shunt capacitor filter 3. Series inductor filter 4. π filter		
Full-wave	1. No filter 2. Shunt capacitor filter 3. Series inductor filter 4. π filter		

2. Input ac voltage, V_m = _____V(rms)
   Peak value, V_m = _____ $\times \sqrt{2}$ = _____V
   Output dc voltage when shunt capacitor filter is used in half-wave
   rectifier circuit   = _____V
   Output dc voltage when shunt capacitor filter is used in full-wave
   rectifier circuit   = _____V

RESULTS:

1. With the use of shunt filter in half-wave and full-wave rectifier circuits, ripple voltages are very much reduced.
2. When a π filter is used, output of half-wave and full-wave rectifier is almost a pure dc.

Transistors

OBJECTIVES: After completing this unit, you will be able to: ○ Explain the construction of a transistor. ○ Explain the action of transistor on the basis of current flow due to the movement of electrons and holes. ○ Explain the flow of leakage current in a transistor in CB configuration. ○ Draw the symbols of *NPN* and *PNP* transistors. ○ Mark the direction of different currents in the symbols of *NPN* and *PNP* transistors. ○ Explain the effect of temperature on leakage current. ○ Connect the external batteries and ac input signal to a transistor in its three configurations (CB, CE and CC). ○ Explain the meaning of α (alpha), β (beta), I_{CBO}, I_{CEO}, input dynamic resistance and output dynamic resistance. ○ Draw the input and output characteristics of a transistor in CB and CE configurations. ○ Calculate transistor parameters from characteristics. ○ Derive the relationship between alpha and beta of a transistor. ○ Compare the CB and CE configurations. ○ Explain the superiority of CE configuration over CB configuration in amplifier circuits. ○ Write code numbers (or type numbers) of at least five commonly used Ge and Si transistors manufactured in India. ○ Draw the circuit diagram of a basic transistor amplifier in CE configuration. ○ Draw the dc loadline on the output characteristics, for the given amplifier circuit. ○ Calculate the current gain, voltage gain, and the power gain for a simple amplifier circuit, by using the dc load line. ○ Explain the phase reversal of the signal when it is amplified by CE amplifier. ○ Explain the basic construction of alloy junction transistors and silicon planar transistors. ○ Refer to the transistor data sheet for a given transistor. ○ Explain the phenomenon of thermal runaway of a transistor. ○ Explain the use of heat sinks in power transistors. ○ Explain the simple structure of a junction field-effect transistor (JFET). ○ Explain the static drain characteristics of a JFET. ○ Define the three parameters (μ, r_d, g_m) of a JFET.

5.1 INTRODUCTION

The transistor was invented in 1948 by John Bardeen, Walter Brattain and William Shockley at Bell Laboratory in America. They were awarded the Nobel Prize in recognition of their contributions to Physics. This invention completely revolutionized the electronic industry. Since then, there has been a rapidly expanding effort to utilize and develop many types of semiconductor devices such as FET, MOSFET, UJT, SCR, etc.

Transistors have replaced bulky vacuum tubes (discussed in the next chapter) in performing many jobs. Transistors offer several advantages over tubes. A few of them are:

(i) No heater or filament is required; hence no heating delays, and no heating power needed.

(ii) They are much smaller in size and are light in weight.

(iii) Very low operating voltages can be used.

(iv) They consume little power, resulting in greater circuit efficiency.

(v) They have long life with essentially no ageing effect.

(vi) They are essentially shock-proof.

5.2 JUNCTION TRANSISTOR STRUCTURE

A transistor is basically a silicon or germanium crystal containing three separate regions. It can either be *NPN*-type or *PNP*-type. Figure 5.1*a* shows an *NPN* transistor. It has three regions. The middle region is called the *base* and the two outer regions are called the *emitter* and the *collector*. Although the two outer regions are of the same type (*N*-type), their functions cannot be interchanged. The two regions have different physical and electrical properties. In most transistors, the collector region is made physically larger than the emitter region since it is required to dissipate more heat. The base is very *lightly doped,* and is very *thin*. The emitter is heavily doped. The doping of the collector is between the heavy doping of the emitter and the light doping of the base. The function of the emitter is to *emit* or inject electrons (holes in case of a *PNP* transistor) into the base. The base passes most of these electrons (holes in case of *PNP*) onto the collector. The collector has the job of *collecting* or gathering these electrons (holes in case of a *PNP*) from the base.

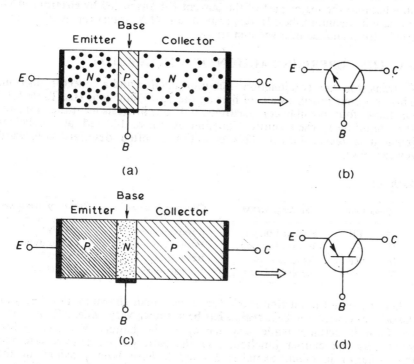

Fig. 5.1   Junction transistor: (*a*) *NPN*-type; (*b*) *NPN*-transistor symbol
(*c*) *PNP*-type;   (*d*) *PNP* transistor symbol

A transistor has two *PN*-junctions. One junction is between the emitter and the base, and is called the emitter-base junction, or simply the *emitter junction*. The other junction is between the base and the collector, and is called collector-base junction, or simply *collector junction*. Thus, a transistor is like two *PN*-junction diodes connected back-to-back. The *PN*-junction theory, learnt in the last chapter, will be used to discuss the action of a transistor.

Figure 5.1*b* shows the symbol for an *NPN* transistor. Note that in the symbol, the emitter (not the collector) has an arrowhead. The arrowhead points in the direction of the conventional emitter current (from *P* region to *N* region).

Figure 5.1*c* shows the structure of a *PNP* transistor and Fig. 5.1*d* shows its symbol. Note the direction of the arrowhead in the emitter. In a *PNP* transistor, the conventional emitter current will flow from the emitter to the base. That is why the direction of arrowhead is inward (from *P* region to *N* region).

Both types (*PNP* and *NPN*) of transistors are widely used; sometimes together in the same circuit. We shall study both the types. However, to avoid confusion, the discussion in this chapter will concentrate on the *NPN* type. Since a *PNP* transistor is the *complement* of an *NPN* transistor, it is merely necessary to read hole for electron, electron for hole, negative for positive, and positive for negative, for the corresponding operation of a *PNP* transistor. The choice of an *NPN* transistor is found to be more suitable because the major part of the current is transported by electrons as is the case in vacuum tubes. In this respect, an *NPN* transistor is the semiconductor analogue of a vacuum triode.

5.3 THE SURPRISING ACTION OF A TRANSISTOR

A transistor has two junctions—emitter junction and a collector junction. There are four possible ways of biasing these two junctions (see Table 5.1). Of these four possible combinations, only one interests us at the moment. It is condition I, where emitter junction is forward-biased and collector junction is reverse-biased. This condition is often described as forward reverse (FR).

Table 5.1

Condition		Emitter junction	Collector junction	Region of operation
I.	FR	Forward-biased	Reverse-biased	Active
II.	FF	Forward-biased	Forward-biased	Saturation
III.	RR	Reverse-biased	Reverse-biased	Cutoff
IV.	RF	Reverse-biased	Forward-biased	Inverted

Let us connect a junction transistor in the circuit shown in Fig. 5.2. For the sake of clarity, the base region has been shown very wide. (Remember, the base is actually made very narrow.) The battery V_{EE} acts to forward bias the emitter junction, and the battery V_{CC} acts to reverse-bias the collector junction. Switches S_1 and S_2 have been provided in the emitter and collector circuits. When the two switches are open, the two junctions are unbiased. We thus have depletion or space-charge regions at the two junctions.

If we close the switch S_1 and keep the switch S_2 open, the emitter junction will be forward biased as shown in Fig. 5.3. The barrier at the emitter junction is reduced. Since, emitter and base regions are just like those in a *PN* diode, we can expect a large current due to forward biasing. This current consists of majority carriers diffusing across the junction. Electrons diffuse from the emitter to the base, and holes from the base to the emitter. *The total current flowing across the junction is the sum of the electron diffusion current and the hole diffusion current.* In a transistor, the base

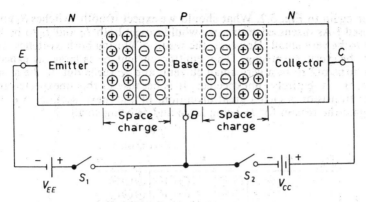

Fig. 5.2 Biasing an *NPN* transistor for active operation

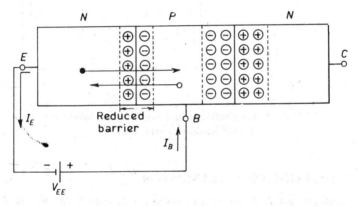

Fig. 5.3 Only emitter junction is forward-biased —
a large current flows

region is deliberately doped very lightly compared to the emitter region. Because of this, there are very few holes in the base region. As a result, over 99 % of the total current is carried by the electrons (diffusing from the emitter to the base). The emitter current I_E and the base current I_B in Fig. 5.3 are quite large. The two currents must be equal ($I_E = I_B$). The collector current I_C is zero.

Next, we close switch S_2 and keep the switch S_1 open in Fig. 5.2. This situation is shown in Fig. 5.4. The collector junction is reverse-biased. Very small current flows across this reverse-biased junction. The reverse leakage current is due to the movement of minority carriers. These carriers are accelerated by the potential barrier. Just as in the *PN*-junction diode, this leakage current is very much temperature dependent. The current flows into the collector lead and out of the base lead. There is no emitter current ($I_E = 0$). The small collector current is called the collector leakage current. It is given a special symbol, I_{CBO}. The subscript *CBO* in this symbol signifies that it is a current between Collector and Base, when the third terminal (i.e. emitter) is Open.

Refer again to Fig. 5.2. What should we expect if both switches S_1 and S_2 are closed ? As discussed above, we would expect both I_E and I_B to be large and I_C to be very small. However, the result of closing both switches turns out to be very surprising. The emitter current I_E is large, as expected. But I_B turns out to be a very small current, and I_C turns out to be a large current. It is entirely unexpected. It is because of this unexpected result that the transistor is such a great invention. In the next section, we shall investigate the reason for I_C being large and I_B being small.

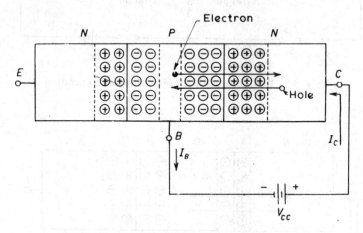

Fig. 5.4   Only collector junction is reverse biased —
a small leakage current flows

5.4   THE WORKING OF A TRANSISTOR

Let us consider an *NPN* transistor biased for active operation. As shown in Fig. 5.5, the emitter-base junction is forward biased by V_{EE}, and the collector-base junction is reverse-biased by V_{CC}. The directions of various currents that flow in the transistor are also indicated in Fig. 5.5. As is the usual convention, the direction of current flow has been taken opposite to the direction of electron movement. To understand the action of the transistor, we have numbered some of the electrons and holes. This will simplify the description.

The emitter junction is forward-biased (may be, by a few tenths of a volt). The barrier potential is reduced. The space-charge region at the junction also becomes narrow. As such, majority charge carriers diffuse across the junction. The resulting current consists of electrons travelling from the emitter to the base, and holes passing from the base to the emitter. As will soon be evident, only the electron current is useful in the action of the transistor. Therefore, the electron current is made much larger than the hole current. This is done by doping the base region more lightly than the emitter region. In Fig. 5.5, we have shown electrons 1, 2, 3 and 4 crossing from the emitter to the base, and hole 7 from the base to the emitter. The total sum of these charge-carrier movements constitutes the emitter current I_E. Only a portion of this current is due to the movement of electrons 1, 2, 3 and 4. These are the electrons injected by the emitter into the base. The ratio of the electron current to the total emitter current is known as *emitter*

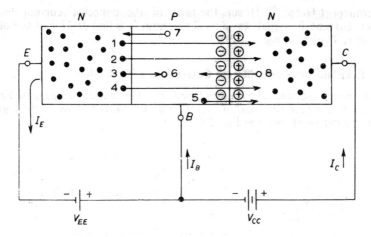

Fig. 5.5 An *NPN* transistor biased for active operation

injection ratio, or the *emitter efficiency.* This ratio is denoted by symbol γ (greek letter *gamma*). Typically, γ is equal to 0.995. This means that only 0.5 % of the emitter current consists of the holes passing from the base to the emitter.

Once the electrons are injected by the emitter into the base, they become minority carriers (in the base region). These electrons do not have separate identities from those which are thermally generated in the base region itself. (Note that these electrons are emitted by the emitter, and are in addition to the thermally generated minority carriers in the base region). The central idea in transistor action is that the base is made very narrow (about 25 μm) and is very lightly doped. Because of this, most of the minority carriers (electrons) travelling from the emitter end of the base region to its collector end do not recombine with holes in this journey. Only a few electrons (like 3) may recombine with holes (like 6). The ratio of the number of electrons arriving at collector to the number of emitted electrons is known as the *base transportation factor.* It is designated by symbol* β'. Typically, $\beta' = 0.995$.

Refer to Fig. 5.5. Movement of hole 8 from the collector region and electron 5 from the base region constitutes leakage current, I_{CBO}. Movement of electron 3 and hole 7 constitute a part of emitter current I_E. These two currents are not equal. Actually, the number of electrons (like 3) and holes (like 7) crossing the emitter-base junction is much more than the number of electrons (like 5) and holes (like 8) crossing the collector-base junction. The difference of these two currents in the base region makes the base current I_B.

The collector current is less than the emitter current. There are two reasons for this. Firstly, a part of the emitter current consists of holes that do not contribute to the collector current. Secondly, not all the electrons injected into the base are successful in reaching the collector. The first factor is represented by the emitter injection ratio γ; and the second, by the

---

*We are using the symbol β' to represent base transport factor, so as not to confuse it with the β of the transistor. The β of a transistor stands for its short-circuit current gain in CE mode.

base transport factor β'. Hence, the ratio of the collector current to the emitter current is equal to $\beta'\gamma$. This ratio is called dc *alpha* (α_{dc}) of the transistor. Typically, $\alpha_{dc} = 0.99$.

5.4.1 Relations between Different Currents in a Transistor

Let us now examine the role played by the batteries V_{EE} and V_{CC} in Fig. 5.5. These batteries help in maintaining the current flow in the transistor. To understand this, see Fig. 5.6.

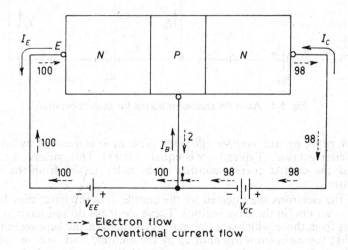

--→ Electron flow

—→ Conventional current flow

Fig. 5.6 Relationship between different transistor currents

We have seen that the emitter region emits a large number of electrons into the base region. Also, some holes diffuse from base to the emitter region. These holes recombine with electrons available in the emitter region. This way, the emitter region becomes short of electrons temporarily. This shortage is immediately made up by the battery V_{EE}. The negative terminal of this battery supplies electrons to the emitter region. After all, the batteries is a storehouse of charge; they can supply as much charge as needed. To make matters simple, let us assume that 100 electrons are supplied by the negative terminal of the battery V_{EE}. (In actual practice, the electrons that flow are very large in number.) These 100 electrons enter the emitter region and constitute the current I_E in the emitter terminal. The conventional current (flow of positive charge or holes) flows in a direction opposite to that of the electron flow. The current I_E is shown coming out of the emitter terminal. This is why the symbol of an *NPN* transistor has an arrow in the emitter lead, pointing outward (Fig. 5.1b).

What happens to the 100 electrons that enter the emitter region? The majority of these electrons (say 99 electrons) are injected into the base region. One electron is lost in the emitter region because of the recombination with a hole that has diffused from the base region. Out of the 99 electrons injected into the base, say, only one recombines with a hole; the rest of them (98 electrons) reach the collector region. This happens because of the special properties of the base region (it is lightly doped and very thin).

In this manner, the base region loses only two holes (one diffuses to the emitter region and the other is lost in the base region itself, due to recombination). The loss of these two holes is made up by creation of two fresh holes in the crystal near the base terminal. In the process of creation of holes, two electrons are generated. These two electrons flow out of the base terminal and constitute the base current. The conventional base current I_B flows into the transistor and is very small in magnitude.

The 98 electrons reaching the collector region experience an attractive force due to the battery V_{CC}. They travel out of the collector terminal and reach the positive terminal of the battery V_{CC}. The conventional collector current I_C (due to the flow of 98 electrons) flows into the transistor. The current I_C is almost equal to, but slightly less than the emitter current I_E.

The negative terminal of the battery V_{CC} gives out as many electrons as are received by its positive terminal. These 98 electrons from the negative terminal of V_{CC} and the 2 electrons from the base terminal combine together (at the junction) to make up a total of 100 electrons. These 100 electrons reach the positive terminal of the battery V_{EE}. The circuit is thus complete. The battery V_{EE} had given out 100 electrons from its negative terminal.

The total current flowing into the transistor must be equal to the total current flowing out of it. Hence, *the emitter current is equal to the sum of the collector and base currents.* That is,

$$I_E = I_C + I_B \qquad (5.1)$$

This equation is a simple statement of what we have discussed up to now; the emitter current distributes itself into the collector current and base current.

There is another point. From the discussions above we can state that the collector current is made up of two parts: (i) The fraction of emitter current which reaches the collector; (ii) The normal reverse leakage current I_{CO}. In equation form, we can write

$$I_C = \alpha_{dc} I_E + I_{CO} \qquad (5.2)$$

where α_{dc} is the fraction of the emitter current I_E that reaches the collector.

5.4.2 DC Alpha (α_{dc})

We can solve Eq. 5.2 for α_{dc} and write

$$\alpha_{dc} = \frac{I_C - I_{CO}}{I_E}$$

Usually, the reverse leakage current I_{CO} is very small compared to the total collector current. Neglecting this current, the above equation can be written as

$$\alpha_{dc} = \frac{I_C}{I_E} \qquad (5.3)$$

Here it is simply given as the ratio of dc collector current to dc emitter current in the transistor. If, for instance, in a transistor, we measure an I_C of 4.9 mA and an I_E of 5 mA, its dc alpha will be

$$\alpha_{dc} = \frac{4.9}{5} = 0.98$$

The thinner and more lightly doped the base is, the greater is the value of α_{dc}. But dc alpha of a transistor can never exceed unity. Many transistors have α_{dc} greater than 0.99, and almost all have α_{dc} greater than 0.95.

Example 5.1 A certain transistor has α_{dc} of 0.98 and a collector leakage current I_{CO} of 1 μA. Calculate the collector and the base currents, when $I_E = 1$ mA.

Solution: With $I_E = 1$ mA, we can use Eq. 5.2 to calculate the collector current.

$$I_C = \alpha_{dc}I_E + I_{CO}$$
$$= 0.98 \times 1 \times 10^{-3} + 1 \times 10^{-6} = 0.981 \times 10^{-3}$$
$$= \textbf{0.981 mA}$$

Now, using Eq. 5.1, the base current can be calculated as

$$I_B = I_E - I_C$$
$$= 1 \times 10^{-3} - 0.981 \times 10^{-3} = 0.019 \times 10^{-3} = 0.019 \text{ mA}$$
$$= \textbf{19 } \boldsymbol{\mu} \textbf{A}$$

Note that I_C and I_E are almost equal and I_B is very small.

5.4.3 Sign Conventions

The sign convention for the currents and voltages in a transistor is the same as followed in a two-port network. Figure 5.7a shows a general two-port network. The port on the left (with terminals 1 1') is the input port and the one on the right (with terminals 2 2') is the output port. Usually, one terminal is made common to the input and the output, and is often grounded. Figure 5.7a also shows the reference directions of input and output currents as well as voltages.

A transistor is a three-terminal device. If one of the terminals is considered common to input and output, a transistor becomes a two-port device. Figure 5.7b shows an *NPN* transistor as a two-port network in which base is made common to the input and the output. Similarly, Fig. 5.7c shows a *PNP* transistor.

As a standard convention, *all the currents entering into the transistor are taken to be positive*. A current flowing out is negative. In other words, if the actual conventional current flows in the outward direction, a negative sign is included along with the magnitude. Hence, for an *NPN* transistor (see Fig. 5.7b), the emitter current I_E is negative, whereas both the base current and the collector current are positive. In a *PNP* transistor (Fig. 5.7c) the emitter current is positive, but the base and collector currents are negative. In many textbooks, however, to avoid confusion, the actual direction of current flow is indicated in the diagrams.

In Figs. 5.7b and c, the transistors are connected in common-base configuration. The base is common to the input and the output. The potential (or voltages) of the emitter and collector terminals are written with reference to the common terminal (here, base). Thus, voltage V_{EB} is the voltage of emitter with respect to base. The reference direction of the voltage is indicated by a single-ended arrow (as in Fig. 5.7b), or by a double-ended arrow with a plus and a minus sign (as in Fig. 5.7c). The voltage V_{CB} represents

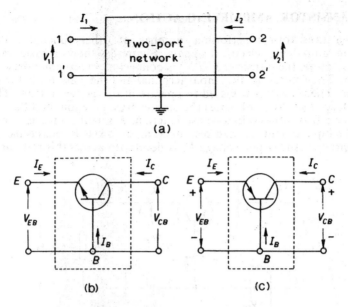

Fig. 5.7 **Sign convention for currents and voltages:**
(a) Two-port network; (b) *NPN* transistor
(c) *PNP* transistor

the voltage of the collector with respect to the base. In case, the common (reference) terminal is at higher potential, the voltage is given negative sign. For an *NPN* transistor, biased to operate in active region (as is done in Fig. 5.6), the voltage V_{EB} is negative and voltage V_{CB} is positive (since the battery V_{EE} sets the emitter at lower potential and the battery V_{CC} sets the collector at higher potential with respect to the base).

5.4.4 Other Conditions of Operation

To understand the operation of a transistor completely, we should briefly discuss other conditions of operation given in Table 5.1. Condition II has FF bias (both the junctions are forward biased). The transistor is in *saturation*. The collector current becomes almost independent of the base current. The transistor acts like a closed switch.

Condition III has RR bias, and it represents *cut-off* operation. In this condition, both junctions are reverse-biased. The emitter does not emit carriers into the base. There are no carriers to be collected by the collector. The collector current is thus zero (except a little current because of thermally generated minority carriers). The transistor acts like an open switch in this condition. We will talk more about saturation and cut-off when we discuss transistor characteristics.

Condition IV has RF bias, and it leads to *inverted* operation. This operation is quite different from the normal operation (condition I, active). Since the emitter and the collector are not doped to the same extent, they cannot be interchanged. The RF bias will result in very poor transistor action, and is rarely used.

5.5 TRANSISTOR AMPLIFYING ACTION

Though a transistor can perform a number of other functions, its main use lies in amplifying electrical signals. Figure 5.8 shows a basic transistor amplifier. Here, the transistor (NPN) is connected in common-base configuration. The emitter is the input terminal and the collector is the output terminal. The transistor is biased to operate in the active region. That is, the battery V_{EE} forward-biases the emitter-base junction, and the battery V_{CC} reverse-biases the collector-base junction. A signal source v_s is connected in the input circuit. A load resistance R_L of 5 kΩ is connected in the output circuit. An output voltage V_o is developed across this resistor.

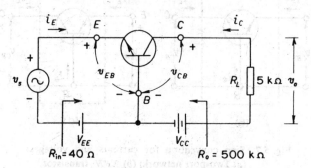

Fig. 5.8   A basic transistor amplifier in common-base
configuration

When the signal V_s is superimposed on the dc voltage V_{EE}, the emitter voltage V_{EB} varies with time. As a result, the emitter current I_E also varies with time. Since the collector current is a function of the emitter current, a similar variation occurs in the collector current. This varying current passes through the load resistor R_L and a varying voltage is developed across it. This varying voltage is the output voltage V_o.

The output signal V_o is many times greater than the input signal voltage V_s. To understand how the signal voltage is magnified (or amplified), let us consider how the transistor responds to the ac signal. Since the emitter-base junction is forward biased, it offers very low impedance to the signal source V_s. In the common-base configuration, the input resistance typically varies from 20 Ω to 100 Ω. The output junction (the collector-base junction) being reverse-biased, offers high resistance. Typically, the output resistance may vary from 100 kΩ to 1 MΩ. Assume that the input signal voltage is 20 mV (rms or effective value). Using an average value of 40 Ω for the input resistance, we get the effective value of the emitter-current variation as

$$I_e = \frac{20 \times 10^{-3}}{40} = 0.5 \text{ mA}$$

Since the collector current is almost the same as the emitter current (in fact it is slightly less), the effective value of the collector current variation is

$$I_e \cong I_c = 0.5 \text{ mA}$$

Now, the output resistance of the transistor is very high (say, 500 kΩ) and the load resistance is comparatively low (5 kΩ). The output side of the transistor acts like a constant current source; almost all the current I_c passes through the load resistance R_L.

Therefore, the output voltage is

$$V_o = I_c R_L$$
$$= (0.5 \times 10^{-3}) \times (5 \times 10^3) = 2.5 \text{ V}$$

The ratio of the output voltage V_o to the input voltage V_s is known as the *voltage amplification* or *voltage gain* A_v of the amplifier. For the amplifier in Fig. 5.8,

$$A_v = \frac{V_o}{V_s}$$

$$= \frac{2.5}{20 \times 10^{-3}} = 125$$

The transistor's amplifying action is basically due to its capability of *transferring* its signal current from a *low resistance* circuit to *high resistance* circuit. Contracting the two terms *transfer* and *resistor* results in the name *transistor*; that is,

*trans*fer + re*sistor* → *transistor*

5.5.1 Standard Notation for Symbols

When a transistor (or a vacuum tube) is used in a circuit, we talk of various quantities to explain its working. A standard notation of symbols to denote these quantities has been adopted by the Institution of Electrical and Electronics Engineers (IEEE). The notation is summarized as follows:

1. Instantaneous values of quantities which vary with time are represented by lower case (small) letters (for example, i for current, v for voltage, and p for power).
2. Upper case (capital) letters are used to indicate either the dc values or the effective (rms) values of ac.
3. Average (or dc) values and instantaneous total values are indicated by the capital subscripts of the proper electrode symbol (E for emitter, C for collector, and B for base; K for cathode, P for plate and G for grid in case of vacuum triode; and S for source, D for drain, and G for gate in case of FET).
4. Time varying components (ac components) are indicated by the small letter subscripts of the proper electrode symbol.
5. The current reference direction is indicated by an arrow. The voltage reference polarity is indicated by plus and minus signs or by an arrow that points from the negative to the positive terminal. For example, in Fig. 5.8, instantaneous total value of the emitter-to-base voltage is written as v_{EB}, but if the base terminal is understood to be common and grounded, we may shorten the symbol v_{EB} to simply v_E. Here, the voltage v_E is the voltage of emitter (with respect to the common terminal base) and is negative.
6. The conventional current flow into an electrode from the external circuit is taken as positive.
7. The magnitude of dc supply is indicated by using double subscripts of the proper electrode symbol. For instance, in Fig. 5.8, V_{CC} represents the magnitude (the sign is taken care of separately) of the dc supply in the collector circuit.

For better understanding of the above rules, let us examine the input circuit of Fig. 5.8. Before the signal voltage v_s is connected to the input circuit, the emitter-to-base voltage is V_{EB}, (or simply V_E). This voltage is the same as the dc supply voltage V_{EE} with a negative sign. That is, $V_{EB} = V_E = -V_{EE}$. See Fig. 5.9a. The variation of signal voltage v_s with time is shown in Fig. 5.9b. When this signal voltage is connected, the total instantaneous value of the emitter voltage (with respect to the base) v_E varies with time as shown in Fig. 5.9c, since

$$v_E = -V_{EE} + v_s$$

At any instant t_1, the instantaneous value of the ac component (v_e) of the voltage (v_E) is also shown. In the figure, note that the voltage v_e is the same as signal voltage v_s.

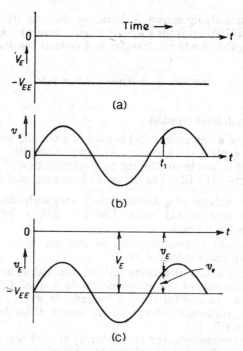

Fig. 5.9 When signal voltage v_s is connected in the input circuit, the instantaneous value of the emitter voltage changes with time

5.6 THREE CONFIGURATIONS

In the previous section, we have seen how a transistor amplifies ac signals when connected in common-base configuration. Is common-base (CB) the only configuration in which a transistor can work as an amplifier ? No. In fact, a transistor can be used as an amplifier in any one of the three configurations. Any of its three electrodes can be made common to input and output. (This common terminal is usually grounded or connected to the chassis.) The connection is then described in terms of the common electrode. For example, in the circuit of Fig. 5.8, the base terminal has been made common to both input and output. This connection is called common-base connection.

The input signal is fed between the emitter and the base. The output signal is developed between the collector and the base. By making the emitter or the collector common, we can have what are known as common-emitter (CE) or common-collector (CC) configurations, respectively. In all the configurations, *the emitter-base junction is always forward-biased and the collector-base junction is always reverse-biased.*

Figure. 5.10 shows three configurations from the ac (signal) point of view None of the configurations shows dc biasing. But it is understood that in all the three configurations, the transistor is working in the active region (i.e. it has FR bias). In common-emitter configuration (see Fig. 5.10*b*) the base is the input terminal and the collector is the output terminal. The input signal is connected between the base and the emitter and the load resistor is connected between the collector and the emitter. The output appears across this load resistor.

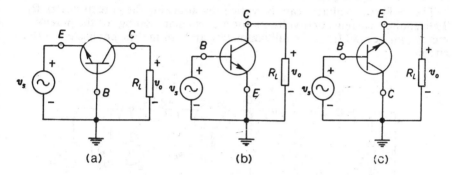

(a)   (b)   (c)

Fig. 5.10 Three configurations in which a transistor may be connected

Figure. 5.10*c* shows common-collector (CC) configuration. Here, the input signal is connected between the base and the collector. The output appears between the emitter and the collector. This circuit is popularly known as *emitter follower.* The voltage gain of this amplifier is poor (it never exceeds unity). But it has got an important characteristics of having very high input resistance and very low output resistance. This property of the emitter follower makes it very useful in certain applications.

5.7 TRANSISTOR CHARACTERISTICS

Knowing α_{dc} (dc alpha) of a transistor does not describe its behaviour. Many more details about a transistor can be studied with the help of curves that relate transistor currents and voltages. These curves are known as *static characteristic curves.* Though many sets of characteristic curves can be plotted for a given configuration, two of them are most important. In fact, these two sets of characteristics completely describe the static operation of the transistor. One is the *input characteristic* and the other is the *output characteristic.* Each curve of the input characteristic relates the input current with the input voltage, for a given output voltage. The output characteristic curve relates the output current with the output voltage, for a given input current.

Although, it is possible to draw the CC characteristics of a transistor, usually they are not needed. The common-collector configuration can be treated as a special case of common-emitter configuration (with feedback applied). The details (or the design) of a CC amplifier can be known from the CE characteristics. The static characteristics of a transistor in CC configuration are therefore not required. This is why the CC characteristics are not discussed in this book.

5.7.1 Common-Base (CB) Configuration

The circuit arrangement for determining CB characteristics of a transistor (here, we have taken *PNP* type) is shown in Fig. 5.11. The emitter-to-base voltage V_{EB} can be varied with the help of a ponentiometer R_1. Since the voltage V_{EB} is quite low (less than one volt) we include a series resistor R_S (say, 1 kΩ) in the emitter circuit. This helps in limiting the emitter current I_E to a low value; without this resistor, the current I_E may change by large amount even if the potentiometer (R_1) setting is moved slightly.

The collector voltage can be varied by adjusting the potentiometer R_2. The required currents and voltages for a particular setting of the potentiometers can be read from the milliammeters and voltmeters connected in the circuit.

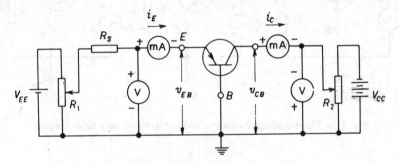

Fig. 5.11 Circuit arrangement for determining the static characteristics of a *PNP* transistor in CB configuration

Input CB Characteristics The common-base input characteristics are plotted between emitter current i_E and the emitter-base voltage v_{EB}, for different values of collector–base voltage V_{CB}. Figure 5.12 shows typical input characteristics for a *PNP* transistor in common-base configuration.

For a given value of V_{CB}, the curve is just like the diode characteristic in forward-bias region. Here, the emitter-base is the *PN*-junction diode which is forward-biased. This junction becomes a better diode as V_{CB} increases. That is, there will be a greater i_E for a given v_{EB} as V_{CB} increases, although the effect is very small.

For a diode, we had seen that its dynamic resistance is calculated from the slope of its forward characteristic curve. In a similar way, from the slope of the input characteristic we can get the *dynamic input resistance* of the transistor:

$$r_i = \frac{\Delta v_{EB}}{\Delta i_E}\bigg|_{V_{CB} = \text{const.}} \tag{5.4}$$

The dynamic input resistance r_i is very low (20 to 100 Ω). Since the curve is not linear, the value of r_i varies with the point of measurement. As the

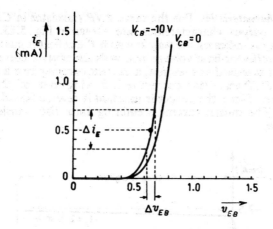

Fig. 5.12 Common-base input characteristics for a
typical *PNP* silicon transistor

emitter-base voltage increases, the curve tends to become more vertical. As a result, r_i decreases.

The input characteristics of an *NPN* transistor are similar to those in Fig. 5.12, differing only in that both i_E and v_{EB} would be negative and V_{CB} would be positive.

Example 5.2 The input characteristics of a *PNP* transistor in common-base configuration are given in Fig. 5.12. Determine the dynamic input resistance of the transistor at a point where $I_E = 0.5$ mA and $V_{CB} = -10$ V.

Solution: Around $I_E = 0.5$ mA, we take a small change Δi_E. Let

$$\Delta i_E = 0.7 - 0.3 = 0.4 \text{ mA}$$

From the curve for $V_{CB} = -10$ V (see Fig. 5.12), the corresponding change Δv_{EB} in emitter-base voltage is

$$\Delta v_{EB} = 0.70 - 0.62 = 0.08 \text{ V}$$

The dynamic input resistance is

$$r_i = \frac{\Delta v_{EB}}{\Delta i_E}\bigg|_{(V_{CB} = -10 \text{ V})} = \frac{0.08}{0.4 \times 10^{-3}}$$

$$= 200 \ \Omega$$

This value of the input resistance is somewhat higher than what is expected. When the transistor is operated as an amplifier, the emitter current may be a few milliamperes. For higher values of emitter currents the input characteristic curve becomes steeper. The input resistance r_i decreases to a very low value (say, 20 Ω).

Output CB Characteristics For the same *PNP* transistor in *CB* configuration, a set of output characteristics are shown in Fig. 5.13. The output characteristic curve indicates the way in which the collector current i_C varies with change in collector-base voltage v_{CB}, with the emitter current I_E kept constant. As per standard convention, a current entering into a transistor is positive. For a *PNP* transistor, current i_C is flowing out of the transistor and is negative. Since the collector junction is reverse biased, the voltage v_{CB} is negative. The emitter current is entering into the transistor, and is taken as positive.

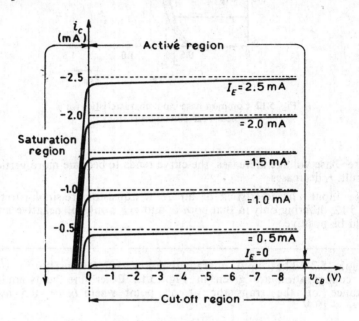

Fig. 5.13 Common-base output characteristics for a
PNP transistor

A close look at the output characteristics of Fig. 5.13 reveals the following interesting points:

(i) The collector current I_C is approximately equal to the emitter current I_E. This is true only in the *active* region, where collector-base junction is reverse-biased.

(ii) In the active region, the curves are almost flat. This indicates that i_C (for a given I_E) increases only slightly as v_{CB} increases. Is it not what happens in a constant current source? The transistor characteristic (in *CB* configuration) is similar to that of the current source. It means that the transistor should have high output resistance (r_o).

(iii) As v_{CB} becomes positive (the collector-base junction becomes forward-biased), the collector current i_C (for a given I_E) sharply decreases. This is the *saturation* region. In this region, the collector current does not depend much upon the emitter current.

(iv) The collector current is not zero when $I_E = 0$. It has a very small value. This is the reverse leakage current I_{CO}. The conditions that exist when $I_E = 0$ for *CB* configuration is shown in Fig. 5.14. The

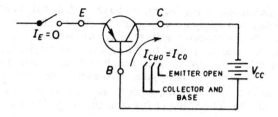

Fig. 5.14 Reverse leakage current in CB configuration

notation most frequently used for I_{CO} is I_{CBO}, as indicated in the figure. In this notation, the subscript CBO means that it is the current between the collector and base when the third terminal (the emitter) is open. Mind you, the current I_{CBO} is like the reverse saturation current for a diode. This too is temperature sensitive. At room temperature, the typical values of I_{CBO} ranges from $2~\mu A$ to $5~\mu A$ for germanium transistors, and $0.1~\mu A$ to $1~\mu A$ for silicon transistors.

From the output characteristics of Fig. 5.13, we can determine a number of important transistor parameters, such as dynamic output resistance (r_o), dc current gain (α_{dc}), and ac current gain (α). The dynamtic output resistance is defined as

$$r_o = \frac{\Delta v_{CB}}{\Delta i_c}\bigg|_{I_E = \text{const.}} \qquad (5.5)$$

where Δv_{CB} and Δi_c are small changes in collector voltage and collector current around a given point on the characteristic curve (for given I_E). Since the output curves are very flat, for a given Δv_{CB}, the Δi_c is very small. It means the output resistance is very high (of the order of 1 MΩ).

The dc alpha of the transistor is defined as

$$\alpha_{dc} = \frac{I_C}{I_E} \qquad (5.6)$$

The characteristics tell us that at any point (in the active region) on the curve, I_C is less than I_E and the difference is very small. The value of α_{dc} is less than, but very close to unity. A typical value is 0.98.

A transistor is used as an amplifier. The amplifier handles ac (varying) signals. Under such a condition, we are more interested in the small changes in the voltages and currents rather than their absolute (dc) values. Specifically, we would like to know what change occurs in collector current for a given change in emitter current. This information is given by a parameter called ac current gain or ac alpha (α or h_{fb})*. It is defined as

$$h_{fb} \text{ or } \alpha = \frac{\Delta i_c}{\Delta i_E}\bigg|_{V_{CB} = \text{const.}} \qquad (5.7)$$

---

*The symbol h_{fb} comes from the analysis based on h-parameters or *hybrid* parameters. The letter f in the subscript stands for *forward*, and the letter b indicates a common-*base* connection. For details about h-parameters, the reader is advised to see Unit **8** on 'Small-Signal Amplifiers' (Sec. 8.3.2).

In the above definition of h_{fb} (or α), we have stated that $V_{CB} = $ constant See Fig. 5.8. When you apply an ac signal to the input, the current will change, and so will the collector voltage. The only way to keep V_{CB} constant (even when the ac signal is applied to the input) is to short-circuit the load resistor R_L. Therefore, the current gain h_{fb} should have been more appropriately called *short-circuit current gain*. However, very often, h_{fb} is simply referred to as *current gain*, with the understanding that it is defined under short-circuit condition. The value of h_{fb} is in the range from 0.95 to 0.995.

Summarizing the common-base configuration, we can say that the current gain h_{fb} (or α) is less than unity (typical value is 0.98), dynamic input resistance r_i is very low (typical value is 20 Ω), and dynamic output resistance is very high (typical value is 1 MΩ). The leakage current I_{CBO} is quite low (typically, 4 μA for germanium and 0.02 μA for silicon transistors). This current is temperature dependent.

Example 5.3 In a certain transistor, a change in emitter current of 1 mA produces a change in collector current of 0.99 mA. Determine the short circuit current gain of the transistor.

Solution : The short-circuit current gain of the transistor is given as

$$\alpha \text{ or } h_{fb} = \frac{\Delta i_c}{\Delta i_E}$$

$$= \frac{0.99 \times 10^{-3}}{1 \times 10^{-3}} = 0.99$$

5.7.2 Common-Emitter (CE) Configuration

In CE configuration, the emitter is made common to the input and the output. The signal is applied between the base and emitter, and the output is developed between the collector and emitter. Whether the transistor works in CB or CE configuration, it is to be ensured that it works in the active region. It means that the emitter-base junction is forward-biased and the collector-base junction is reverse-biased. Such biasing (FR biasing) is achieved in CE configuration by connecting the batteries V_{BB} and V_{CC} as shown in Fig. 5.15a. Here an *NPN* transistor is used. The emitter-base junction is forward-biased by the battery V_{BB}. This forward biasing needs a very small voltage (say, 0.6 V). The battery V_{CC} (say, 9 V) is connected between emitter and collector. Since the base is at $+V_{BB}$ potential with respect to the emitter, and the collector is at $+V_{CC}$ potential with respect to the emitter, the net potential of the collector with respect to the base is $V_{CC} - V_{BB}$. The collector-base junction is reverse-biased by this potential. Since V_{CC} is much larger than V_{BB}, the reverse-bias voltage may be taken as merely V_{CC}.

In Fig. 5.15b, the transistor is replaced by its symbol. The directions of actual currents are also marked in the figure.

Current Relations in CE Configuration We have seen that in CB configuration I_E is the input current and I_C is the output current. These currents are related through Eqs. 5.1 and 5.2 (rewritten below for convenience) :

$$I_E = I_C + I_B \tag{5.1}$$

$$I_C = \alpha_{dc} I_E + I_{CBO} \tag{5.2}$$

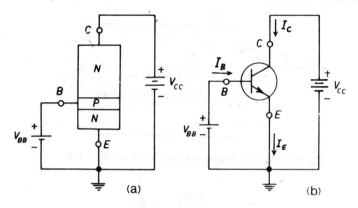

Fig. 5.15 FR biasing of an *NPN* transistor in common emitter
(CE) configuration

In CE configuration, I_B becomes the input current and the I_C is the
output current. We are interested in knowing how the output current I_C is
related with the input current I_B. That is, we should find a relation such as

$$I_C = f(I_B) \tag{5.8}$$

To obtain this relation, we simply substitute the expression of I_E from
Eq. 5.1 into Eq. 5.2, so that

$$I_C = \alpha_{dc}(I_C + I_B) + I_{CBO}$$

or $\qquad (1 - \alpha_{dc})I_C = \alpha_{dc}I_B + I_{CBO}$

or $\qquad I_C = \dfrac{\alpha_{dc}}{1 - \alpha_{dc}} I_B + \dfrac{1}{1 - \alpha_{dc}} I_{CBO} \tag{5.9}$

In this equation, I_C is given in terms of I_B. The equation can be simplified
somewhat by defining

$$\beta_{dc} = \frac{\alpha_{dc}}{1 - \alpha_{dc}} \tag{5.10}$$

and

$$I_{CEO} = \frac{I_{CBO}}{1 - \alpha_{dc}} \tag{5.11}$$

Thus, Eq. 5.9 becomes

$$I_C = \beta_{dc}I_B + I_{CEO} \tag{5.12}$$

This equation states that I_C is equal to β_{dc} multiplied by the input current I_B,
plus a leakage current I_{CEO}. This leakage current is the current which would
flow between the collector and emitter, if the third terminal (base) were
open. This is illustrated in Fig. 5.16. The magnitude of I_{CEO} is much larger
than that of I_{CBO}, as indicated by Eq. 5.11. For example, if $\alpha_{dc} = 0.98$, the
value of I_{CEO} is fifty times that of I_{CBO}. For silicon transistors, I_{CEO} would
typically be a few microamperes, but it may be a few hundred microamperes
for germanium transistors.

The factor β_{dc} is called the common-emitter dc current gain. It relates
the dc output current I_C to the input current I_B. Eq. 5.10 indicates that β
can be very large. For example, if $\alpha_{dc} = 0.98$, the value of β_{dc} is

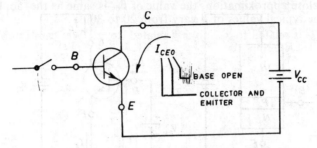

Fig. 5.16 Reverse leakage current in CE configuration

$$\beta_{dc} = \frac{0.98}{1-0.98} = 49$$

Typically β_{dc} can have values in the range from 20 to 300.

If we solve Eq. 5.12 for β_{dc}, we obtain

$$\beta_{dc} = \frac{I_C - I_{CEO}}{I_B} \tag{5.13}$$

If I_{CEO} is very small compared to I_C (as is the case usually) then

$$\beta_{dc} = \frac{I_C}{I_B} \tag{5.14}$$

Thus, β_{dc} is the ratio of dc collector current to dc base current.

How Beta of a Transistor is Related to Its Alpha The dc current gain of a transistor when connected in CE configuration, is β_{dc}. It is defined by Eq. 5.14. The same transistor connected in CB configuration gives a dc current gain of α_{dc}. Therefore, there is nothing surprising if beta (β_{dc}) of a transistor is related to its alpha (α_{dc}). This relation is given by Eq. 5.10. If the value of α_{dc} of a transistor is known, its β_{dc} can be calculated. Manipulating Eq. 5.10, we get

$$\beta_{dc} = \frac{\alpha_{dc}}{1-\alpha_{dc}}$$

or

$$\beta_{dc} - \beta_{dc}\alpha_{dc} = \alpha_{dc}$$

or

$$\beta_{dc} = \alpha_{dc}(1+\beta_{dc})$$

or

$$\alpha_{dc} = \frac{\beta_{dc}}{\beta_{dc}+1} \tag{5.15}$$

Thus, knowing the value of β_{dc}, we can calculate α_{dc} using the above equation.

When a transistor is used as an amplifier, we are more interested in knowing the ratio of *small changes* in the collector and base currents, rather than the ratio of their absolute values. This ratio is called ac or dynamic beta (β_{ac} or simply β). Thus, the ac beta is

$$\beta = \frac{\Delta i_C}{\Delta i_B}\bigg|_{V_{CE}=\text{const.}} \tag{5.16}$$

To a very close approximation, the value of β_{dc} is same as the ac beta (β). Like β_{dc}, the typical values of β vary from 20 to 300.

Just as β_{dc} is related to α_{dc}, so is β related to α. We can establish this relation by considering that

$$I_E = I_C + I_B$$

If we let I_C and I_B change by small amounts Δi_C and Δi_B, so that I_E changes by Δi_E, we would still have

$$\Delta i_E = \Delta i_C + \Delta i_B$$

Dividing the above equation by Δi_C and rearranging the terms, we get

$$\frac{\Delta i_E}{\Delta i_C} = 1 + \frac{\Delta i_B}{\Delta i_C}$$

or

$$\frac{1}{\alpha} = 1 + \frac{1}{\beta}$$

or

$$\beta = \frac{\alpha}{1-\alpha} \qquad (5.17)$$

Example 5.4 When the emitter current of a transistor is changed by 1 mA, its collector current changes by 0.995 mA. Calculate (a) its common-base short circuit current gain α, and (b) its common-emitter short circuit current gain β.

Solution : (a) Common-base short circuit current gain is given by

$$\alpha = \frac{\Delta i_C}{\Delta i_E} = \frac{0.995 \times 10^{-3}}{1 \times 10^{-3}} = 0.995$$

(b) Common-emitter short circuit current gain is

$$\beta = \frac{\alpha}{1-\alpha} = \frac{0.995}{1-0.995} = 199$$

Example 5.5 The dc current gain of a transistor in common-emitter configuration is 100. Find its dc current gain in common-base configuration.

Solution : We can use Eq. 5.15 to calculate the dc current gain in common-base configuration

$$\alpha_{dc} = \frac{\beta_{dc}}{\beta_{dc}+1} = \frac{100}{100+1} = 0.99$$

Input CE Characteristics In CE configuration, i_B and v_{BE} are the input variables. The output variables are i_C and v_{CE}. We can use the circuit arrangement of Fig. 5.17 to determine the input characteristics of a *PNP* transistor (for an *NPN* transistor, terminals of all the batteries, milli-ammeters and voltmeters will have to be reversed). Typical input characteristics are shown in Fig. 5.18. They relate i_B to v_{BE} for different values of V_{CE}. These curves are similar to those obtained for CB configuration (Fig. 5.12). Note that the change in output voltage V_{CE} does not result in a large devia-tion of the curves. In fact, for the commonly used dc voltages, the effect of changing V_{CE} on input characteristics may be ignored.

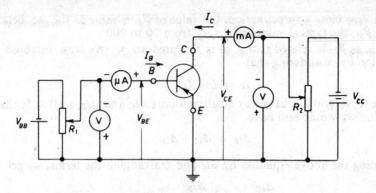

Fig. 5.17 Circuit arrangements for determining the static characteristics of a *PNP* transistor, in CE configuration

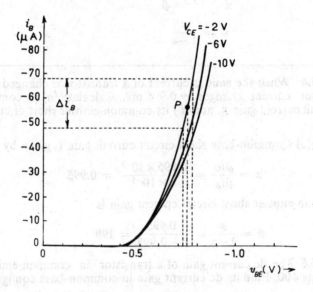

Fig. 5.18 Common-emitter input characteristics of a *PNP* transistor

We can find out the *dynamic input resistance* of the transistor at a given voltage V_{BE}, from Fig. 5.18. It is given by the reciprocal of the slope of the curve at that point. That is,

$$r_i = \frac{\Delta v_{BE}}{\Delta i_B}\bigg|_{V_{CE}=\text{const.}} \tag{5.18}$$

For example, the input resistance of the transistor at the point

$$V_{BE} = -0.75 \text{ V, and } V_{CE} = -2 \text{ V}$$

is calculated from Fig. 5.18, as follows·

$$r_i = \frac{\Delta v_{BE}}{\Delta i_B}\bigg|_{V_{CE}=-2\text{V}} = \frac{0.78-0.72}{(68-48)\times 10^{-6}} = \frac{0.06}{20\times 10^{-6}} = 3 \text{ k}\Omega$$

The value of r_i is typically 1 kΩ, but can range from 800 Ω to 3 kΩ.

Output CE Characteristics From the circuit arrangement of Fig. 5.17, we can also determine the output characteristics of a *PNP* transistor. Figure 5.19 shows typical output characteristics of a *PNP* transistor. They relate the output current i_C, to the voltage between collector and emitter, v_{CE}, for various values of input current, I_B. Note that the quantities v_{CE}, i_C and I_B are all negative for a *PNP* transistor. If the transistor is *NPN* type, we reverse the terminals of the batteries V_{CC} and V_{BB}, so that v_{CE}, i_C and I_B become positive.

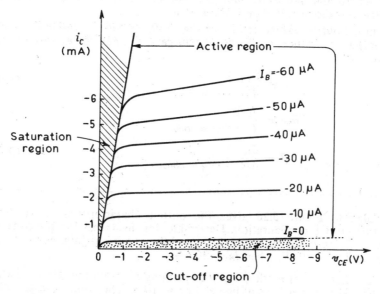

Fig. 5.19 Common-emitter output characteristics of a *PNP* transistor

A study of these output characteristics reveals following interesting points:

(i) In the active region, i_C increases slowly as v_{CE} increases. The slope of these curves is somewhat greater than the CB output characteristics (see Fig. 5.13). We know that β_{dc} is equal to the ratio I_C/I_B. For each curve of Fig. 5.19, the input current I_B is constant, but current i_C increases with v_{CE}. This indicates that β_{dc} increases with v_{CE}.

(ii) When v_{CE} falls below a few tenths of a volt, i_C decreases rapidly as v_{CE} decreases. This occurs as v_{CE} drops below the value of V_{BE}; the collector-base junction then becomes forward-biased. In this condition, both junctions of the transistor are forward-biased. The transistor is working in the *saturation region*. It is called saturation region, because the current I_C no longer depends upon the input current I_B.

(iii) In the active region, the collector current is β_{dc} times greater than the base current. Thus, small input current, I_B, produces a large output current I_C.

(iv) The collector current is not zero when I_B is zero. It has a value of I_{CEO}, the reverse leakage current. The current I_{CEO} is related with

I_{CBO} by Eq. 5.11. Using Eq. 5.10, we can write Eq. 5.11 in another form:

$$I_{CEO} = \frac{1}{1-\alpha_{dc}} I_{CBO}$$

$$= \frac{1}{1-\dfrac{\beta_{dc}}{1+\beta_{dc}}} I_{CBO}$$

or $\qquad I_{CEO} = (1+\beta_{dc})I_{CBO} \qquad\qquad (5.19)$

For a germanium transistor, I_{CEO} may typically have a value of 500 μA. For silicon transistor, it is only about 20 μA.

From the output characteristics of Fig. 5.19, we can determine the dynamic output resistance, r_o, the dc current gain, β_{dc}, and the ac current gain β, as follows:

$$r_o = \frac{\Delta v_{CE}}{\Delta i_C}\bigg|_{I_B=\text{const.}} \qquad\qquad (5.20)$$

$$\beta_{dc} = \frac{I_C}{I_B}\bigg|_{V_{CE} = \text{const.}} \qquad\qquad (5.21)$$

$$\beta = \frac{\Delta i_C}{\Delta i_B}\bigg|_{V_{CE} = \text{const.}} \qquad\qquad (5.22)$$

Example 5.6 Figure 5.20 gives the output characteristics of an *NPN* transistor in CE configuration. Determine, for this transistor, the dynamic output resistance, the dc current gain and the ac current gain, at an operating point $V_{CE} = 10$ V, when $I_B = 30\ \mu$A.

Solution: Let us first mark the given operating point on the given characteristics. We draw a vertical line at $V_{CE} = 10$ V. The point of intersection of this line with the characteristic curve for $I_B = 30\ \mu$A gives the operating point. The collector current at this point is 3.6 mA.

To determine the dynamic output resistance of the transistor we take a small change of collector voltage around the operating point. Let the voltage v_{CE} change from 7.5 V to 12.5 V. For a constant base current of 30 μA, the corresponding change in collector current may be seen to be from 3.5 mA to 3.7 mA. Therefore, the dynamic output resistance is given as

$$r_o = \frac{\Delta v_{CE}}{\Delta i_C}\bigg|_{I_B = 30\ \mu\text{A}} = \frac{12.5-7.5}{(3.7-3.5)\times 10^{-3}} = \frac{5}{0.2\times 10^{-3}}$$

$$= 25\ \text{k}\Omega$$

To find β_{dc} we should know the value of dc collector current corresponding to $I_B = 30\ \mu$A. From the characteristics it can be seen that $I_C = 3.6$ mA at this point. Therefore,

$$\beta_{dc} = \frac{I_C}{I_B} = \frac{3.6\ \text{mA}}{30\ \mu\text{A}} = 120$$

In order to calculate ac current gain (β), a vertical line corresponding to $V_{CE} = 10$ V is drawn. From the given characteristics it is clear that when

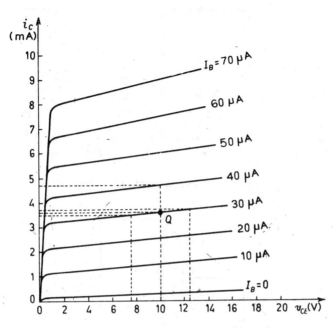

Fig. 5.20 Determination of dynamic output resistance, dc beta,
and ac beta, of an *NPN* transistor in CE mode

base current changes from 30 μA to 40 μA, the collector current changes
from 3.6 mA to 4.7 mA. Therefore, the ac current gain is given as

$$\beta = \frac{\Delta i_c}{\Delta i_B}\bigg|_{V_{CE} = 10\ V} = \frac{4.7\ mA - 3.6\ mA}{40\ \mu A - 30\ \mu A} = \frac{1.1 \times 10^{-3}}{10 \times 10^{-6}}$$

$$= 110$$

5.7.3 Common-Collector (CC) Configuration

In CC configuration, we make the collector common to the input and the
output. This is shown in Fig. 5.21a. The same circuit can be drawn in a
different way (Fig. 5.21b). Here the transistor is shown in the conventional
manner (the collector terminal at the upper end and the emitter terminal at
the lower end). Now, do you see some similarity between this circuit and
that of CE configuration (Fig. 5.10b). The two circuits look alike, except
for the fact that in the CC configuration the output is taken at the emitter
rather than the collector. Also, the load resistance R_L is connected between
the emitter and the ground.

The biasing arrangement for the CC configuration is shown in Fig. 5.22.
The battery V_{BB} forward-biases the base-emitter junction. The battery V_{CC}
has large voltage so that the collector-junction is reverse biased. If a *PNP*
transistor is used in place of the *NPN*, the polarities of the batteries V_{BB}
and V_{CC} are reversed. Note that the load resistor is connected to the emitter
terminal. Quite often, we name it R_L. You will see later that this circuit is
also called *emitter follower*.

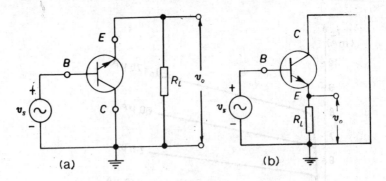

Fig. 5.21 Transistor connected in CC configuration

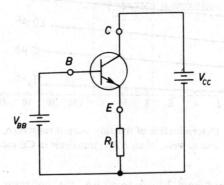

Fig. 5.22 Biasing arrangement for an *NPN*
transistor connected in common-
collector configuration,

Current Relations in CC Configuration In CC configuration, the base current is the input current, and the emitter current is the output current. The output current is dependent on the input current. That is,

$$I_E = f(I_B) \tag{5.23}$$

To find this functional relationship, we start with the basic current relations of a transistor (see Eqs. 5.1 and 5.2):

$$I_E = I_B + I_C$$

and

$$I_C = \alpha_{dc} I_E + I_{CBO}$$

Since the colletor is the common terminal, we are not interested in the value of collector current I_C. We, therefore, eliminate the collector current I_C from the above two equations. Substituting the expression for collector current from the second equation into the first equation, we get

$$I_E = I_B + \alpha_{dc} I_E + I_{CBO}$$
$$(1 - \alpha_{dc}) I_E = I_B + I_{CBO}$$

or

$$I_E = \frac{1}{1 - \alpha_{dc}} I_B + \frac{1}{1 - \alpha_{dc}} I_{CBO}$$

or

Since

$$\frac{1}{1-\alpha_{dc}} = \beta_{dc}+1$$

Therefore,

$$I_E = (\beta_{dc}+1)I_B + (\beta_{dc}+1)I_{CBO} \qquad (5.24)$$

If we neglect the leakage current I_{CBO}, then

$$I_E = (\beta_{dc}+1)I_B$$

or

$$\frac{I_E}{I_B} = (\beta_{dc}+1) \qquad (5.25)$$

Equation 5.25 shows that the dc current gain (sometimes designated as γ_{dc}) of this configuration is maximum. It is equal to $(\beta_{dc}+1)$. The leakage current in this configuration is as high as in CE configuration.

5.8 COMPARISON BETWEEN THE THREE CONFIGURATIONS

We have seen that a transistor can be connected in any one of the three configurations. It behaves differently in different configurations. In which configuration should we connect a transistor? This depends upon the particular application we desire. A configuration may be suitable for some application, whereas it may not be suitable for the other. What are the important parameters that govern the suitability of the configuration? We should know the input dynamic resistance, output dynamic resistance, dc current gain, ac current gain, ac voltage gain and leakage current of the transistor in a given configuration.

Out of the three configurations, the common-collector configuration has maximum input dynamic resistance. So we use this configuration where high input resistance is of prime importance, even though its voltage gain is less than unity. The decreased voltage gain can be compensated by subsequently using the CE configuration. We do not study the CC configuration separately as an independent circuit. It is usual practice to consider the CC configuration as a special case of the CE configuration*. We shall therefore consider and compare only the CB and CE configurations. Table 5.2 gives the typical values of the important parameters in the two configurations.

Table 5.2 COMPARISON BETWEEN CB AND CE CONFIGURATIONS

Parameters	Common-base configuration	Common-emitter configuration
1. Input dynamic resistance	Very low (20 Ω)	Low (1 kΩ)
2. Output dynamic resistance	Very high (1 MΩ)	High (10 kΩ)
3. Current gain	Less than unity (0.98)	High (100)
4. Leakage current	Very small (5 μA for Ge, 1 μA for Si)	Very large (500 μA for Ge, 20 μA for Si)

---

*This circuit, also called emitter follower, is discussed in Unit 12 on "Feedback in Amplifiers".

5.8.1 Input Dynamic Resistance

The input dynamic resistance of the CB configuration is much lower than that of the CE configuration. This fact can be seen from the definition of the input dynamic resistance of the two configurations:

(i) r_i for CB configuration $= \dfrac{\Delta v_{EB}}{\Delta i_E}\bigg|_{V_{CB}\,=\,\text{const.}}$

(ii) r_i for CE configuration $= \dfrac{\Delta v_{BE}}{\Delta i_B}\bigg|_{V_{CE}\,=\,\text{const.}}$

The numerators of the above two expressions are the same. But, the denominator of the first is of the order of a few milliamperes, whereas that of the second is of the order of a few microamperes. Hence, r_i for the CB configuration is much lower than that for the CE configuration.

5.8.2 Output Dynamic Resistance

Let us look at the output static characteristics of the two configurations (see Figs. 5.13 and 5.19). We note that the output characteristics of the CB configuration (Fig. 5.13) are almost horizontal. There is hardly any change in the collector current for a given variation in collector-to-base voltage. This means that the output resistance

$$r_o = \dfrac{\Delta v_{CB}}{\Delta i_c}\bigg|_{I_E\,=\,\text{const.}}$$

is very high (since Δi_c is very small for a certain value of Δv_{CB}). Now see Fig. 5.19. These curves are not so horizontal. As we increase v_{CB}, the collector current is seen to increase by an appreciable amount. This shows that the output dynamic resistance

$$r_o = \dfrac{\Delta v_{CE}}{\Delta i_c}\bigg|_{I_B\,=\,\text{const.}}$$

of the common-emitter configuration is not very high. It is of the order of 10 kΩ. Note that the slope of the output characteristic curve is not the same everywhere. It is for this reason that the value of the output dynamic resistance of the transistor depends upon the point around which the variations are taken.

5.8.3 Current Gain

The current gain (both dc as well as ac) of CB configuration is less than unity. It is typically 0.98. The closer its value to unity, the better is the transistor. A transistor having a low value of alpha (say, less than 0.95) will not make a good amplifier. Such transistors are rejected during manufacture.

The current gain of the CE configuration is quite high. It is typically 100, and it may be as high as 250. Such high current gain in the CE configuration makes it possible to have quite high voltage gain as well as high power gain.

5.8.4 Leakage Current

The leakage current in the CB configuration is very low (of the order of only a few μA). In the CE configuration, it is quite high (a few hundred μA). The leakage current in a transistor is due to the flow of minority carriers. The concentration of these minority carriers is very much dependent on temperature. Thus, the leakage current is temperature dependent. As the temperature rises, the leakage current rises. This may lead to what is known as thermal runaway of the transistor. The high value of the leakage current (and its rapid increase with temperature) in CE configuration is its great disadvantage. Since, silicon transistors have much less leakage current as compared to germanium transistors, we prefer to use silicon transistors.

5.9 WHY CE CONFIGURATION IS WIDELY USED IN AMPLIFIER CIRCUITS

The main utility of a transistor lies in its ability to amplify weak signals. The transistor alone cannot perform this function. We have to connect some passive components (such as resistors and capacitors) and a biasing battery. Such a circuit is then called an *amplifier*. Thus, an amplifier is an electronic circuit that is capable of amplifying (or increasing the level of) signals.

Very often, a single transistor amplifying stage is not sufficient. In almost all applications we use a number of amplifier stages, connected one after the other. The signal to be amplified is fed to the input of the first stage. The output of the first stage is connected to the input of the second stage. The second stage feeds the third stage, and so on. Ultimately, the output appears across the load connected to the output of the final stage. Such a connection of amplifier stages is known as *cascaded amplifier*.

Figure 5.23 shows a cascaded amplifier having two stages. The first stage is energized by a signal source having voltage v_s and internal resistance R_s. The load is connected to the output of the second stage at terminals A_3B_3. If this cascaded amplifier is to work properly, certain conditions must be satisfied. The working of one stage should not adversely affect the performance of the other.

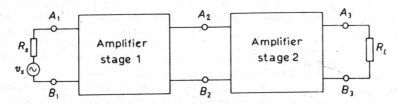

Fig. 5.23 Two amplifier stages cascaded to increase amplifying action

First of all, we would want the whole of (if not whole, then atleast most of) the signal voltage v_s to reach the input of the first stage. This can happen only when the input resistance of this stage, at terminals A_1B_1 is high (compared to source resistance R_s). Recall that a source works as a good voltage source when the load resistance is much greater than the source resistance. Here, the input resistance of the first stage acts as the load resistance to the source. Secondly, it is desirable that the performance of first stage is not disturbed when we connect the second stage at terminals $A_2 B_2$. For this, the output resistance of the first stage should be low. Also, the input resistance of the second stage (which comes in parallel with the load resistance of the first stage) should be high. You may recollect that connecting a high resistance in parallel with a low resistance element of a circuit does not much affect the working of the circuit. The resistance of the parallel combination will almost be the same as the low resistance itself.

Moreover, the first stage serves as the voltage source for the second stage. The input resistance of second stage acts as the load resistance for the voltage source (i.e. the first stage). The output resistance of the first stage is the internal resistance of the voltage source. The internal resistance of the source must be low compared to load resistance. Again, the second amplifier stage will deliver more power to the load R_L (this load may be a loudspeaker) only if its output resistance is low. Thus, we find that *a good amplifier stage is one which has high input resistance and low output resistance*.

A transistor in CB configuration has a very low input resistance ($\simeq 20 \ \Omega$) and a very high output resistance ($\simeq 1 \ M\Omega$). It is just the reverse of what we desire (high input resistance and low output resistance). That is why the CB configuration is unpopular. Comparatively, the CE configuration is much better, as regards its input and output resistances. Its input resistance is about 1 $k\Omega$ and output resistance about 10 $k\Omega$. A transistor in the CE configuration makes a much better amplifier. Furthermore, the current gain, voltage gain and power gain of CE is much greater than those of CB.

From the point of view of cascading of amplifier stages, the CC configuration would have been the best. Its input resistance is very high ($\simeq 150 \ k\Omega$) and output resistance is quite low ($\simeq 800 \ \Omega$). However, unfortunately the voltage gain of the CC amplifier is low (less than unity). Therefore, we use CC amplifier only in such applications where the requirement of high input resistance is of prime importance.

Thus we see that CE configuration is best suited for most of the amplifier circuits. We shall study this circuit in some detail.

5.10 BASIC CE AMPLIFIER CIRCUIT

Figure 5.24 shows a basic CE amplifier circuit*. Here, we have used an *NPN* transistor. The battery V_{BB} forward-biases the emitter junction. The series resistance R_B is meant to limit the base current within certain specified values. The battery V_{CC} is a relatively high-voltage battery (9 V). It reverse-biases the collector junction. The resistor R_C in the collector circuit is the load resistance. The amplified ac voltage appears across it.

The signal to be amplified is represented by voltage source v_s. The signal is applied to the base through the coupling capacitor C_{C1}. The capacitor permits only ac to pass through. It blocks dc voltage. The dc base current flows only through resistor R_B, and not through the voltage source v_s. Similarly, the coupling capacitor C_{C2} blocks dc from reaching the output terminals. Only ac signal voltage appears at the output v_o.

To observe the performance of the amplifier circuit, we take the help of a dc load line.

---

*This is not a practical circuit. In practice, we use only one battery (say, V_{CC}) for biasing both the collector-junction as well as the emitter-junction. We shall study such practical circuits later.

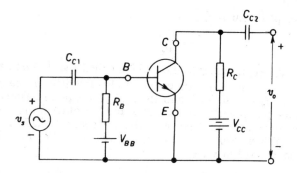

Fig. 5.24 Basic CE amplifier circuit

5.10.1 DC Load Line

Let us consider the amplifier circuit of Fig. 5.24, when no signal is applied to its input. This condition (of having no input signal) is described as a *quiescent condition*. The circuit then reduces to the one shown in Fig. 5.25. The battery V_{CC} sends current I_C through the load resistor R_C and the transistor. There is some voltage drop across the load resistor R_C due to the flow of current I_C. The polarity of this voltage drop $I_C R_C$ is shown in the figure. The remaining voltage drops across the transistor. This voltage is written as V_{CE}. Applying Kirchhoff's voltage law to the collector circuit, we get

$$V_{CC} = I_C R_C + V_{CE} \qquad (5.26)$$

We can rearrange the terms of the above equation and put it as

$$I_C = \left(-\frac{1}{R_C}\right) V_{CE} + \frac{V_{CC}}{R_C} \qquad (5.27)$$

We have rewritten Eq. 5.26 in above form, because we wanted to put it in the form

$$y = mx + c \qquad (5.28)$$

which is the equation of a straight line. If Eq. 5.27 is plotted on the transistor's output characteristics (i.e. the curves between v_C and i_{CE}), we get a straight line. Comparison of Eq. 5.27 with Eq. 5.28 indicates that the slope of this line is

$$m = -\frac{1}{R_C} \qquad (5.29)$$

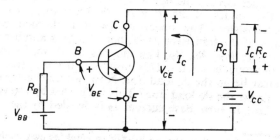

Fig. 5.25 CE amplifier in quiescent condition

and its intercept on the ic axis is

$$c = \frac{V_{CC}}{R_C} \qquad (5.30)$$

The straight line represented by Eq. 5.27 is called the *dc load line*.

Plotting of the dc load line on collector characteristics is easy. Find any two points satisfying Eq. 5.27, and then join these points. The simplest way, then, is to take one point on the v_{CE} axis and the other on ic axis. On the v_{CE} axis, the current I_C must be zero. Hence, from Eq. 5.27, we should have $V_{CE} = V_{CC}$. When $V_{CE} = 0$, Eq. 5.27 gives $I_C = V_{CC}/R_C$. Thus, the two points on the dc load line are

(i) $V_{CE} = V_{CC}$; $I_C = 0$

(ii) $V_{CE} = 0$; $I_C = \dfrac{V_{CC}}{R_C}$.

These two points can be located on the collector characteristics. See Fig. 5.26. Join these two points. This is the dc load line. The slope of this line is $(-1/R_C)$ and is decided by the value of resistor R_C. Since this resistance is the dc load* of the amplifier, we call the line as dc load line.

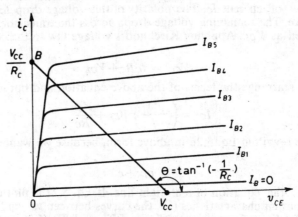

Fig. 5.26 Plotting of dc load line on collector characteristics

In an amplifier circuit, the operating conditions of the transistor are described by the values of its V_{CE} and I_C. These values fix up the *operating point* of the transistor. The operating point is decided not only by the characteristics of the transistor itself, but also by a number of other factors. These factors are V_{CC}, R_C, R_B, V_{BE} and V_{BB}. First, we fix the values of V_{CC} and R_C in an amplifier circuit. This ensures that the operating point of the transistor must lie on the dc load line. Now, where exactly does the operating point lie on the dc load line? This is decided by the value of the base current I_B. And, in turn, base current I_B is decided by the values of V_{BE} (of

---

*Later we shall learn that the ac load of an amplifier may be different from its dc load.

the transistor), R_B and V_{BB}. Applying Kirchhoff's voltage law to the base circuit of the transistor, we get

$$V_{BB} = I_B R_B + V_{BE}$$

or
$$I_B = \frac{V_{BB} - V_{BE}}{R_B} \cong \frac{V_{BB}}{R_B} \qquad (5.31)$$

Knowing the values of V_{BB}, R_B and V_{BE} (value of V_{BE} is 0.7 V for Si transistors and 0.3 V for Ge transistors), the above equation gives the value of base current I_B. Corresponding to this base current, there will be a collector characteristic curve. If by chance this curve is not present on the characteristics, we can plot the curve (see Example 5.7). The exact operating point will lie at the intersection of this curve and the dc load line. This point is called *quiescent operating point* or simply Q point.

Example 5.7 A silicon transistor is used in the circuit of Fig. 5.25, with $V_{CC} = 12$ V, $R_C = 1$ kΩ, $V_{BB} = 10.7$ V, and $R_B = 200$ kΩ. The collector characteristics of the transistor are given in Fig. 5.27. Determine the Q point.

Solution: First we plot the dc load line on the output characteristic curves. Two points on the dc load line are $(V_{CC}, 0)$ and $(0, V_{CC}/R_C)$. Here, $V_{CC} = 12$ V and $R_C = 1$ kΩ. Therefore, the two points are (12 V, 0) and (0, 12 mA). In other words, the load line cuts the v_{CE} axis at 12 V and the i_C axis at 12 mA. We join these two points to get the dc load line.

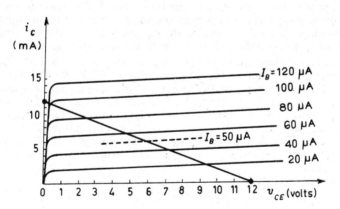

Fig. 5.27 Fixing the Q point of the transistor on its output characteristics

The operating point may lie anywhere on this dc load line. To fix the Q point, we will determine the base current I_B. Applying Kirchhoff's voltage law (KVL) to the input circuit gives

$$V_{BB} = I_B R_B + V_{BE}$$

or
$$I_B = \frac{V_{BB} - V_{BE}}{R_B}$$

Here, $V_{BB} = 10.7$ V, and $R_B = 200$ kΩ. For a silicon transistor, $V_{BE} = 0.7$ V. Therefore, the base current is

$$I_B = \frac{10.7 - 0.7}{200 \times 10^3} = 50 \ \mu A$$

However, it is seen from Fig. 5.27 that the curve for $I_B = 50 \ \mu A$ is not given. We draw this curve between the curves for $I_B = 40 \ \mu A$, and $I_B = 60 \ \mu A$. This curve is shown dotted. The point of intersection of this curve and the dc load line gives the Q point. At this point

$$V_{CE} = 6 \text{ V}$$

and

$$I_C = 6 \text{ mA}$$

5.10.2 Amplifier Analysis Using DC Load Line

A transistor can amplify ac signals only after its dc operating point is suitably fixed. We have seen in the last section how to fix the Q point on the output characteristics. The Q point should preferably lie in the middle portion of the active region of the characteristics. This helps the transistor to amplify ac signals faithfully, i.e. without distorting its waveshape.

Under quiescent condition, the base current has a constant dc value. It is determined from the Q point. Now, we apply the ac signal to the input of the amplifier circuit (see Fig. 5.24). The base voltage varies as per the signal voltage v_s. As a result, the base current will also vary. As the base current varies, the instantaneous operating point of the transistor moves along the dc load line. Thus, the instantaneous values of collector current and voltage also vary according to the input signal. The variation in collector voltage is many times larger than the variation of the input signal. The collector-voltage variation reaches the output terminals through capacitor C_{C2}. The output is therefore many times larger than the input.

Let us take an illustrative example. See the amplifier circuit in Fig. 5.24. As in Example 5.7, let us assume that $V_{CC} = 12$ V, $R_C = 1$ kΩ, $V_{BB} = 10.7$ V and $R_B = 200$ kΩ. In this circuit, the dc base current is found to be 50 μA. The collector dc current and dc voltage are 6 mA and 6 V, respectively. Let us now apply a small ac signal voltage, say, 7 mV to the input. This voltage will have about 20 mV peak-to-peak variation. If the input dynamic resistance r_i (or h_{ie}) of the transistor is assumed to be 1 kΩ, the input voltage will produce a peak-to-peak variation of 20 μA in base current, since

$$\Delta i_B = \frac{20 \text{ mV}}{1 \text{ k}\Omega} = 20 \ \mu A$$

This variation in base current takes place around its quiescent value of 50 μA. As the base current varies, the instantaneous operating point moves along the dc load line between the points A ($I_B = 60 \ \mu A$) and B ($I_B = 40 \ \mu A$). To show the variation in I_B on the collector characteristics of the transistor, we draw a line perpendicular to the dc load line and passing through the Q point. This line is taken as ωt axis, and then, variation in I_B (assumed sinusoidal) is plotted (Fig. 5.28).

As the instantaneous operating point moves along dc load line between the points A and B, both the collector current and collector voltage vary. The current i_C varies between the points A_1 ($I_C = 7.3$ mA) and B_1 ($I_C = 4.8$ mA). This variation is shown on the left side of the characteristics. The

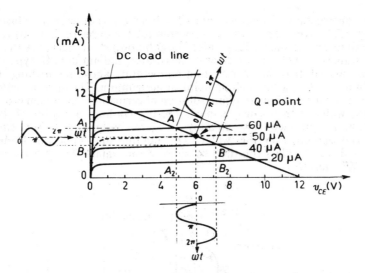

Fig. 5.28 Variation in base current produces variation in collector
current and voltage in a CE amplifier

voltage v_{CE} varies between points A_2 ($V_{CE} = 4.9$ V) and B_2 ($V_{CE} = 7.1$ V).
The collector-voltage variation is shown at the bottom of the characteristics.

The current gain, voltage gain and the power gain of the amplifier can
now be computed. We shall make the calculations on the basis of peak-to-
peak variation:

(i) Current gain, $A_i = \dfrac{\Delta i_C}{\Delta i_B} = \dfrac{(7.3 - 4.8)10^{-3}}{(60 - 40) \times 10^{-6}} = 125$

(ii) Voltage gain, $A_v = \dfrac{\Delta v_{CE}}{\Delta v_{BE}} = \dfrac{7.1 - 4.9}{20 \times 10^{-3}} = 110$

(iii) Power gain $= \dfrac{\text{output ac power}}{\text{input ac power}} = \dfrac{I_c V_{ce}}{I_b V_{be}}$

$= A_i \times A_v = 125 \times 110 = 13\,750$

We find that the CE amplifier has sufficiently large values of current gain,
voltage gain, and power gain.

5.11 CONSTRUCTION OF TRANSISTORS

In recent years, the construction of transistors has undergone a great
many changes and improvements. A number of different methods of manufacturing transistors have been developed since the invention of transistor
in 1948. A description of all the methods is outside the scope of this book.
The most commonly used types of transistors are alloy junction transistor
and silicon planar transistor. We shall discuss these two types here.

5.11.1 Alloy Junction Transistor

The alloy junction transistor is one of the earliest types of transistor that is
still in use. It is relatively inexpensive and provides high current gain. It
can be constructed to operate at high current and power levels.

The construction of a germanium alloy junction transistor is illustrated in Fig. 5.29. We start with a very thin (of the order of 250 μm) N-type germanium crystal wafer. This wafer is lightly doped and serves as the base of the transistor. On the two sides of this wafer, indium dots (P-type impurity) are placed. It is then heated to a temperature above the melting point of indium and below the melting point of germanium. The indium melts and dissolves the germanium. A liquid solution of germanium in indium is obtained. The wafer is then slowly cooled. During cooling, a region of P-type germanium is produced and an alloy of germanium and indium (mainly indium) is deposited on the wafer. The emitter and collector leads are connected to this alloy (on the two sides of the wafer). The process results in a PNP transistor.

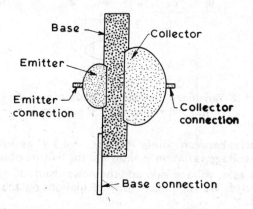

Fig. 5.29 Construction of an alloy junction transistor

5.11.2 Silicon Planar Transistor

The construction of a silicon planar transistor is shown in Fig. 5.30. The important feature of this type of transistor is that the PN-junctions are buried under a layer of silicon dioxide. This layer protects the PN-junctions from impurities.

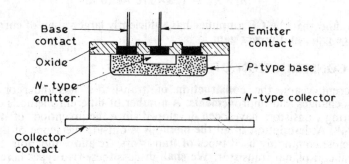

Fig. 5.30 Construction of a silicon planar transistor

The steps involved in the manufacture of a silicon planar transistor are illustrated in Fig. 5.31. To make an NPN transistor, we start with an N-type silicon wafer, which would ultimately make the collector. The top surface of this wafer is oxidized to a depth of approximately $1\,\mu\text{m}$ (Fig. 3.31a). SiO_2 is an insulating material which cannot be penetrated by impurities. To make the base region, we diffuse acceptor-type impurity (e.g. boron) into the wafer. However, because the SiO_2 film checks impurity diffusion, we must remove the film from those areas on the wafer where the base is to be diffused. This is done by etching away the SiO_2 from that area, with a masked photo-resist process (Fig. 5.31b). The wafer is now exposed to a vapour of boron (P-type impurity) and the impurity is allowed to diffuse into the wafer to a predecided depth. Now, another layer of SiO_2 is grown over the entire wafer (Fig. 5.31c). A part of the SiO_2 film is again etched away by the photo-resist process using another mask (Fig. 5.31d). The wafer is now exposed to a vapour of donor-type impurity (e.g. phosphorus) and is also reoxidized again (Fig. 5.31e). The wafer now contains a layer of P-type material that makes the base of the transistor and a layer of N-type material that makes the emitter. SiO_2 is again etched away from the surface of the wafer to separate the base and emitter regions (Fig. 5.31f). Finally, metal contacts are made onto the etched areas (Fig. 5.31g). The wafer is now cut to the required size. It is mounted on a suitable collector contact. Leads are then connected to the base and emitter contacts.

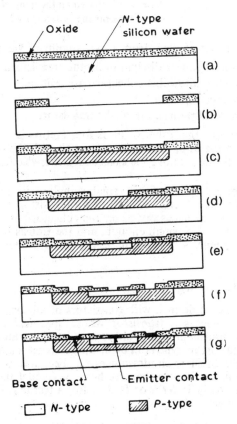

Fig. 5.31 Various stages in the manufacture of a silicon planar *NPN* transistor

5.12 TRANSISTOR DATA SHEETS

To analyse or to design a transistor circuit, one must have sufficient information about the transistor. This information is obtained from the manufacturer's data sheets. These sheets describe the transistor. Sometimes the outline and dimensions are also given. The lead orientation is also identified here. Commonly, the lead orientation is as shown in Fig. 5.32. A red dot is placed near one of the terminals. This represents the collector lead. Now put the transistor such that the leads are facing you, as in Fig. 5.32b. The central terminal is the base. The third one is the emitter. However, a word of caution is necessary. The convention described above for recognizing the three leads is not a standard one. Different manufacturers use different conventions for this purpose.

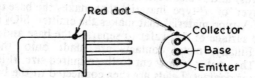

Fig. 5.32 Orientation of leads in a transistor.
A red dot is placed on the body of the
transistor near the collector lead

The important set of data, from a user's point of view are as follows:

1. The maximum power dissipation in the transistor at 25 °C.
2. The maximum allowable collector-base voltage.
3. The current gain β or h_{fe}.
4. The transition frequency f_T of the transistor.

Generally, one or the other of these four factors is of prime importance, depending upon the application. In no case should the maximum ratings, given in items 1 and 2 above, be exceeded. Otherwise the transistor may be damaged.

If a transistor is required for a small-signal audio frequency amplifier, the most important factor in selecting a transistor is its current gain. In some cases, it may be necessary to see the collector-base voltage. Since the power involved will be small enough, and the transistor is not required to handle high frequencies, it is not necessary to consider the factors at 1 and 4.

5.12.1 Transistor Testing

Today, the market is flooded with transistors of all sorts and makes. Very often, we come across a transistor whose specifications are not known. Sometimes, the transistor type number may be obliterated from its body. Even if the transistor type is known, the reference data-book may not be readily available. In these circumstances, it becomes necessary to test a transistor. In this section we shall see how to conduct the test to determine whether the transistor is *NPN* or *PNP*. Also, a test is given to identify the transistor terminals.

Test to Distinguish between PNP and NPN Transistors Figure 5.33a shows a simple circuit for testing a transistor for its nature (*PNP* or *NPN*). In this circuit, two germanium rectifier diodes and two LEDs (light-emitting diodes) are used. A resistor R_L is also placed in series so as to prevent a heavy current from flowing in the circuit. The two leads of the tester are marked x and y. If a resistor R is placed between these terminals, the current passes for both the halves of the input wave. In the positive half,

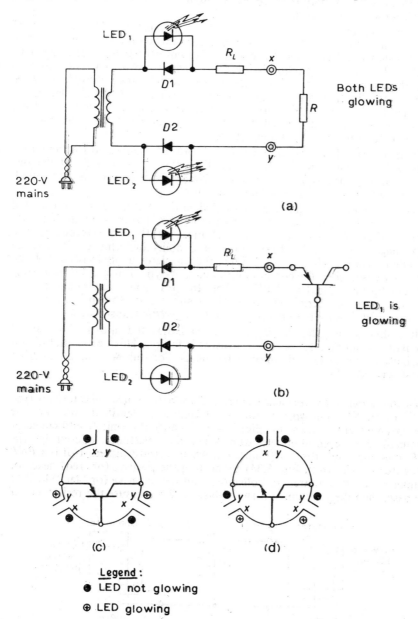

Fig. 5.33 *PNP/NPN* check for a transistor:
(*a*) Test circuit; (*b*) E-B junction connected to test leads; (*c*) Condition of LEDs for different pairs of terminals in a *PNP* transistor; (*d*) Condition of LEDs for different pairs of terminals in a *NPN* transistor.

current flows through LED_1, R_L, R, and $D2$. The diode $D1$ will not conduct during this half. The forward voltage drop of 0.3 V across $D2$ will prevent LED_2 from glowing in this half. During the negative half cycle, the current flows through LED_2, R, R_L and $D1$. During this period, LED_2 will glow, while LED_1 will not. Thus both the LEDs will glow alternatively. As the frequency of supply is 50 Hz (quite high) we shall observe both the LEDs glowing continuously.

Now, consider the case when one of the junctions (say E-B junction) of a transistor (say, PNP-type) is connected across the test leads x and y. This is shown in Fig. 5.33b. In this case, current cannot flow for those half-cycles when the E-B junction is reverse-biased. However, current flows in the direction from emitter to base (from P to N) during those half cycles when the E-B junction is forward biased. As such, only LED_1 will glow. This test indicates that the terminal connected to the lead x is P-type (and that connected to lead y is N-type). Thus, this simple circuit identifies P- and N-type terminals of a PN-junction.

None of the LEDs will glow when test leads are connected to the terminals of the same type (emitter and collector) of the transistor. Under this condition, the base is open circuited. No current (except a very small leakage current) flows through the transistor.

As seen earlier, only one LED glows when the test leads are connected across an E-B junction or across a C-B junction. That is, between those two pair of terminals, the common terminal must be the base terminal. The remaining two must be the emitter and collector terminals. In case the common terminal (the base) is P-type, the transistor is obviously NPN. In the other case, when the common terminal is N-type, it is a PNP transistor. The conditions of the two LEDs when the test leads are connected to different pairs of terminals of a PNP transistor are shown in Fig. 5.33c. Figure 5.33d shows the same procedure for an NPN transistor.

It is obvious that the glowing of both the LEDs indicate a short-circuited pair of terminals, i.e. the transistor is faulty. This test cannot distinguish between emitter and collector terminals. For this, we conduct another test described below.

Identification of Emitter and Collector Terminals   Once the transistor type (PNP or NPN) is known, and the base terminal is identified, we can use the arrangement shown in Fig. 5.34 to identify the emitter and collector terminals. We use an ohmmeter to measure the resistance offered by the E-B junction and the C-B junction, when forward-biased. If it is a PNP transistor (as shown in Fig. 5.34) connecting the positive (or red) lead of ohmmeter to the emitter (or collector), and the negative (or black) lead to the base, then the junction is forward-biased. We measure the resistance of

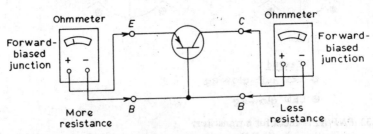

Fig. 5.34 Test arrangement for the identification of emitter and collector terminals

one junction, and then using the same ohmmeter we measure the resistance of the other junction. *The measurement that results in the higher resistance reading indicates the emitter terminal.* The other terminal is obviously the collector terminal.

5.13 THERMAL RUNAWAY AND HEAT SINK

If the temperature of the collector-base junction increases, the collector leakage current I_{CBO} increases. Because of this, collector current increases. The increase in collector current produces an increase in the power dissipated at the collector junction. This, in turn, further increases the temperature of the junction and so gives further increase in collector current. The process is cumulative. It may lead to the eventual destruction of the transistor. This is described as the *thermal runaway* of the transistor. In practice, thermal runaway is prevented in a well designed circuit by the use of stabilization circuitry.

For transistors handling small signals, the power dissipated at the collector is small. Such transistors have little chances of thermal runaway. However in power transistors, the power dissipated at the collector junction is larger. This may cause the junction temperature to rise to a dangerous level. We can increase the power handling capacity of a transistor if we make a suitable provision for rapid conduction of heat away from the junction. This is achieved by using a sheet of metal called *heat sink*. As the power dissipated within a transistor is predominantly the power dissipated at its collector-base junction, sometimes the collector of the power transistor is connected to its metallic case. The case of the transistor is then bolted on to a sheet of metal as shown in Fig. 5.35a. This sheet serves as the heat sink.

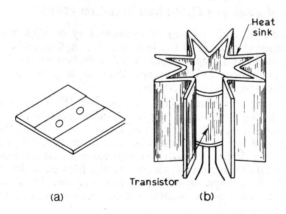

Fig. 5.35 Two kinds of heat sinks used with power transistors

Connecting a heat sink to a transistor increases the area from which heat is to be transferred to the atmosphere. Heat moves from the transistor to the heat sink by conduction, and then it is removed from the sink to the ambient by convection and radiation.

Another type of heat sink is shown in Fig. 5.35b. It consists of a push-fit clip. This clip is pushed on to the transistor. To increase the surface area of the heat sink, it is usually given a ribbed structure. Because of this structure, the heat sink does not occupy much space within the equipment.

For maximum efficiency, a heat sink should (i) be in good thermal contact with the transistor case, (ii) have the largest possible surface area, (iii) be painted black, and (iv) be mounted in a position such that free air can flow past it.

5.14 FIELD-EFFECT TRANSISTOR (FET)

The field-effect transistor (FET), developed in the early 1960s, is emerging as an important member of the semiconductor family. Today, the junction FET (JFET) and insulated gate FET (IGFET) or metal oxide semiconductor FET (MOSFET) are rapidly replacing both vacuum tubes and the junction transistors in applications requiring high input impedance.

The advantages FETs have over the vacuum tubes* used earlier are their small size, low power consumption and lack of a filament. The advantages of FETs over junction transistors are that they have high input impedanc: and high power gain. FETs are comparatively easier to fabricate. Also, their fabrication process is particularly suited to make ICs.

As we shall see, the FET differs from the conventional junction transistor in that its operation depends on the flow of majority carriers only. That is why it is sometimes called a *unipolar transistor*. In contrast, the conventional transistor depends on both the majority and minority carriers for its operation. Hence, it is called a *bipolar junction transistor* (BJT).

In the sections that follow, we shall see how the construction and characteristics of an FET differ from those of a BJT. As we shall find, the electrical characteristics of a FET are similar to those of a vacuum pentode tube. That is why many FET circuits resemble their vacuum tube counterparts.

5.14.1 Structure of a Junction Field-Effect Transistor (JFET)

A JFET can be of *N-channel* type or of *P-channel* type. (The meaning of *channel* will be made clear later in the section.) We shall describe the structure of an *N*-channel JFET. The structure of a *P*-channel JFET is similar to that of an *N*-channel JFET, except that in its structure, *N*-type is replaced by *P*-type and *P*-type by *N*-type.

In its simplest form, the structure of an *N*-channel JFET starts with nothing more than a bar of *N*-type silicon. This bar behaves like a resistor between its two terminals, called *source* and *drain* (Fig. 5.36a). We introduce heavily doped *P*-type regions on either side of the bar. These *P* regions are called *gates* (Fig. 5.36b). Usually, the two gates are connected together (Fig. 5.36c). The gate terminal is analogous to the base of a BJT. This is used to control the current flow from source to drain. Thus, source and drain terminals are analogous to emitter and collector terminals respectively, of a BJT.

In Fig. 5.36d, the bar of the JFET has been placed vertically. The circuit symbol of *N*-channel JFET is shown in Fig. 5.36e. Note that the arrow is put in the gate terminal (and not in the source terminal, though source is analogous to emitter in a BJT). The gate arrow points into the JFET. (In a *P*-channel JFET, the gate arrow would point out of the JFET).

Let us now see why the *N*-type bar is called a channel. Normally, we operate an *N*-channel JFET by applying positive voltage to the drain with respect to the source (Fig. 5.37a). Due to this voltage, the majority carriers in the bar (electrons in this case) start flowing from the source to the drain.

---

*The vacuum tubes are described in the next unit.

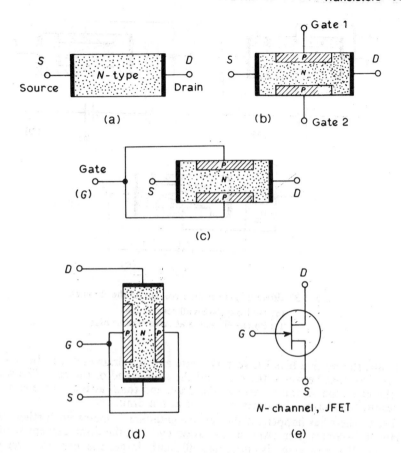

Fig. 5.36 Junction field-effect transistor (*N*-channel type)

This flow of electrons makes the drain current I_D. The current I_D is analogous to the collector current I_C in a BJT. The electrons in the bar have to pass through the space between the two P regions. As we shall see, the width of this space between the P regions can be controlled by varying the gate voltage. That is why this space is called a channel.

To see how the width of the channel changes by varying the gate voltage, let us consider Fig. 5.37b. Here we have applied a small reverse bias to the gate. Because of the reverse bias, the width of the depletion increases. Since the N-type bar is lightly doped compared to the P regions, the depletion region extends more into the N-type bar. This reduces the width of the channel. Recall that the depletion regions do not contain any charge carriers. The electrons have to pass through the channel of reduced width. Reduction in the width of the channel (the conductive portion of the bar) increases its resistance. This reduces the drain current I_D.

See Fig. 5.37b carefully. There is one important point about the channel shape. It is narrower at the drain end. This happens because the amount of reverse-bias is not same throughout the length of the PN-junction. When current flows through the bar, a potential drop occurs across its length. As

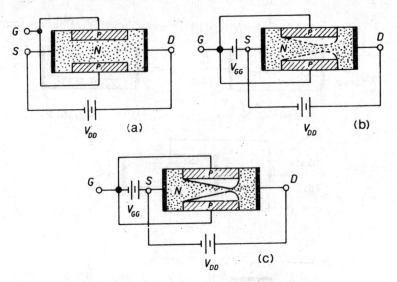

Fig. 5.37 Effect of gate-source voltage on the channel:
(*a*) No bias; (*b*) Small reverse bias
(*c*) Pinch-off occurs at large reverse bias

a result, the reverse bias between the gate and the drain end of the bar is more than that between the gate and the source end of the bar. The width of the depletion region is more at the drain end than at the source end. As a result, the channel becomes narrower at the drain end.

Let us see what happens if the reverse gate-bias is increased further. The channel becomes narrower at the drain end and the drain current further reduces. If the reverse bias is made sufficiently large, the depletion regions will extend into the channel and meet. This *pinches off* all the current flow (Fig. 5.37c). The gate-source voltage at which pinch-off occurs is called *pinch-off voltage* V_P.

You may think that the channel completely closes at the drain end when the gate-source voltage reaches the pinch-off value. But in practice it does not happen, simply because it cannot happen. Suppose, if it were possible, the channel completely closes at the drain end. The drain current would then reduce to zero. As a result, there would be no voltage drop along the length of the channel. The amount of reverse-bias would become uniformly same throughout the length. The wedge shaped depletion region would try to become straight (rectangular shaped). The channel would then open at the drain end. The drain current flows.

When the gate-source voltage reaches the pinch-off value, the channel width reduces to a constant minimum value. The drain current flows through this constricted channel.

Some important terminology regarding a JFET:

Source : The source is the terminal through which the majority carriers (electrons in case of *N*-channel FET, and holes in case of *P*-channel FET) enter the bar.

Drain : The drain is the terminal through which the majority carriers leave the bar

Gate : On both sides of the N-type bar, heavily doped P regions are formed. These regions are called gates. Usually, the two gates are joined together to form a single gate.

Channel : The region between the source and drain, sandwiched between the two gates, is called channel. The majority carriers move from source to drain through this channel.

5.14.2 JFET Characteristics

As a BJT has static collector characteristics, so does a JFET have static drain characteristics. Such characteristics are shown in Fig. 5.38. For each curve, the gate-to-source voltage V_{GS} is constant. Each curve shows the variation of drain current i_D versus drain-to-source voltage v_{DS}.

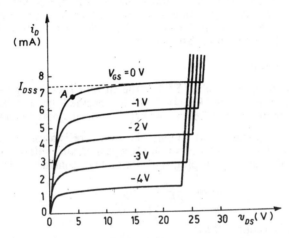

Fig. 5.38 Typical drain characteristics of an
N-channel JFET

Let us consider first, the curve for zero gate bias. For this curve, $V_{GS} = 0$. When v_{DS} is zero, the channel is entirely open. But the drain current is zero, because the drain terminal does not have any attractive force for the majority carriers. For small applied voltage v_{DS}, the bar acts as a simple resistor. Current i_D increases linearly with voltage v_{DS}. This region (to the left of point *A*) of the curve is called *ohmic region*, because the bar acts as an ohmic resistor.

Ohmic voltage drop is caused in the bar due to the flow of current i_D. This voltage drop along the length of the channel reverse-biases the gate junction. The reverse biasing of the gate junction is not uniform throughout. The reverse bias is more at the drain end than at the source end of the channel. So, as we start increasing v_{DS}, the channel starts constricting more at the drain end. The channel is eventually *pinched off*. The current i_D no longer increases with the increase in v_{DS}. It approaches a constant saturation value. The voltage v_{DS} at which the channel is "pinched off" (that is, all the free charges from the channel are removed), is called *pinch-off voltage*, V_P. Note that the voltage V_P is not sharply defined on the curve. The region of the curve to the right of point *A* is called *pinch-off region*.

A special significance is attached to the drain current in the pinch-off region when $V_{GS} = 0$. It is given the symbol I_{DSS}. It signifies the drain source current at pinch-off, when the gate is shorted to the source. It is measured well into the pinch-off region. In this case, $I_{DSS} = 7.4$ mA.

Further increase in voltage v_{DS} increases the reverse bias across the gate junction. Eventually, at high v_{DS} breakdown of the gate junction occurs. The drain current i_D shoots to a high value. Of course, when we use a JFET in a circuit, we avoid the gate junction breakdown.

If the gate reverse-bias is increased (say, $V_{GS} = -1$ V), the curve shifts downward. The pinch-off occurs for smaller value of v_{DS}. The maximum saturation drain current is also smaller, because the conducting channel now becomes narrower.

For an increased reverse bias at the gate, the avalanche breakdown of the gate junction occurs at lower value of v_{DS}. This happens because the effective bias at the gate junction (at the drain end) is the voltage V_{GS} plus voltage V_{DS}. The greater the value of V_{GS}, the lower the value of V_{DS} required for the junction to breakdown.

5.14.3 JFET Parameters

An important parameter of a JFET is the current I_{DSS}. It signifies the drain saturation current when $V_{GS} = 0$. It is specified by the manufacturer. Besides this, there are the following three important parameters of a JFET.

Dynamic drain resistance (r_d) : Dynamic drain resistance, at an operating point, is defined as the ratio of small change in drain voltage to the small change in drain current, keeping the gate voltage constant. That is

$$r_d = \left.\frac{\Delta v_{DS}}{\Delta i_D}\right|_{V_{GS} = \text{const.}} \tag{5.32}$$

Typically, r_d is about 400 kΩ.

Mutual conductance or transconductance (g_m) : The mutual conductance, at an operating point, is defined as the ratio of small change in drain current to the small change in gate voltage, keeping the drain voltage constant. That is

$$g_m = \left.\frac{\Delta i_D}{\Delta v_{GS}}\right|_{V_{DS} = \text{const.}} \tag{5.33}$$

It is measured in siemens (S). Typically, its value ranges from 150 μS to 250 μS.

Amplification factor (μ) : It is defined as the ratio of small change in drain voltage to the small change in gate voltage, when current I_D is kept constant. That is

$$\mu = \left.\frac{\Delta v_{DS}}{\Delta v_{GS}}\right|_{I_D = \text{const.}} \tag{5.34}$$

Since μ is a ratio of two voltages, it does not have any units. The amplification factor of a FET can be as high as 100.

The above three parameters of a JFET are related as :

$$\mu = r_d g_m \tag{5.35}$$

Thus, if any two parameters are known, the third can be computed.

Example 5.8 For a JFET type BFW10 (made by BEL, Bangalore), the typical values of amplification factor and transconductance are specified as 80 and 200 μS, respectively. Calculate the dynamic drain resistance of this JFET.

Solution : The three parameters of a JFET are related by the formula

$$\mu = r_d g_m$$

Here, $\mu = 80$, and $g_m = 200 \ \mu S = 200 \times 10^{-6}$ S. Therefore, the dynamic drain resistance is given as

$$r_d = \frac{\mu}{g_m} = \frac{80}{200 \times 10^{-6}}$$
$$= 4 \times 10^5 \ \Omega = 400 \ k\Omega$$

REVIEW QUESTIONS

5.1 Draw a sketch showing the structure of an *NPN*-junction transistor. Label the emitter, base and collector regions. Also label the emitter-base and collector-base junctions.

5.2 Repeat the above for a *PNP*-junction transistor.

5.3 Draw the circuit symbol of an *NPN* transistor and indicate the reference directions, according to standard convention, for the three currents.

5.4 Explain why an ordinary junction transistor is called bipolar.

5.5 Show the biasing arrangement for a *PNP* transistor in CB configuration so that it works in active region.

5.6 Explain the function of the emitter in the operation of a junction transistor.

5.7 What is done to the base region of a transistor to improve its operation ?

5.8 Though the collector-base junction of a transistor operating in active region is reverse-biased, the collector current is still quite large. Explain briefly, say, within 10 lines.

5.9 What do you understand by the collector reverse-saturation current ? In which configuration (CB or CE) does it have a greater value ?

5.10 Besides the active region of operation of a transistor, what are the other possible conditions of operation of a transistor ? Give the biasing conditions of each.

5.11 What causes collector current to flow when the emitter current is zero ? What is this collector current called ?

5.12 Explain the reason why the base current in a transistor is usually much smaller than I_E or I_C in active operation.

5.13 For a *PNP* transistor in the active region, what is the sign (positive or negative) of I_E, I_B, I_C, V_{EB} and V_{CF} ?

5.14 Draw an *NPN* transistor in the CB configuration biased for operation in active region.

5.15 What is considered the input terminal, and what is the output terminal in the CB configuration ?

5.16 Sketch typical CB input characteristic curves for an *NPN* transistor. Label all variables. Explain how you will calculate the input dynamic resistance of the transistor from these curves.

5.17 Sketch typical output characteristic curves for a *PNP* transistor in CB configuration. Label all variables and indicate active, cut-off and saturation regions.

5.18 What are the input and output terminals in the CE configuration ?

5.19 An *NPN* transistor is to be used in common-emitter configuration. Show how you will connect the external batteries so that the transistor works in the active region.

5.20 Sketch typical CE input characteristics for an *NPN* transistor. Label all variables. Outline the procedure of calculating the input dynamic resistance of the transistor at a given point from these curves.

5.21 Sketch typical CE output characteristic curves for an *NPN* transistor. Label all variables. Explain in brief how you will compute the beta of the transistor from these characteristic curves ?

5.22 Derive the relationship between the beta and alpha of a transistor.

5.23 Compare the relative values of input and output resistances for the common-base and common-emitter configurations. Give their typical values.

5.24 Explain why CE configuration is most popular in amplifier circuits.

5.25 Draw the circuit diagram of a simple transistor amplifier in CE configuration. Write down the equation of a dc load line.

5.26 Explain how you will determine the voltage gain of the CE amplifier by plotting the dc load line on the output characteristics of the transistor.

5.27 Describe briefly the procedure for manufacturing alloy junction transistors.

5.28 Explain the important steps in making a silicon planar transistor. Why has this technology of manufacturing transistors become so popular recently ?

5.29 You are given a transistor. Somehow the red dot on its body has been obliterated Explain how you will determine its three terminals.

5.30 What is meant by thermal runaway in a transistor? Explain.

5.31 What is a heat sink ? Draw a typical heat sink. List the factors which determine its efficiency.

5.32 State the order of magnitude of the collector reverse saturation current I_{CBO} for (a) germanium transistor, (b) silicon transistor. How does it vary with temperature in each case?

5.33 Sketch the basic structure of *N*-channel JFET.

5.34 Draw the circuit symbols of (a) an *N*-channel JFET, (b) a *P*-channel JFET.

5.35 Show the biasing arrangement of an *N*-channel JFET.

5.36 Draw typical drain characteristics curves of a JFET. Explain the shape of these curves qualitatively.

5.37 What do you understand by the term 'channel' in a JFET ?

5.38 Define all the important parameters of a JFET.

OBJECTIVE-TYPE QUESTIONS

I. Here are some incomplete statements. Four alternatives are provided below each. Tic the alternative that completes the statement correctly:

1. In a *PNP* transistor with normal bias
   (a) only holes cross the collector junction
   (b) only majority carriers cross the collector junction
   (c) the collector junction has a low resistance
   (d) the emitter-base junction is forward-biased and the collector-base junction is reverse-biased.

2. In a transistor with normal bias, the emitter junction
   (a) has a high resistance
   (b) has a low resistance
   (c) is reverse-biased
   (d) emits such carriers into the base region which are in majority (in the base)

3. For transistor action

  (a) the collector must be more heavily doped than the emitter region

  (b) the collector-base junction must be forward-biased

  (c) the base region must be very narrow

  (d) the base region must be N-type material

4. The symbol I_{CBO} signifies the current that flows when some dc voltage is applied

  (a) in the reverse direction to the collector junction with the emitter open-circuited

  (b) in the forward direction to the collector junction with the emitter open-circuited

  (c) in the reverse diretion to the emitter junction with the collector open-circuited

  (d) in the forward direction to the emitter junction with the collector open-circuited

5. The current I_{CBO}

  (a) is generally greater in silicon than in germanium transistors

  (b) depends largely on the emitter-base junction bias

  (c) depends largely on the emitter doping

  (d) increases with an increase in temperature

6. The main current crossing the collector junction in a normally biased NPN transistor is

  (a) a diffusion current

  (b) a drift current

  (c) a hole current

  (d) equal to the base current

7. In a PNP transistor, electrons flow

  (a) out of the transistor at the collector and base leads

  (b) into the transistor at the emitter and base leads

  (c) into the transistor at the collector and base leads

  (d) out of the transistor at the emitter and base leads

8. The current I_{CBO} flows in

  (a) the emitter, base, and collector leads

  (b) the emitter and base leads

  (c) the emitter and collector leads

  (d) the collector and base leads

9. The emitter region in the PNP junction transistor is more heavily doped than the base region so that

  (a) the flow across the base region will be mainly because of electrons

  (b) the flow across the base region will be mainly because of holes

  (c) recombination will be increased in the base region

  (d) base current will be high

10. For a given emitter current, the collector current will be higher if

  (a) the recombination rate in the base region were decreased

  (b) the emitter region were more lightly doped

  (c) the minority-carrier mobility in the base region were reduced

  (d) the base region were made wider

11. The arrowhead on the transistor symbol always points in the direction of

  (a) hole flow in the emitter region

  (b) electron flow in the emitter region

  (c) minority-carrier flow in the emitter region

  (d) majority-carrier flow in the emitter region

12. A small increase in the collector reverse-bias will cause
    (a) a large increase in emitter current
    (b) a large increase in collector current
    (c) a large decrease in collector current
    (d) very small change in collector reverse saturation current

13. One way in which the operation of an *NPN* transistor differs from that of a *PNP* transistor is that
    (a) the emitter junction is reverse-biased in the *NPN*
    (b) the emitter injects minority carriers into the base region of the *PNP* and majority carriers in the base region of the *NPN*
    (c) the emitter injects holes into the base region of the *PNP* and electrons into the base region of the *NPN*
    (d) the emitter injects electrons into the base region of the *PNP* and holes into the base region of the *NPN*

14. The emitter current in a junction transistor with normal bias
    (a) may be greatly increased by a small change in collector bias
    (b) is equal to the sum of the base current and collector current
    (c) is approximately equal to the base current
    (d) is designated as I_{co}

15. In CB configuration, the output volt-ampere characteristics of the transistor may be shown by plots of
    (a) v_{CB} versus i_C for constant values of I_E
    (b) v_{CB} versus i_B for constant values of I_E
    (c) v_{CE} versus i_E for constant values of I_B
    (d) v_{CE} versus i_C for constant values of I_B

16. A transistor-terminal current is considered positive if
    (a) the electrons flow out of the transistor at the terminal
    (b) the current is due to the flow of holes only
    (c) the current is due to the flow of electrons only
    (d) the electrons flow into the transistor at the terminal

17. A transistor-terminal voltage is considered positive if
    (a) the terminal is more negative than the common terminal
    (b) the terminal is more positive than the common terminal
    (c) the terminal is the output terminal
    (d) the terminal is connected to *P*-type material

18. The current I_{CEO} is
    (a) the collector current in the common-emitter connected transistor with zero base current.
    (b) the emitter current in the common-collector connected transistor with zero base current
    (c) the collector current in the common-emitter connected transistor with zero emitter current
    (d) the same as I_{CBO}

19. The common-emitter input volt-ampere characteristics may be shown by plots of
    (a) v_{CB} versus i_C for constant values of I_E
    (b) v_{CF} versus i_C for constant values of I_B
    (c) v_{CF} versus i_E for constant values of V_{LB}
    (d) v_{BE} versus i_B for constant values of V_{CE}

20. In CE configuration, the output volt-ampere characteristics may by shown by plots of
   (a) v_{CB} versus i_C for constant values of I_E
   (b) v_{CE} versus i_C for constant values of I_B
   (c) v_{CE} versus i_E for constant values of V_{EB}
   (d) v_{BE} versus i_B for constant values of V_{CE}

21. The beta (β) of a transistor may be determined directly from the plots of
   (a) v_{CB} versus i_C for constant values of I_E
   (b) v_{EC} versus i_E for constant values of I_B
   (c) v_{CE} versus i_C for constant values of I_B
   (d) v_{BE} versus i_B for constant values of V_{CE}

22. The most noticeable effect of a small increase in temperature in the common emitter connected transistor is
   (a) the increase in the ac current gain
   (b) the decrease in the ac current gain
   (c) the increase in output resistance
   (d) the increase in I_{CEO}

23. When determining the common-emitter current gain by making small changes in direct currents, the collector voltage is held constant so that
   (a) the output resistance will be high
   (b) the transistor will not burn out
   (c) the change in emitter current will be due to a change in collector current
   (d) the change in collector current will be due to a change in base current

24. The high resistance of the reverse-biased collector junction is due to the fact that
   (a) a small change in collector bias voltage causes a large change in collector current
   (b) a large change in collector bias voltage causes very little change in collector current
   (c) a small change in emitter current causes an almost equal change in collector current
   (d) a small change in emitter bias voltage causes a large change in collector current

25. A transistor connected in common-base configuration has
   (a) a low input resistance and high output resistance
   (b) a high input resistance and a low output resistance
   (c) a low input resistance and a low output resistance
   (d) a high input resistance and a high output resistance

26. Compared to a CB amplifier, the CE amplifier has
   (a) lower input resistance
   (b) higher output resistance.
   (c) lower current amplification
   (d) higher current amplification

27. A transistor, when connected in common-emitter mode, has
   (a) a high input resistance and a low output resistance
   (b) a medium input resistance and a high output resistance
   (c) very low input resistance and a low output resistance
   (d) a high input resistance and a high output resistance

28. The input and output signals of a common-emitter amplifier are
   (a) always equal
   (b) out of phase
   (c) always negative
   (d) in phase

29. A transistor is said to be in a quiescent state when

    (a) no signal is applied to the input
    (b) it is unbiased
    (c) no currents are flowing
    (d) emitter-junction bias is just equal to collector-junction bias

30. When a positive voltage signal is applied to the base of a normally biased *NPN* common-emitter transistor amplifier

    (a) the emitter current decreases
    (b) the collector voltage becomes less positive
    (c) the base current decreases
    (d) the collector current decreases

31. A field-effect transistor (FET)

    (a) has three *PN*-junctions
    (b) incorporates a forward-biased junction
    (c) depends on the variation of a magnetic field for its operation
    (d) depends on the variation of the depletion-layer width with reverse voltage, for its operation

32. The operation of a JFET involves

    (a) a flow of minority carriers
    (b) a flow of majority carriers
    (c) recombination
    (d) negative resistance

33. A field-effect transistor (FET)

    (a) uses a high-concentration emitter junction
    (b) uses a forward-biased *PN*-junction
    (c) has a very high input resistance
    (d) depends on minority-carrier flow

II. Indicate which of the following statements pertain to *NPN* transistors and which pertain to *PNP* transistor :

    1. The emitter injects holes into the base region.
    2. When biased in the active region, current flows into the emitter terminal.
    3. The electrons are the minority carriers in the base region.
    4. The collector is biased negatively relative to the base for active operation.
    5. The principal current carriers are electrons.
    6. The *E-B* junction is forward-biased for active operation.
    7. The base is made by doping the intrinsic semiconductor with indium.

Ans. I. 1. *d*; 2. *b*; 3. *c*; 4. *a*; 5. *d*; 6. *b*; 7. *c*; 8. *d*; 9. *b*; 10. *a*; 11. *a*; 12. *d*; 13. *c*; 14. *b*: 15. *a*; 16. *a*; 17. *b*; 18. *a*; 19. *d*; 20. *b*; 21 *c*; 22. *d*; 23. *d*; 24. *b*; 25. *a*; 26. *d*; 27. *b*; 28. *b*; 29. *a*; 30. *b*; 31. *d*; 32. *b*; 33. *c*.

II. 1. *PNP*; 2. *PNP*: 3. *NPN*; 4. *PNP*; 5. *NPN*; 6. both *PNP* and *NPN*; 7. *NPN*.

TUTORIAL SHEET 5.1

1. For a certain transistor $\alpha_{dc} = 0.98$ and emitter current $I_E = 2$ mA. Calculate the values of collector current I_C and base current I_B.

   [**Ans.** $I_C = 1.96$ mA, $I_B = 40$ μA]

2. The collector current $I_C = 2.9$ mA in a certain transistor circuit. If base current $I_B = 100$ μA, calculate α_{dc} of the transistor. [**Ans.** $\alpha_{dc} = 0.97$]

3. The emitter current I_E in a transistor is 2 mA. If the leakage current I_{CBO} is 5 μA and $\alpha_{dc} = 0.985$, calculate the collector and base currents.

   [**Ans.** $I_C = 1.975$ mA, $I_B = 25$ μA]

4. In an *NPN* silicon transistor, $\alpha_{dc} = 0.995$, $I_E = 10$ mA, leakage current $I_{CO} = 0.5$ μA. Determine I_C, I_B, β_{dc} and I_{CEO}.
   [**Ans.** $I_C = 9.9505$ mA, $I_B = 49.5$ μA, $\beta_{dc} = 199$, $I_{CEO} = 100$ μA]
5. A transistor is supplied with dc voltages so that $I_B = 40$ μA. If $\beta_{dc} = 80$ and leakage current is 5 μA, what is the value of emitter current I_E?
   [**Ans.** $I_E = 3.645$ mA]

TUTORIAL SHEET 5.2

1. In a transistor circuit, $I_E = 5$ mA, $I_C = 4.95$ mA and $I_{CEO} = 200$ μA. Calculate β_{dc} and leakage current I_{CBO}. [**Ans.** $\beta_{dc} = 99$, $I_{CBO} = 2$ μA]
2. Collector current in a BC107 transistor is 5 mA. If $\beta_{dc} = 140$ and base current is 35 μA, calculate the leakage current I_{CO}. [**Ans.** $I_{CO} = 0.71$ μA]
3. A transistor is connected in CB configuration. When the emitter voltage is changed by 200 mV, the emitter current changes by 5 mA. During this variation, collector-to-base voltage is kept fixed. Calculate the dynamic input resistance of the transistor.
   [**Ans.** $r_i = 40$ Ω]
4. A variation of 5 μA in the base current produces a change of 1.2 mA in the collector current. Collector-to-emitter voltage remains fixed during this variation. Calculate the current amplification factor β_{dc}. [**Ans.** $\beta_{dc} = 240$]

TUTORIAL SHEET 5.3

1. Table T.5.3.1 gives values of the collector current and collector voltage for a series of base current values in a transistor in the CE configuration. Plot these characteristics and hence find (a) the current gain when the collector voltage is 6 V, (b) the output resistance for a base current of 45 μA. [**Ans.** (a) 40.25; (b) 13.33 kΩ]

Table T.5.3.1

V_{CE} (V)	Collector current (mA)			
	$I_B = 25$ μA	$I_B = 45$ μA	$I_B = 65$ μA	$I_B = 85$ μA
3	0.91	1.59	2.25	3.00
5	0.92	1.69	2.45	3.20
7	0.96	1.84	2.65	3.50
9	0.99	2.04	2.95	4.00

2. Table T.5.3.2 gives the data of a transistor which is used in a common-emitter amplifier. Plot the output characteristics assuming them to be linear between the values indicated. The collector supply voltage is 10 V, and the collector load resistance is 1.2 kΩ. Draw the load line and choose a suitable operating point. Use this load line to calculate (a) the voltage gain and (b) the current gain, when a 12 μA peak signal is applied at the base. Assume the dynamic input resistance of the transistor to be 1.8 kΩ. [**Ans.** (a) 41.61; (b) 62.5]

Table T.5.3.2

V_{CE} (V)	I_C (mA)		
	$I_B = 40\ \mu A$	$I_B = 60\ \mu A$	$I_B = 80 \mu A$
1	3	4.5	6.0
3	3.4	5.0	6.5
5	3.8	5.5	7.0
9	4.2	6.0	7.6
11	4.6	6.5	8.2

3. In a basic transistor amplifier shown in Fig. T.5.3.1a an *NPN* transistor is used. The output characteristics of this transistor are shown in Fig. T.5.3.1b. Draw the dc lo⸱ ᵌ line on the characteristics and locate the Q point. (a) Write down the coordinates of the Q point. (b) Determine the current gain of this amplifier.

[**Ans.** (a) 7.0 V, 2.7 mA; (b) 16]

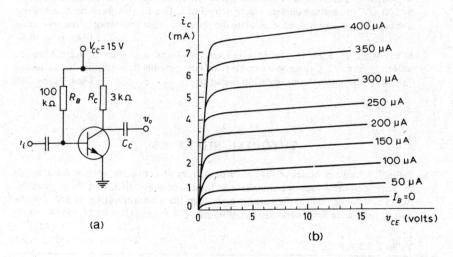

(a)

(b)

Fig. T.5.3.1

EXPERIMENTAL EXERCISE 5.1

TITLE: Common-base transistor characteristics.

OBJECTIVES: To,

1. trace the given circuit;
2. measure emitter current for different values of emitter-base voltage keeping collector-base voltage constant;
3. calculate the input dynamic resistance from the input characteristic at a given operating point;
4. plot the output characteristics (graph between the collector current and collector-to-base voltage, keeping emitter current fixed) for the given transistor;

5. calculate the output dynamic resistance r_o, α_{dc} and α at a given operating point.

APPARATUS REQUIRED: Experimental board, transistor (or IC) power supply, two milliammeters (0 to 50 mA), two electronic multimeters.

CIRCUIT DIAGRAM: The circuit diagram is shown in Fig. E.5.1.1.

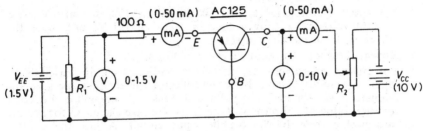

Fig. E.5.1.1

BRIEF THEORY: A transistor is a three-terminal active device. The three terminals are emitter, base and collector. In common-base configuration, we make the base common to both input and output. For normal operation, the emitter-base junction is forward-biased and the collector-base junction is reverse-biased.

The input characteristic is a plot between i_E and v_{EB} keeping voltage V_{CB} constant. This characteristic is very similar to that of a forward-biased diode. The input dynamic resistance is calculated using the formula

$$r_i = \frac{\Delta v_{EB}}{\Delta i_E}\bigg|_{V_{CB} = \text{const.}}$$

The output characteristic curves are plotted between i_C and v_{CB}, keeping I_E constant. These curves are almost horizontal. This shows that the output dynamic resistance, defined below, is very high.

$$r_o = \frac{\Delta v_{CB}}{\Delta i_C}\bigg|_{I_E = \text{const.}}$$

The collector current I_C is less than, but almost equal to the emitter current. The current I_E divides into I_C and I_B. That is,

$$I_E = I_C + I_B$$

When the output side is open (i.e. $I_E = 0$), the collector current is not zero, but has a small (a few μA) value. This value of collector current is called collector reverse saturation current, I_{CBO}.

At a given operating point, we define the dc and ac current gains (alpha) as follows:

$$\text{dc current gain, } \alpha_{dc} = \frac{I_C}{I_E}$$

$$\text{ac current gain, } \alpha = \frac{\Delta i_C}{\Delta i_E}\bigg|_{V_{CB} = \text{const.}}$$

PROCEDURE :

1. From the experimental board, note down the type number of the transistor. Note the important specifications of the transistor from the data book. Identify the terminals of the transistor. Trace the circuit.

2. Make the circuit connections as shown in Fig. E.5.1.1. Use milli-ammeters of proper range.

3. For input characteristics, first fix the voltage V_{CB}, say, at 6 V. Now vary the voltage v_{EB} slowly (say, in steps of 0.1 V) and note the current i_E for each value of v_{EB}.

4. Repeat the above for another value of V_{CB} say, 10 V.

5. For output characteristics, first fix the collector voltage, say, at 4 V. Open the input circuit. Note the collector current by using a micro-ammeter. Vary the collector voltage in steps and note collector current for each value of collector voltage. This will give the curve for reverse saturation current. Now, close the input circuit. Adjust the emitter current I_E to, say, 1 mA with the help of potentiometer R_1. Again vary the voltage V_{CB} in steps. Note current I_C for each. Repeat this process for 3 to 4 different values of emitter current (say, 2 mA 3 mA, 4 mA, etc.). See to it that you do not exceed the maximum ratings of the transistor.

6. Plot the input and output characteristics by using the readings taken above.

7. Select a suitable operating point well within the active region (say, $V_{CB} = 6$ V, $I = 3$ mA). At this operating point, draw a tangent to the curve of input characteristics (you should have the curve for the selected value of V_{CB}). The slope of this curve will give the input dynamic resistance. Similarly, by drawing tangent to the output characteristic curve gives the output dynamic resistance.

8. To determine dc alpha, simply divide the dc collector current (at the selected operating point) by the dc emitter current.

9. To determine ac alpha, draw a vertical line through the selected operating point on the output characteristics. Take a small change in i_E (say, 1 mA) around the operating point and read from the graph, the corresponding change in i_C. Divide the change in i_C by the change in i_E to get ac alpha.

OBSERVATIONS:

1. *Type number of the transistor* = _____

2. *Information from data book* :

    (*a*) Maximum collector current rating = _____ mA

    (*b*) Maximum collector-to-emitter voltage rating = _____V

    (*c*) Maximum collector dissipation power = _____W

3. *Input characteristics*:

Sr. No.	$V_{CB} = 6$ V		$V_{CB} = 10$ V	
	v_{FB} *in* mV	I_E *in* mA	v_{EB} in V	I_E *in* mA
1				
2.				
3.				

4. *Output characteristics*:

S. No.	$I_E = 0$		$I_E = 1$ mA		$I_E = 2$ mA		$I_E = 3$ mA	
	v_{CB} (V)	I_C (mA)	v_{CB} (V)	I_C (mA)	v_{CB} (V)	I_C (mA)	v_{CB} (V)	I_C (mA)
1.								
2.								
3.								

CALCULATIONS:

1. Input dynamic resistance,

$$r_i = \left. \frac{\Delta v_{EB}}{\Delta i_E} \right|_{V_{CB} = \rule{1cm}{0.4pt} V} = \rule{1cm}{0.4pt} = \rule{1.5cm}{0.4pt} \, \Omega$$

2. Output dynamic resistance,

$$r_o = \left. \frac{\Delta v_{CB}}{\Delta i_C} \right|_{I_E = \rule{1cm}{0.4pt} mA} = \rule{1cm}{0.4pt} = \rule{1.5cm}{0.4pt} \, k\Omega$$

3. DC current gain, $\alpha_{dc} = \dfrac{I_C}{I_E} = \rule{1cm}{0.4pt} = \rule{1.5cm}{0.4pt}$

4. AC current gain, $\alpha = \left. \dfrac{\Delta i_C}{\Delta i_E} \right|_{V_{CB} = \rule{1cm}{0.4pt} V} = \rule{1.5cm}{0.4pt}$

RESULTS:

1. Input and output characteristics are plotted on the graph.
2. The transistor parameters are given below:

Parameter	Value determined
1. r_i	_____ Ω
2. r_o	_____ $k\Omega$
3. α_{dc}	_____
4. α	_____

EXPERIMENTAL EXERCISE 5.2

TITLE: Transistor characteristics in common-emitter configuration.

OBJECTIVES: To,

1. trace the given circuit;
2. plot the input characteristics (graph between the base current i_B and base-to-emitter voltage v_{BE}, keeping collector-to-emitter voltage V_{CE} constant);
3. calculate the input dynamic resistance from the input characteristic at a given operating point;

4. plot the output characteristics (graph between i_C and v_{CE}, for fixed values of I_B);

5. calculate the output ac resistance (r_o), the dc beta (β_{dc}), and ac beta at a given operating point.

APPARATUS REQUIRED: Experimental board, transistor (or IC) power supply, one milliammeter, (0-50 mA), one microammeter (0-50 μA), two electronic multimeters.

CIRCUIT DIAGRAM: The circuit diagram is shown in Fig. E.5.2.1.

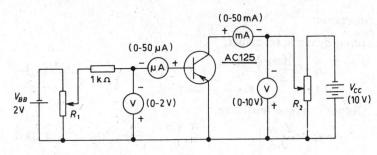

Fig. E.5.2.1

BRIEF THEORY: In Experimental Exercise 5.1, we had drawn the transistor characteristics in CB configuration. In CE configuration, we make the emitter terminal common to the input and output. Whether the transistor is connected in CB or CE, the E-B junction is forward biased and the C-B junction is reverse biased.

For CE configuration, we defined the important parameters as follows:

1. Input dynamic resistance, $r_i = \dfrac{\Delta v_{BE}}{\Delta i_B}\bigg|_{V_{CE}\,=\,\text{const.}}$

2. Output ac resistance, $r_o = \dfrac{\Delta v_{CE}}{\Delta i_C}\bigg|_{I_B\,=\,\text{const.}}$

3. DC current gain, $\beta_{dc} = \dfrac{I_C}{I_B}$

4. AC current gain, $\beta = \dfrac{\Delta i_C}{\Delta i_B}\bigg|_{V_{CE}\,=\,\text{const.}}$

PROCEDURE:

1. Note down the type number of the transistor used in the experimental board. Find the important specifications of the transistor from the data book. Identify the terminals of the transistor and trace the circuit.

2. Make the circuit connections as shown in Fig. E.5.2.1. Use meters with proper range.

3. For input characteristic, first fix the voltage V_{CE}, say, at 9 V. Vary the voltage v_{BE} slowly, in steps. Note the value of current i_B at each step.

4. For output characteristics, first open the input circuit. Vary the collector voltage v_{CE} in steps and note the collector current. This current is the reverse saturation current I_{CEO}, and the magnitude will be small. Now close the input circuit and fix the base current I_B at, say, 10 μA. For this you can use the potentiometer R_1. Vary the voltage v_{CE} with the help of potentiometer R_2 in steps. Note current i_C for each step. Repeat the process for other values of I_B (say, 20 μA, 30 μA, 40 μA, etc.). Be careful not to go beyond the maximum ratings of the transistor.

5. Plot the input and output characteristics by using the readings taken above.

6. Select a suitable operating point in the linear portion of the characteristics. Determine the slope of the input characteristic curve at this operating point. This gives the input dynamic resistance. Similarly, using the definition given above (in brief theory), calculate the output ac resistance r_o, dc beta and ac beta.

OBSERVATIONS:

1. *Type number of the transistor* = _____.
2. *Information from the data book*:

    (*a*) Maximum collector current rating = _____ mA

    (*b*) Maximum collector voltage rating = _____ V

    (*c*) Maximum collector dissipation
        power rating   = _____ W

3. *Input characteristics*:

S. No.	$V_{CB} =$ ___ V		$V_{CE} =$ ___ V	
	v_{BE} (in V)	i_E (in μA)	v_{BF} (in V)	i_B (in μA)
1.				
2.				
3.				

4. *Output characteristics*:

S. No.	$I_B = 0$		$I_B = 10$ μA		$I_B = 20$ μA		$I_B = 30$ μA	
	v_{CE} (V)	i_C (mA)	v_{CE} (V)	i_C (mA)	v_{CE} (V)	i_C (mA)	v_{CB} (V)	i_C (mA)
1.								
2.								
3.								

CALCULATIONS:

1. Input dynamic resistance,

$$r_i = \left.\frac{\Delta v_{BE}}{\Delta i_B}\right|_{V_{CE} = \underline{\hspace{1cm}} V} = \frac{}{} = \underline{\hspace{1cm}} \text{ k}\Omega$$

2. Output ac resistance,

$$r_o = \frac{\Delta v_{CE}}{\Delta I_C}\bigg|_{I_B = \underline{\quad} \text{mA}} = \underline{\quad\quad} = \underline{\quad\quad}\text{k}\Omega$$

3. DC current gain,

$$\beta_{dc} = \frac{I_C}{I_B}\bigg|_{V_{CE} = \underline{\quad}\text{V}} = \underline{\quad\quad} = \underline{\quad\quad}$$

4. AC current gain,

$$\beta = \frac{\Delta i_C}{\Delta i_B}\bigg|_{V_{CE} = \underline{\quad}\text{V}} = \underline{\quad\quad} = \underline{\quad\quad}$$

RESULTS:

1. Input and output characteristics are plotted on the graph.
2. The parameters of the transistor in CE mode are given below:

Parameter	Value determined
1. r_i	_____ Ω
2. r_o	_____ kΩ
3. β_{dc}	_____
4. β	_____

EXPERIMENTAL EXERCISE 5.3

TITLE: FET characteristics.

OBJECTIVES: To,

1. trace the given circuit of FET;
2. plot the static drain characteristics of FET;
3. calculate the FET parameters (drain dynamic resistance r_d, mutual conductance g_m, and amplification factor μ) at a given operating point.

APPARATUS REQUIRED: Experimental board, transistor (or IC) power supply, milliammeter (0 to 25 mA), two electronic multimeters.

CIRCUIT DIAGRAM: The circuit diagram is shown in Fig. E.5.3.1.

BRIEF THEORY: Like an ordinary junction transistor, a field effect transistor is also a three terminal device. It is a unipolar device, because its function depends only upon one type of carrier. (The ordinary transistor is bipolar, hence it is called bipolar-junction transistor). Unlike a BJT, an FET has high input impedance. This is a great advantage.

A field-effect transistor can be either a JFET or MOSFET. Again, a JFET can either have N-channel or P-channel. An N-channel JFET, has an N-type semiconductor bar; the two ends of which make the drain and source

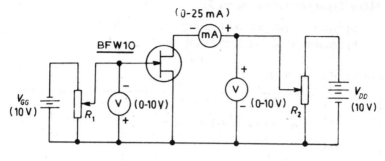

Fig. E.5.3.1

terminals. On the two sides of this bar, PN-junctions are made. These P regions make gates. Usually, these two gates are connected together to form a single gate. The gate is given a negative bias with respect to the source. The drain is given positive potential with respect to the source. In case of a P-channel JFET, the terminals of all the batteries are reversed.

The important parameters of a JFET are defined below:

1. *Drain dynamic resistance,* $r_d = \left.\dfrac{\Delta v_{DS}}{\Delta I_D}\right|_{V_{GS} = \text{const.}}$

2. *Mutual conductance,* $g_m = \left.\dfrac{\Delta i_D}{\Delta v_{GS}}\right|_{V_{DS} = \text{const.}}$

3. *Amplification factor,* $\mu = \left.\dfrac{\Delta v_{DS}}{\Delta v_{GS}}\right|_{I_D = \text{const.}}$

These parameters are related by the equation

$$\mu = r_d g_m$$

PROCEDURE:

1. Note the type number of FET connected in the experimental board. See its specifications from the data book. Identify its terminals. Trace the circuit.
2. Make the circuit connections as shown in E.5.3.1. Use milliammeter and electronic voltmeter in suitable range.
3. First, fix V_{GS} at some value, say 0 V. Increase the drain voltage v_{DS} slowly in steps. Note drain current i_D for each step. Now, change V_{GS} to another value and repeat the above. This way, take readings for 3 to 4 gate-voltage values.
4. Plot the drain characteristics (graph between i_D and v_{DS} for fixed values of V_{GS}).
5. Use the definitions given in brief theory to calculate the FET parameters, from the characteristics.

OBSERVATIONS:

1. *Type number of the FET* = _____

2. *Information from the data book:*
   (*a*) Maximum drain current rating =_____ mA
   (*b*) Maximum drain voltage rating =_____ V

3. *Drain characteristics:*

S. No.	v_{DS} (in V)	Drain current i_D (in mA)				
		$V_{GS} = 0V$	$V_{GS} = -1$ V	$V_{GS} = -2$ V	$V_{GS} = -3$ V	$V_{GS} = -4$ V
1.						
2.						
3.						

CALCULATIONS:

A suitable operating point is selected, say at $V_{DS} = 8$ V, $V_{GS} = -3$ V. At this operating point, the parameters are calculated as follows:

1. $r_d = \dfrac{\Delta v_{DS}}{\Delta i_D}\bigg|_{V_{GS} = -3\text{ V}}$ = ——— = _____ kΩ

2. $g_m = \dfrac{\Delta i_D}{\Delta V_{GS}}\bigg|_{V_{DS} = 8\text{ V}}$ = ——— = _____ mS

3. $\mu = \dfrac{\Delta v_{DS}}{\Delta v_{GS}}\bigg|_{I_D = \text{——— mA}}$ = ——— = _____

RESULTS:

1. The drain characteristics of the FET are plotted on the graph.
2. The parameters of FET determined from the drain characteristics are given below.

Parameter	Value determined
1. r_d	_____ kΩ
2. g_m	_____ mS
3. μ	_____

Vacuum Tubes

OBJECTIVES: After completing this unit, you will be able to: ○ State a few applications where vacuum tubes have not yet been replaced by their solid-state counterparts. ○ Draw the schematic symbols of diode, triode, tetrode and pentode tubes. ○ Explain the phenomenon of thermionic emission. ○ Explain the construction and working of a vacuum diode. ○ Explain the space-charge limited and temperature-limited regions in the V-I characteristics of the diode. ○ Explain the meaning of directly heated cathodes and indirectly heated cathodes in vacuum tubes. ○ Name some cathode materials. ○ Explain the function of control grid in a vacuum triode. ○ Explain the plate characteristics and mutual characteristics of a triode. ○ Define the triode parameters (μ, r_p, and g_m) and state their typical values. ○ Calculate the triode parameters from its static plate characteristics. ○ Explain the function of screen grid in a tetrode. ○ Explain why the negative resistance region exists in the plate characteristics of a tetrode. ○ Explain the pentode plate characteristics. ○ Explain why a pentode can be used as a constant current source. ○ Compare the values of pentode parameters with triode parameters.

6.1 WHY VACUUM TUBES ARE STILL USED

This is the age of solid-state electronics. Since 1948, the year transistors were invented, there has been a rapid development in this field. Today we use ICs and microprocessors. But, before transistors came into vogue, we had big-sized *vacuum tubes* (also called *thermionic valves*). A vacuum tube has two or more electrodes contained in an evacuated glass (or metal) envelope.

Transistors have not fully replaced vacuum tubes. There are still some applications where the use of vacuum tubes is essential. Visit a radio transmitting station in your city. You will find large vacuum tubes inside the transmitter room. These tubes are used to generate high-power radio waves. Wherever high power is involved, we have to use vacuum tubes. The transistor cannot handle high power (may be, after a few years, high power transistors are developed).

There are other applications where we have to use vacuum tubes. In modern TV sets, generally, transistors and ICs are used. As you already know, the transistors and ICs are minute components. Then, why is the TV set so large in size. Using ICs and transistors, the complete circuit of a TV could be wired on a printed board Had it not been for the picture tube, the TV set could be made in the form of a plane board. For the picture display in TV, we use a vacuum tube. Even with the state-of-the art* solid-state technology, it has not yet been possible to find the substitute of the picture tube by its solid-state counterpart.

---

*If the research work done recently in the field of solid-state devices is any indication, it may be possible within a couple of years to have a plane-board TV set. Such TV sets could then be hung on the wall just like a calendar.

Thus we find that we have not done away with vacuum tubes completely. It is worthwhile to know the basics of vacuum-tube devices.

6.2 VACUUM DIODE

It is the simplest of all vacuum tubes. It has only two electrodes. They are called cathode and anode (or plate). The operation of a vacuum diode (and also of other vacuum tubes) depends on the emission of electrons from the cathode. In almost all cases, this emission is achieved by heating the cathode. Such emission of electrons is called *thermionic emission*.

6.2.1 Thermionic Emission

"To emit" means "to throw out". The act of throwing free electrons out of a metallic surface is known as electron emission. If this emission is caused due to the heating of the metal, it is called thermionic emission.

In a metal like copper or tungsten, there are plenty of free electrons. At room temperature, these electrons wander randomly in the atomic structure, but they cannot leave the metallic surface. At room temperature, ordinary metals do not lose their electrons. This means that a force must exist, which prevents electrons from leaving the metallic surface permanently. To understand what this force is, and how it is created, let us assume that due to its random motion, an electron leaves the surface. Immediately after it leaves the surface, the metal gains a positive charge (losing a negatively charged electron is equivalent to gaining a positive charge). This positive charge exerts a force of attraction on the emitted electron. This force pulls the electron back to the metal. For an electron to *escape* from the metal surface, it must have sufficient kinetic energy to overcome this force. This force is described as *surface barrier*.

The surface barrier is analogous to the gravitational pull of the earth. If a body is thrown upwards, it comes back to the earth because of the gravitational force. For a rocket or spaceship to come out of the earth's attracting field, it has to be launched with a velocity greater than a particular value. This value of minimum velocity is called *escape velocity*. Similarly, an electron can come out of the metallic surface permanently, only if its velocity (or kinetic energy) is more than a particular value. Modern physics tells us that even at the absolute zero of temperature, the velocity (or kinetic energy) of all the electrons does not reduce to zero; there are many electrons that possess appreciable energy. The highest energy that an electron in a metal has, at the absolute zero of temperature is called the *Fermi level of energy*. It is designated as E_F. For emission to take place, we have to supply additional energy from outside. This additional energy needed for emission is called *work function* of the metal. It is expressed as E_W. If E_B is the total barrier an electron has to overcome for coming out of the surface, the additional energy needed is

$$E_W = E_B - E_F \tag{6.1}$$

The work function is usually expressed in terms of the energy unit, *electron volt* (abbreviated as eV)*.

---

*One electron volt (eV) is that amount of energy which an electron acquires, if it is moved through a potential difference of one volt.

Just as the escape velocity is different for different planets, so is the work function for different metals. It also depends to some extent on the condition of the surface.

There are many ways to supply the additional energy needed for electron emission. For example, we can supply the additional energy in any of the following ways:

(i) Heat
(ii) Electric field
(iii) Light
(iv) Thrust produced by bombarding the surface with electrons

According to the way we choose to supply additional energy, the electron emission is classified as

(i) Thermionic emission
(ii) High-field emission
(iii) Photoelectric emission
(iv) Secondary emission

Out of these, thermionic emission is most common.

6.2.2 Cathode Materials

The lower the work function of a metal, easier will be the electron emission, i.e. we have to supply less amount of external energy. Unfortunately, metals having low work function also have low melting point. As their temperature is raised, they start melting before any electron emission can take place. The cathode material should have low work function, but at the same time its melting point should also be high.

Three different materials are used for the cathodes of thermionic tubes. The choice of the material depends upon the range of potential and power the anode is expected to handle. The important properties of these three materials are given in Table 6.1.

Table 6.1 CATHODE MATERIALS

Emitter material	Work function E_W (in eV)	Operating temperature (in K)	Usual plate voltage (in V)
1. Oxide-coated	1.0	1000	Below 1000
2. Thoriated tungsten	2.63	2000	1000–5000
3. Pure tungsten	4.52	3000	5000–20 000

Oxide-coated cathodes are used for the tube designed to handle only *small power*, up to about 300 W. These tubes can have voltages up to about 1000 V applied to their anodes. All the tubes, used in radio receivers have oxide-coated cathodes. An oxide-coated cathode consists of a sleeve of *konel* (an alloy of cobalt, nickel, iron and titanium) or some other metal. This sleeve is coated with barium oxide and strontium oxide. Their operating temperature is about 1000 K. As they have very low work functions (only 1 eV) their emission efficiency is high. The emission efficiency of a cathode is defined as the amount of emission current per watt of heating power.

Thoriated tungsten is used in *medium power* valves. These tubes can handle power upto 1000 W. A maximum of about 5000 V can be applied to their anodes. The emission efficiency of these cathodes is less than that of the oxide-coated cathodes. But a thoriated tungsten cathode gives sufficient electron emission if it is heated to about 2000 K. An oxide-coated cathode cannot be used if the voltage at the anode is as high as 500 V. The severity of bombardment of the cathode by ionized atoms (positive ions of the atoms of the gas moving towards the cathode) depends upon the anode voltage. If the anode voltage is high, the bombardment of the cathode by positive ions will be more severe. This results in rapid disintegration of the oxide-coated cathode.

When the tube is required to handle very large power, say, up to 100 kW, thoriated tungsten too becomes impracticable. These tubes have voltage upto about 20 000 V at their plates. The resulting ionic bombardment of the cathode becomes very severe. In such cases, we use pure tungsten for the cathode, although its emission efficiency is very low. It gives large emission current when heated to about 3000 K.

6.2.3 Types of Cathodes

In their physical form, cathodes may be of two types: (i) directly heated, and (ii) indirectly heated. See Fig. 6.1. In directly heated tubes, the heater current passes directly through the cathode. In other words, the heater filament itself works as the cathode. A directly heated cathode may be made of tungsten, thoriated tungsten, or tungsten coated with strontium oxide.

In the case of indirectly heated tubes, the heater current is passed through a resistance wire mounted inside a hollow cylindrical cathode. That is, the heater filament is separate from the cathode. The heater filament is made of tungsten. The cathode is invariably the oxide coated type.

Although a directly heated cathode utilizes heating power in much better way, it has some disadvantage. When current (ac or dc) passes through the filament, there is a voltage drop across the length of the wire. All the points of the filament are not at the same potential. This may adversely affect the working of the tube. On the other hand, the potential at every point on the surface of an indirectly heated cathode is the same.

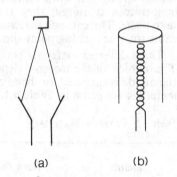

(a)   (b)

Fig. 6.1 Thermionic cathode
construction:
(a) Directly heated cathode
(b) Indirectly heated cathode

6.2.4 Construction of a Diode

Figure 6.2 illustrates the construction of an indirectly heated diode. It has two electrodes*, namely, cathode and anode (or plate). Usually, the cathode is

---

*The heater filament is not considered as an electrode in vacuum tubes, as its function is limited only to the heating of the cathode.

a cylindrical sleeve made of nickel coated with barium oxide and strontium oxide. Inside the cathode, the heater filament is inserted. This is made of tungsten. Surrounding the cathode is another cylinder made of nickel. This serves as the anode. Often the anode is fitted with cooling fins. This helps to dissipate the heat produced at the anode. For the same reason, the surface of the anode is blackened and roughened. The assembly of heater filament, cathode and anode is enclosed in a glass envelope. The pin connections from the electrodes and the heater filament are brought out at the bottom of the tube. The glass envelope is then evacuated. Figure 6.2 also shows the symbol of the diode.

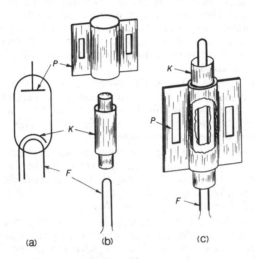

(a) (b) (C)

Fig. 6.2 An indirectly heated cathode:
(a) Symbol; (b) Assembly components
(c) The whole assembly

6.2.5 *V-I* Characteristics of Diode

What happens when the cathode of a valve is heated? Does it go on emitting electrons continuously? Consider Fig. 6.3. The cathode is heated up by the heater filament. The anode (or plate) is not connected to any external circuit. Some electrons are emitted from the surface of the hot cathode. These electrons travel a short distance due to their initial velocity and they form a cluster around the cathode. This cloud or cluster of electrons is known as *space charge*. Obviously the space charge is negative. As electrons leave the cathode surface, it becomes positively charged. As more electrons come out of the surface, they experience a retarding force. They are returned to the cathode. There are two reasons for this retarding force. Firstly, the outgoing electrons experience an attracting force from the

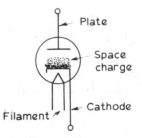

Fig. 6.3 Formation of space charge around the cathode in a diode

positively charged cathode surface. Secondly, the outgoing electrons experience a repulsive force from the negative space charge.

Now let us see what happens to the space charge if the anode is given a positive voltage (with respect to the cathode). This is shown in Fig. 6.4. The filament current is supplied from the low tension (6.3 V) winding of a transformer. A variable dc voltage is connected between the plate and cathode.

Let us increase the plate voltage slightly. Some electrons from the space charge are now attracted by the plate. This results in a flow of plate current. The electrons collected by the plate reach the positive terminal of the battery. The negative terminal of the battery pumps electrons into the cathode of the diode. Remember, some electrons were taken away from the space-charge by the plate. This had created a deficiency of electrons in the space charge. Now, the cathode emits the same number of electrons so that the space charge is maintained. Thus, a continuous flow of current is maintained in the plate circuit.

As the plate voltage is increased, more electrons are attracted by it. The plate current increases. As a result, more number of electrons emitted from the cathode are permitted to join the space charge. The variation of plate current with the plate voltage is shown in Fig. 6.4b. This plot is called the *V-I characteristic* of the vacuum diode.

It can be seen that at first, the plate current increases rapidly with increase in plate voltage. A stage is reached when further increase in plate voltage does not result in much increase in plate current. Soon, the current reaches its *saturation* value. The voltage beyond which the current starts saturating is called *knee*, or *saturation voltage*.

Up to the saturation point (such as point *A* in Fig. 6.4b) space charge exists around the cathode. The current in this region is limited due to the presence of the space charge. That is why the current in this region is called *space-charge limited current*. When the plate voltage is more than the saturation voltage, almost all the emitted electrons are collected by the plate. The rate of emission becomes maximum, and the rate of collection of electrons by the plate also becomes the same. Therefore, the space charge is exhausted. The plate current remains almost constant whatever is the value of plate voltage. This current is known as *temperature-limited current*.

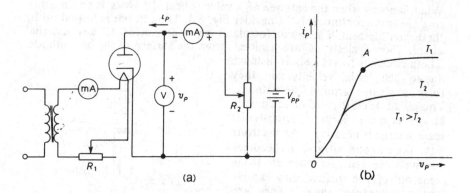

Fig. 6.4 (a) Circuit for determining the *V-I* characteristics of a vacuum diode
(b) Plate characteristics of a vacuum diode

We can reduce the temperature of the cathode by reducing the filament current. At reduced temperature, the cathode cannot supply the electrons at as high a rate as it was doing at higher temperature. Therefore, the saturation current is now lower. If the temperature of the cathode is further reduced, the saturation current further reduces. Note that till the space charge exists around the cathode, the plate current is decided by this space charge. It does not matter what the temperature of the cathode is. The space-charge limited current is given by *Langmuir-Child's three-half power law*:

$$i_P = k\, v_P^{3/2} \qquad\qquad (6.2)$$

where k is a constant of proportionality.

The saturation current depends upon the temperature of the cathode. The current in the temperature-limited region is given by *Richardson-Dushman's equation*:

$$I = A_0 S T^2 \exp(b_0 T) \qquad\qquad (6.3)$$

where A_0 is a constant, S is the area of cathode, T is its temperature in kelvin, and b_0 is a constant that depends upon the work function of the metal.

The reciprocal of the slope of the characteristic curve is the output resistance of the diode. It is called *dynamic plate resistance* and is designated by symbol r_p

$$\text{Dynamic plate resistance,} \quad r_p = \frac{\Delta v_P}{\Delta i_P} \qquad\qquad (6.4)$$

Example 6.1   Measurement on a vacuum diode gave the following observations:

Plate voltage (*in* V)	0	0.5	1.0	1.5	2.0
Plate current (*in* mA)	0	1.6	4.0	6.7	9.4

Plot the characteristic and determine the dynamic plate resistance of the diode at $V_P = 1.25$ V.

Solution: The static characteristic is plotted in Fig. 6.5. Let us take small increments of 0.25 V on either side of the point $V_P = 1.25$ V. Note that the curve is a straight line around the point $V_P = 1.25$ V. Hence there is no need of drawing a tangent to the curve at this point. The corresponding increments in current can be determined from the curve itself. Drawing projections from points A and B, we note the corresponding values of plate current as 6.7 mA and 4 mA. Thus, a small change $\Delta v_P = 0.5$ V in plate voltage produces a small change $\Delta i_P = 2.7$ mA in plate current. Therefore, the dynamic plate resistance is given as

$$r_p = \frac{\Delta v_P}{\Delta i_P} = \frac{0.5}{2.7 \times 10^{-3}} = \textbf{185 } \Omega$$

6.3 USE OF VACUUM DIODES IN RECTIFIERS

We have seen in Unit 4, how semiconductor diodes are used in rectifiers. Similarly, vacuum diodes can also be used in rectifiers. In fact, where very high voltages are involved, vacuum diodes rather than semiconductor diodes

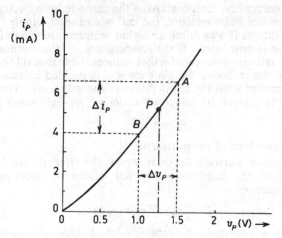

Fig. 6.5 Determination of dynamic plate resistance from the static characteristic curve of a diode

are used. Like a semiconductor diode, a vacuum diode also has the important property of allowing current in one direction only. If a diode is connected in a circuit such that its plate is positive with respect to its cathode, it conducts (Fig. 6.6a), and we say that the diode is *forward-biased*.

On the other hand, if the plate is given a negative voltage with respect to the cathode (as in Fig. 6.6b), it does not conduct. The negative plate cannot attract the electrons from the space charge; it repels them. As electrons do not reach the plate, the diode does not conduct.

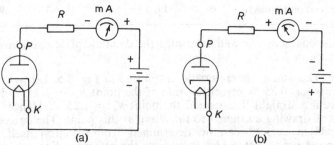

Fig. 6.6 A vacuum tube can be connected in two different ways:
(a) Forward biasing; (b) Reverse biasing

So, we see that vacuum diode permits current to flow in one direction only. It is very much unlike a resistor through which current can flow in either direction with the same ease. This behaviour of a vacuum diode resembles the working of a valve in a water-lifting pump. That is why in the early days of electronics, the vacuum diode was popularly called a "valve".

Note that the characteristic of a vacuum diode is similar to that of a semiconductor diode except for one difference. When a vacuum diode is forward biased, saturation of plate current occurs at high voltages. There is no such saturation in the case of semicoductor diodes. When we use a vacuum diode (or any other tube—triode, pentode, etc.), we avoid this saturation region. The tube always works in the space-charge limited region of its characteristics.

Figure 6.7 illustrates the use of vacuum diodes in rectifiers. As shown in Fig. 6.7*a* and *b*, the transformer has another secondary winding of low voltage. This supplies power to the heating filament. In the full-wave rectifier circuit in Fig. 6.7*b*, the two diodes are shown housed in a single glass envelope. Such tubes are available in market. For the sake of clarity, often we do not show the heater-filament connections, as in bridge rectifier circuit of Fig. 6.7*c*.

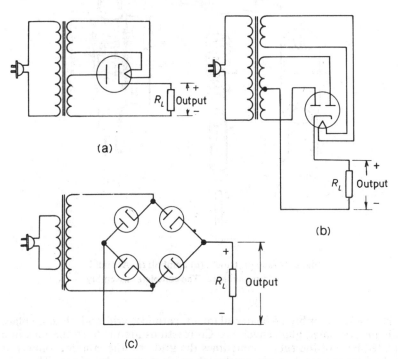

Fig. 6.7 Use of vacuum diodes in (*a*) Half-wave rectifier; (*b*) Full-wave (centre-tap) rectifier; (*c*) Full-wave (bridge) rectifier.

6.4 VACUUM TRIODE

In a vacuum diode, there is only one way of varying plate current, i.e. by changing the voltage of the plate. (Of course, we could change the plate current by varying the filament heating current. But then this change is not followed by the plate current as fast as we desire). If we insert another electrode between the cathode and anode, there is a possibility of varying the plate current by varying the potential of this electrode. We make this third electrode in the form of a mesh or a grid so that it does not obstruct the flow of electrons from the cathode to the plate. The third electrode is kept very near to the cathode. It can *control* the plate current in a much better way than the plate voltage. This electrode is called the *control grid*, or simply *grid*.

Lee De Forest, who observed this phenomenon first, in 1907, called this new device "audion". The device had the ability to amplify small electrical signals (audio signals). This is now known as a *triode tube*. The invention of the triode led to the sensational development of radio communications, broadcasting, and other fields of electronics. Figure 6.8 illustrates the construction of a vacuum triode. The schematic symbol of a triode is also shown in this figure.

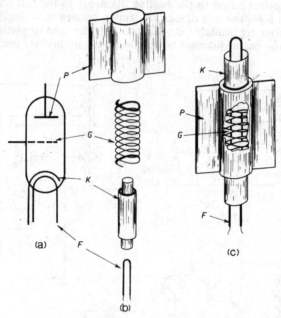

Fig. 6.8 Vacuum triode: (*a*) Symbol; (*b*) Assembly
components; (*c*) The complete assembly

The whole assembly of heater filament, cathode, grid and plate is placed inside an evacuated glass envelope. Connections are made to the pins in an insulated base of the tube. Sometimes the grid, or the anode connections may be brought out to a pin on the top of the glass envelope. The plate (or anode) is provided with cooling fins. This increases the heat dissipation power of the tube.

6.4.1 How the Grid Controls the Plate Current

The control grid is in the form of a mesh of fine wire. Very often, it consists of a fine wire helically wound around the cathode. The grid is normally maintained at a negative potential relative to the cathode. Because of the negative potential of the grid, less number of electrons are able to reach the plate. The electrons experience a retarding force from the control grid. So, only those electrons which have high kinetic energy can reach the plate. As a result, the plate current is reduced. If the control grid is made more negative, still less number of electrons will be able to reach the plate. The plate current is further reduced. Conversely, if the grid is made less negative, more electrons are able to reach the plate. The grid potential is able to control the number of electrons reaching the plate. In other words it can control the plate current (in the same way as plate voltage does).

Here is an important point. The grid is situated much nearer to the cathode as compared to the anode. Therefore, the controlling action per volt change in potential, exerted by the grid, is much greater than that exerted by the anode. And for this control of plate current (by the grid), the grid does not take any current. In this sense, the triode acts as a relay. This is the main feature of this device.

If we continue increasing the negative potential at the grid, the plate current goes on decreasing. Eventually a point is reached where the plate current reduces to zero. For a given plate voltage, there is a particular value of grid voltage at which the plate current is just cut off. This value of grid voltage is called *cut-off bias*.

Usually, the grid is not permitted to become positive relative to the cathode, because this would allow grid current to flow. The power developed at the grid is in the form of heat. This heat must be dissipated to the surroundings. If it is not dissipated, the grid structure would be damaged.

6.4.2 Static Characteristics of a Triode

In a triode, we are interested in three electrical quantities. These are the plate voltage v_P, the plate current i_P, and the grid voltage v_G. They are interrelated. If one quantity varies, the others will be affected. There are three possibilities of showing this interdependence. We can plot the following curves:

(i) Between v_P and i_P for constant value of V_G,—*static plate characteristics*.

(ii) Between v_G and i_P for constant values of V_P—*mutual* or *transfer characteristics*.

(iii) Between v_G and v_P for constant values of I_P—*voltage transfer*, or *constant current characteristics*.

The third set of characteristics is rarely employed in practice. Therefore, we shall not discuss it any further.

Plate Characteristics The plate characteristics are plots of plate current against plate voltage, for constant values of grid voltage. They can be determined experimentally using the circuit shown in Fig. 6.9. The procedure is as follows: First, set the grid voltage v_G at a convenient value (say, 0 V). Now increase the plate voltage v_P from zero in a number of discrete steps. At each step, note the plate current i_P. The grid voltage is then set to a new

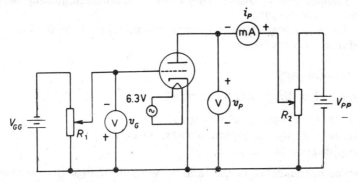

Fig. 6.9 Circuit arrangement for the determination of the static characteristics of a triode

value (say, -1 V). The measurements are again repeated. The plot of these values gives a family of curves, as shown in Fig. 6.10. The curves are seen to be linear over much of their range. The operation of the triode is restricted to this linear part.

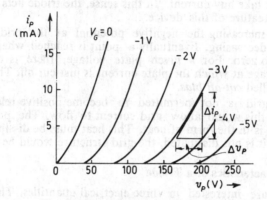

Fig. 6.10 Typical static plate characteristics
of a triode

Mutual Characteristics The mutual or transfer characteristics are the plots of plate current against grid voltage for constant values of plate voltage. This set of characteristics can be determined with the help of the circuit in Fig. 6.9. The procedure of measurement is as follows: First, the plate voltage is set at a convenient value (say, 100 V). The grid voltage is then increased negatively, starting from zero, in steps. At each step, the plate current is noted. The plate voltage is set to other different values and then the procedure is repeated. The plot of these observations gives a family of curves, as shown in Fig. 6.11. For a given curve, the plate voltage is constant. The point at which the curve meets the v_G axis gives the cut-off voltage.

A triode can be used as an amplifier. Usually, the grid and the cathode make the input terminals, and the plate and cathode the output terminals. The curves in Fig. 6.11 relate an input quantity (namely, grid voltage v_G) with an output quantity (namely, plate current i_P). That is why these curves are called *mutual* characteristics. They show the mutual relationship between the input and the output.

To understand the working of a circuit containing a triode, the knowledge of its parameters is necessary.

6.4.3 Triode Parameters

A triode tube has three useful ac parameters :

  (i) Plate ac (or dynamic) resistance, r_p
  (ii) Mutual conductance (or transconductance), g_m
  (iii) Amplification factor, μ.

The plate ac resistance r_p, is the ratio of a small change in plate voltage to the change in plate current produced by it; the grid voltage remaining constant. That is

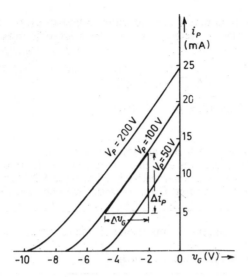

Fig. 6.11 Typical mutual characteristics of a
vacuum triode

$$r_p = \frac{\Delta v_P}{\Delta i_P}\bigg|_{V_G = \text{const.}} \qquad (6.5)$$

It is expressed in kilohms (kΩ). Typically, it ranges from about 8 kΩ to 40 kΩ. The r_p of a triode can be determined from its plate characteristics. See Fig. 6.10. Note that r_p represents the reciprocal of the slope of the plate characteristics curve.

The mutual conductance g_m, is the ratio of a small change in plate current to the change in grid voltage producing it; the plate voltage remaining constant. That is

$$g_m = \frac{\Delta i_P}{\Delta v_G}\bigg|_{V_P = \text{const.}} \qquad (6.6)$$

It is usually expressed in millisiemen (mS). Typically, it ranges from about 1.0 mS to 10.0 mS. The mutual conductance is the slope of mutual characteristic curve as shown in Fig. 6.11.

The amplification factor μ, is the ratio of a small change in plate voltage to the corresponding change in grid voltage, when the plate current remains constant. That is

$$\mu = \frac{\Delta v_P}{\Delta v_G}\bigg|_{I_P = \text{const.}} \qquad (6.7)$$

This, being the ratio of two voltages, does not have any units. It is dimensionless. Typical value of amplification factor of a triode ranges from 16 to 100.

The above three parameters are not independent of each other. We can establish relationship between them in the following way. Multiplying Eqs. 6.5 and 6.6 gives

$$r_p \times g_m = \frac{\Delta v_P}{\Delta i_P} \times \frac{\Delta i_P}{\Delta v_G} = \frac{\Delta v_P}{\Delta v_G}$$

which, from Eq. 6.7, is equal to μ. Therefore,

$$\mu = r_p \times g_m \qquad (6.8)$$

If any two parameters of a triode are known, the third can be computed. The values of these parameters can be determined from either the plate characteristics or the mutual characteristics. Example 6.2 illustrates how the triode parameters are computed from its static plate characteristic curves.

Example 6.2 The measurements made on a triode tube are given in Table 6.2.

Table 6.2 TRIODE CHARACTERISTICS MEASUREMENTS

Plate voltage v_P (in V)	Plate current (in mA)				
	For v_G 0 (in V)	−1	−2	−3	−4
50	3.5	—	—	—	—
100	11.2	4.0	—	—	—
150	20.0	12.4	5.4	—	—
200	—	21.5	14.1	3.4	—
250	—	—	—	12.4	2.5
300	—	—	—	—	11.3

Plot the static plate characteristics. Use the linear parts of these characteristics and determine the values of plate ac resistance, mutual conductance and amplification factor.

Solution: The static plate characteristics are shown in Fig. 6.12. For calculating the plate ac resistance, let us take the curve for $V_G = -1$ V. On this curve, let us take the operating point at $V_P = 150$ V. For this voltage, the plate current can be seen to be $I_P = 12.5$ mA. Let us choose a small increment of 10 V on either side of the operating point P (total increment, $\Delta v_P = 20$ V). This gives points A and B on the curve. We draw projections from these points on the i_P axis. This gives the corresponding increment in plate current as

$$\Delta i_P = 14.0 - 10.7 = 3.3 \text{ mA}$$

Therefore, the plate ac resistance is given by

$$r_p = \frac{\Delta v_P}{\Delta i_P}\bigg|_{V_G = -1 \text{ V}} = \frac{20}{3.3 \times 10^{-3}} = 6.06 \text{ k}\Omega$$

For determining mutual conductance, we have to keep v_P constant. Let $V_P = 150$ V. If we change the grid voltage v_G from -1 V to -2 V, the corresponding change in current is seen from figure as

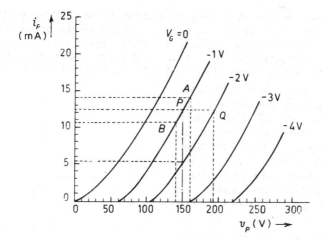

Fig. 6.12 Calculation of triode parameters from its
static plate characteristics

$$\Delta i_P = 12.4 - 5.3 = 7.1 \text{ mA}$$

Therefore, the mutual conductance is

$$g_m = \frac{\Delta i_P}{\Delta v_G}\bigg|_{V_P = 150 \text{ V}} = \frac{7.1 \times 10^{-3}}{1} = 7.1 \text{ mS}$$

We can use Eq. 6.8 to compute amplification factor

$$\mu = r_p g_m = 6.06 \times 10^3 \times 7.1 \times 10^{-3} = 43.03$$

The amplification factor can also be calculated by using its definition given in Eq. 6.7. For this, we take plate current constant at $I_P = 12.4$ mA. Again, let us take a change in v_G from -1 V to -2 V. Corresponding change in plate voltage is found by drawing projections from points P and Q on the v_P axis. This gives

$$\Delta v_P = 192 - 150 = 42 \text{ V}$$

Therefore,

$$\mu = \frac{\Delta v_P}{\Delta v_G}\bigg|_{I_P = \text{const.}} = \frac{42}{1} = 42$$

This value of μ quite tallies with the value earlier calculated.

6.4.4 Limitations of a Triode Tube

The cathode and plate of a triode tube are in the form of two concentric cylinders. These are made of conducting metal and are separated by some distance. In between the cathode and the plate, there is an insulating material (air at very low pressure). This is exactly what we have in the construction of a capacitor. Thus, a capacitance must exist between these two electrodes. This capacitance is designated as C_{pk}. Similarly, capacitances must exist between the plate and grid, and the cathode and the grid. These capacitances are designated as C_{pg} and C_{gk}, respectively. These *inter-electrode capacitances* are shown in Fig. 6.13. All these capacitances are *stray* or *parasitic* capacitances, since they are not deliberately placed there. The value of these capacitances is very small. They may range from about 2 pF to 12 pF.

In what way do these small inter-electrode capacitances put a limitation on the use of the vacuum triode? The triode tube works as an excellent amplifier till the frequencies of the signal are low. But, when the signal frequencies are high, the inter-electrode capacitances affect the working of the tube adversely. The capacitances C_{pk} and C_{gk} are not very important. They can generally be incorporated into the associated circuitry. However, the capacitance C_{pg} between the plate and the grid is of utmost importance. It is most objectionable, since it provides a path linking the plate and grid circuits. Through this linking path, a part of the output signal may be fed

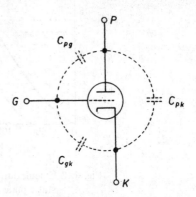

Fig. 6.13 Interelectrode capacitances of a triode tube

back into the input circuit. This unwanted feedback is undesirable. It may make the triode amplifier circuit unstable and may well produce self-oscillations. This is highly objectionable in the working of an amplifier circuit.

6.5 TETRODE TUBE

The tetrode tube was developed in an attempt to reduce the capacitance C_{pg} of a vacuum triode. Between the plate and control grid of a triode tube, another electrode was introduced. This fourth electrode is called *screen grid* (grid 2) (to distinguish it from control grid) (grid 1). This grid, like the control grid, is also made in the form of a fine wire-mesh, so that it does not hinder the flow of electrons from cathode to plate. A plan view of a typical tetrode tube is given in Fig. 6.14a. Note that the control grid is completely enclosed by the screen grid. Figure 6.14b shows the schematic symbol of the tetrode.

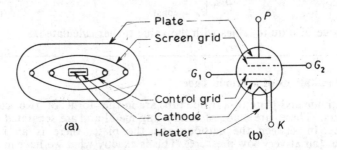

Fig. 6.14 Tetrode tube: (*a*) Construction
(*b*) Schematic symbol

Let us see how the screen grid between the plate and control grid reduces the capacitance C_{pg}. The screen grid is connected, directly or via a resistor, to the HT (high tension) supply and to the earth via a capacitor, as shown in Fig. 6.15a. So far as alternating currents are concerned, the screen grid is at earth potential. It acts as an earthed screen or shield between the plate and the control grid. It effectively divides the capacitance C_{pg} into two capacitances—C_p between the plate and earth, and C_g between grid 1 and earth, as shown in Fig. 6.15b. Ideally, the control grid is now completely isolated from the plate, and C_{pg} is zero. However, in practice, the screening is not perfect. Some capacitance between the plate and control grid still exists. Typically, capacitance C_{pg} for a tetrode is of the order of 0.01 pF.

We have seen how the presence of a screen grid, grounded (from ac point of view) to earth, reduces capacitance C_{pg}. Now the question arises—Why do we have to give a high dc voltage (of say, 100 V) to the screen grid? Could we not connect the screen grid directly to the earth? To answer this, let us assume for a moment that the screen grid is directly connected to the earth (instead of connecting it to earth through a capacitor). Now the electrons emitted from the cathode will not find any attracting force. Even if the plate is given a positive potential, the electrons in the space charge do not experience an attracting force. This is because of the shielding effect of the screen grid. Thus, we find that the electrons do not move from the space charge to the plate. The plate current remains zero even if plate has a high positive potential.

If we give high dc potential to the screen grid, the electrons from the space charge are accelerated. This acceleration is sufficient to take them to the plate. The plate current can flow, if it is given a positive voltage. In a tetrode, the screen grid accelerates the electrons, and the plate collects them.

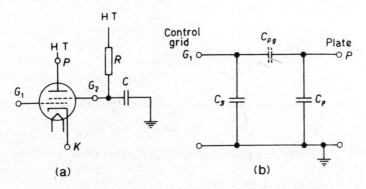

Fig. 6.15 The action of the screen grid in the tetrode

6.5.1 Plate Characteristics of Tetrode

Typical plate characteristics of a tetrode are shown in Fig. 6.16. They are plotted between plate current and plate voltage for fixed values of both screen-grid voltage and control-grid voltage. For all the curves in Fig. 6.16, the screen voltage is fixed (say, at 80 V). There are four important points to be noted about these characteristics:

(i) The plate current rises rapidly from zero as the plate voltage is first increased. This rise in plate current is more rapid than in case of a triode.

(ii) Over a range of plate voltages, from approximately 25 V to 75 V, the plate current decreases with increase in plate voltage. This is somewhat unusual. The tetrode behaves as *negative resistance* in this range.

(iii) The characteristics shown are for one particular value of screen-grid voltage. If we change the screen-grid voltage, the family of characteristics will also change.

(iv) Above approximately 75 V, the plate current increases with increase in plate voltage but the curve soon flattens out. It is this flat part of the characteristic that is useful when the tetrode is used in an amplifier.

The flat part of each curve is almost horizontal. It means that the plate ac resistance (which is equal to the reciprocal of the slope of the curve) is very large. It is typically between 400 kΩ and 1 MΩ. The presence of the screen grid in a tetrode is immaterial, as regards the influence of the control grid over the plate current is concerned. Therefore, the mutual conductance g_m of a tetrode is of the same order as that of a triode. But the amplification factor $(\mu = r_p g_m)$ of a tetrode is much higher than that of a triode.

The disadvantage of the tetrode is that it has "kinks" in its characteristics. The presence of the kinks in the characteristics restricts the plate voltage swing to the higher voltages only. The kink appears because the tetrode behaves as a negative resistance over a range of plate voltages. This negative resistance region is due to a phenomenon called *secondary emission* of electrons from the plate.

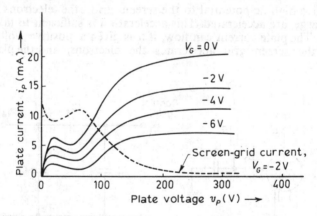

Fig. 6.16 Plate characteristics of a tetrode tube

6.5.2 Secondary Emission

In a tetrode, the electrons emitted from the cathode are accelerated not only by the plate, but also by the positive screen grid. When these electrons arrive at the plate, they gain kinetic energy. This energy usually appears as heat at the plate. The heat has to be dissipated by the plate to the surroundings. When the kinetic energy of the striking electrons is considerably high, some of the electrons in the plate material may be knocked out. These knocked out electrons are called secondary emitted electrons, and this phenomenon is called *secondary emission*.

When the plate voltage is zero, all the electrons emitted from the cathode are collected by the screen grid. Hence the screen-grid current is high. See the dotted curve in Fig. 6.16. It shows the variation of screen-grid current with plate voltage for a grid voltage of -2 V. At zero plate voltage, the plate does not collect any electrons. Hence, the plate current is zero.

As we increase the plate voltage from zero onwards, more number of electrons reach the plate. The plate current increases, and the screen current decreases. The plate current continues to increase till the plate voltage becomes about 25 V. The kinetic energy of the electrons striking the plate (these are called primary electrons) becomes sufficiently high to cause secondary emission. As the screen grid is at higher potential than the plate, it attracts these secondary electrons. Because of this, the screen-grid current increases, and the plate current decreases. When the plate voltage is further increased, the kinetic energy of the primary electrons also increases. Now they are able to cause more secondary emission. So the screen-grid current increases, and plate current continues to decrease. This reduction in plate current with increase in plate voltage continues until the plate voltage is increased to a point where it is sufficiently positive relative to the screen grid. The secondary electrons are then attracted back to the plate. Thereafter, the plate current increases, and the screen current decreases, with increase in plate voltage. Beyond some plate voltage (called "knee" voltage), there is hardly any increase in plate current with increase in plate voltage. The curve becomes almost flat.

Is there no secondary emission in a triode tube?

When primary electrons with high kinetic energy strike the plate, secondary emission is caused. But in the triode there is no other electrode (such as screen grid in tetrode) at higher potential than the plate. So, the secondary electrons again go back to the plate. That is why secondary emission is not observed in the triode characteristics.

The tetrode tube would have served as a good high-frequency amplifier, but for the kink in its characteristics. An effort to remove the kink from the characteristics of the tetrode led to another tube—*pentode*.

6.6 PENTODE TUBE

The pentode tube was originally developed to overcome the effect of secondary emission in the tetrode tube. A pentode has five electrodes. The fifth electrode, called suppressor grid (or grid 3), is placed between the plate and the screen grid. The plan view of a pentode is shown in Fig. 6.17a. Its schematic symbol is shown in Fig. 6.17b.

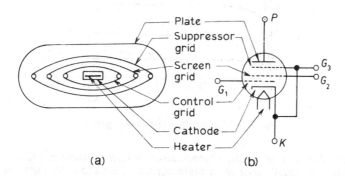

Fig. 6.17 (*a*) Construction of a pentode tube; (*b*) Its symbol

The fifth electrode in a pentode is called the *suppressor grid,* because it suppresses the effects of secondary emission. Let us see how it performs this function. The suppressor grid (like other grids) is in the form of a fine wire-mesh. It does not obstruct the flow of electrons from the cathode to the plate. Generally, the suppressor grid is connected to the cathode. Thus, it is many volts negative with respect to both the plate and screen grid. The electrons emitted from the cathode are accelerated by the screen-grid potential. (The screen grid in a pentode is also kept at high positive potential, say, 80 V.) The electrons attain quite high velocities to enable them to overcome the retarding field of the suppressor grid and reach the anode. The secondary electrons emitted from the plate do not have much initial kinetic energy. They come out of the plate with relatively small velocities. These electrons cannot overcome the retarding field of the suppressor grid. (Remember, the suppressor grid is at a negative potential with respect to the plate.) Hence these electrons are returned to the plate. Thus, the plate current remains unaffected, and the negative resistance region (of the tetrode tube) is removed. The pentode characteristics do not have any kinks or folds.

Besides suppressing the secondary emission, the suppressor grid proves useful in another way. It acts as another screen between the plate and the control grid. The capacitance C_{pg} between the plate and the cathode is further reduced. Typically, it is 0.004 pF only.

6.6.1 Plate Characteristics of Pentode

A typical family of pentode plate characteristics is shown in Fig. 6.18. They are plotted between the plate current and plate voltage for constant values of control-grid voltage. The plate current increases rapidly with the increase in plate voltage in the beginning. Then the curves flatten out. The plate current becomes almost independent of further increase in plate voltage.

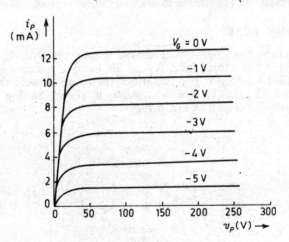

Fig. 6.18 Plate characteristics of a pentode

Over the flat portion of the characteristic, a large change in plate voltage causes only a small change in plate current. This indicates that the plate ac resistance is very high. Typically, it is of the order of 1 MΩ. The mutual conductance of a pentode is approximately the same as that of a triode.

This is so because the value of g_m depends upon the relative position of the control grid, cathode and plate; and for the two tubes (triode and pentode) the position of the three electrodes is the same. With same g_m and high r_p, very large values of amplification factor are common in pentodes. Typically, μ of a pentode may range from 1000 to 10 000.

Compared to a tetrode, a pentode permits greater swings of plate voltage without causing distortion. Therefore, the pentode has replaced the tetrode in almost all applications. The pentode is used in video amplifiers and rf voltage amplifiers. It is also used as a constant-current source, because the plate current is almost constant irrespective of the plate potential.

6.6.2 Variable-mu Pentode

The mutual characteristics of a pentode can be drawn by plotting the variation of plate current with control grid voltage for constant plate voltage (of course, the screen grid voltage also remains constant). If you change the plate voltage to another value, the mutual characteristic curve remains almost the same.

In Fig. 6.19, we have shown the mutual characteristic curves for two pentodes. One is for a pentode type EL84, and the other is for type EF89. The first curve (for tube EL84) is mostly a straight line. If you calculate the value of mutual conductance at different points on this curve, you will get almost the same value. That is, for this tube, the mutual conductance does not depend upon the grid-bias voltage. Also note that on increasing the grid voltage negatively, the cut-off occurs sharply. Hence, the tube is called a *sharp cut-off pentode*.

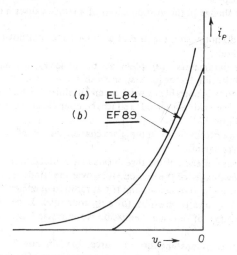

Fig. 6.19 Transfer or mutual characteristic curve
(*a*) EL84 sharp cut-off pentode;
(*b*) EF89 remote cut-off or variable-mu
pentode

The observations made above do not apply to the second curve (tube EF89). For this curve, the value of mutual conductance g_m will be different at different points. At low grid voltage, g_m has high value. As we increase the grid voltage (negatively, of course), the value of g_m reduces. Hence, this type of tubes are called *variable-mu pentodes,* as their mutual conductance can be varied by varying the grid bias. Such a characteristic is obtained by making the grid-cathode spacing, or the spacing between the wires of the grid, or the thickness of the grid wire, non-uniform along the length of the grid structure. These tubes are also called *remote cut-off* tubes, because cut-off is not sharply noticeable.

REVIEW QUESTIONS

6.1 Name some applications where solid-state devices have not been able to replace vacuum tubes as yet.

6.2 In a vacuum diode there are only two electrodes, namely, the cathode and the anode (or plate). But, while connecting it in a circuit, we give connections to four pins of its base. Explain why ?

6.3 There are a large number of free electrons moving randomly in a metal. At room temperature, why do they not come out of the surface of the metal ?

6.4 Explain the phenomenon of "thermionic emisson". Name other types of emission possible.

6.5 What are the three most commonly used cathode materials ? Give important properties of each.

6.6 Explain the meaning of terms *space-charge limited* and *temperature-limited* with reference to a vacuum diode.

6.7 Explain how a vacuum diode can be used to act as a rectifier of alternating current. Compare the relative merits and demerits of vacuum diodes and semiconductor diodes for this purpose.

6.8 Explain why any change in the grid potential of a triode causes a change in its plate current.

6.9 Explain why we do not permit the grid of a triode to be at a positive potential with respect to its cathode.

6.10 Define the triode parameters : (a) plate ac resistance r_p, (b) mutual conductance g_m, and (c) amplification factor μ. Also show that $\mu = r_p g_m$.

6.11 Explain why a triode is not able to work as an amplifier for high-frequency signals.

6.12 How does the screen grid in a tetrode reduce the inter-electrode capacitance between plate and control grid ?

6.13 Why do the kinks or folds occur in the plate characteristics of a tetrode ? In what way are these kinks harmful ?

6.14 Explain the purpose of each of the three grids in a pentode tube.

6.15 What are the advantages of the pentode tube over the triode tube while working as an amplifier (a) at audio frequencies and (b) at radio frequencies ?

6.16 Draw the symbol of (a) a triode tube, (b) a pentode tube. Label each electrode.

6.17 Explain how the effect of secondary emission is suppressed by the suppressor grid in a pentode.

6.18 A pentode can act as a constant current source. Explain how ?

6.19 What is the difference between the mutual characteristics of a remote cut-off pentode and a sharp cut-off pentode ? Which of the two will have variable mutual conductance with grid bias ?

OBJECTIVE-TYPE QUESTIONS

I. Here are some incomplete statements. Four alternatives are provided below each. Tick the alternative that completes the statement correctly:

1. In almost all applications, vacuum tubes have been replaced by solid-state devices. But the vacuum tubes are still in use where
   (a) very long life of the device is a necessary requirement
   (b) very high power is to be handled
   (c) very low-frequency signals are involved
   (d) portability of the electronic system is an important factor

2. In a vacuum tube, the rate at which electrons are collected by the plate is equal to the rate at which electrons are emitted by the cathode, only when tube is working in
   (a) space-charge limited condition only
   (b) temperature-limited condition only
   (c) both the space-charge limited as well as temperature-limited conditions
   (d) none of the two conditions

3. The control grid in a triode is usually given negative potential with respect to the cathode so as to
   (a) reduce the space-charge
   (b) increase the space-charge
   (c) limit the plate current to a safe value
   (d) make the grid current zero

4. A triode tube can be used as an amplifier because
   (a) it has three terminals and any three-terminal device can act as an amplifier
   (b) any small change in grid voltage can cause a large change in plate voltage
   (c) high-power source is available in the plate circuit
   (d) its ac plate resistance is different from the dc plate resistance

5. The function of control grid in a pentode tube is
   (a) to control the secondary emission from the plate
   (b) to accelerate the electrons emitted from the cathode
   (c) to collect electrons from the space-charge
   (d) to control the number of electrons moving from cathode to plate

6. In case of a directly heated cathode
   (a) electrons are emitted from the filament itself
   (b) heating power is supplied to the cathode and electron emission takes place from the filament
   (c) heating voltage is given to the filament and electron emission takes place from the cathode
   (d) the cathode is directly heated by the bombardment of positive ions

7. The heater filament of a vacuum tube is generally supplied with ac voltage (and not dc voltage) for heating because
   (a) it results in a uniform heating of filament, so that the electron emission is also uniform
   (b) it is very easy to obtain this voltage from the ac mains supply
   (c) the dc voltage that would be required for heating has much greater magnitude than the ac voltage used
   (d) when dc is used for heating, a different type of filament is required which is very expensive

8. While doing an experiment to determine the plate characteristics of a triode, a student observed that plate current changes by 1.5 mA when the plate voltage is changed by 20 V. During this experiment, the grid voltage was maintained at —2 V. The plate ac resistance of the triode is

   (a) 7.5 kΩ   (c) 60 kΩ
   (b) 30 kΩ   (d) 13.3 kΩ

9. In tetrode plate characteristics, the negative resistance region comes when

   (a) plate voltage is more than the screen-grid voltage
   (b) plate voltage is less than the screen-grid voltage
   (c) the control grid is at positive potential with respect to the screen grid
   (d) the control grid is at positive potential with respect to the cathode

10. In a tetrode tube, secondary emission means

    (a) the emission of electrons from the filament due to heat energy
    (b) the emission of electrons from the plate due to the bombardment of the fast-moving electrons emitted from the cathode
    (c) emission of high-velocity electrons from the cathode
    (d) emission of those electrons which belong to the second orbit of the atoms of cathode

II. State whether the following statements are TRUE or FALSE ?

1. The space charge formed around the cathode of a vacuum tube is negative. _____

2. When a vacuum diode is working under space-charge limited condition, its dynamic plate resistance is greater at higher plate voltage, than at lower plate voltage.

_____

3. In temperature-limited region of its characteristics, the plate of a diode collects electrons at a faster rate than they are emitted from the cathode. _____

4. Most of the vacuum tubes use oxide-coated cathode because they can be heated to sufficiently high temperature (say, 3000 K) without affecting its physical properties.

_____

5. The cathode of a thermionic tube is always heated by passing an electric current.

_____

6. An increase in the negative potential on a control grid of a triode tube reduces the anode current. _____

7. Typically, the amplification factor of a triode is about 10 000. _____

8. A triode is said to be cut-off when no current flows through its plate circuit, even if the plate has high positive potential on it. _____

9. The main purpose of inserting the screen grid between the control and plate (in a tetrode tube) is to reduce the space-charge around cathode. _____

10. The screen grid of a pentode tube is maintained at a more negative potential than the control grid. _____

11. The inter-electrode capacitor C_{pg} between the plate and control grid of a pentode is reduced to about 4.5 pF. _____

12. The mutual conductance of a pentode is of the same order as that of a triode.

_____

13. Out of the three inter-electrode capacitances in a triode, the most objectionable is the capacitance between the plate and the control grid, i.e. C_{pg}. _____

14. The plate ac resistance of a pentode is of the order of 10 kΩ.

Ans. I.   1. b; 2. c; 3. d; 4. b; 5. d; 6. a; 7. b; 8. d; 9. b; 10. b.
      II.   1. T; 2. F; 3. F; 4. F; 5. T; 6. T; 7. F; 8. T; 9. F; 10. F; 11. F; 12. T;
            13. T; 14. F.

TUTORIAL SHEET 6.1

1. Measurements made on a vacuum diode type EZ81 are tabulated in Table T.6.1.1.

Table T.6.1.1

Plate voltage (in V)	0	2.0	5.0	10.0	12.0
Plate current (in mA)	0	10.0	27.0	60.0	75.0

Plot the V-I characteristic of the diode, and determine the plate dc resistance and plate ac resistance of the diode at a plate voltage of 8 V.

[Ans. 177.8 Ω; 142.85 Ω]

2. Measurements were made on the triode section of the tube ECH81 (made by BEL, Bangalore) to determine its characteristics. The observations are tabulated in Table T.6.1.2.

Table T.6.1.2

Plate voltage, v_p (in V)	Plate current i_P (in mA)					
	$V_G = 0$ V	$V_G = -2$ V	$V_G = -4$ V	$V_G = -6$ V	$V_G = -8$ V	$V_G = 10$ V
0	0	0	0	0	0	0
25	2.2	0.4	0	0	0	0
50	5.4	1.5	0.25	0	0	0
75	9.0	4.2	1.2	0.2	0	0
100	—	7.5	3.5	1.2	0.2	0
125	—	—	6.3	3.0	1.2	0.2
150	—	—	—	5.2	2.7	1.2

Plot the static plate characteristics of the triode. Locate the operating point given by $V_P = 75$ V; $V_G = -2$ V. Determine the parameters r_p, g_m and μ of the triode at this operating point, and hence verify the relation $\mu = r_p\, g_m$.

[Ans. $r_p = 8$ kΩ; $g_m = 2.5$ mS; $\mu = 19$]

3. A pentode tube has a mutual conductance of 2.5 mS and an amplification factor of 2000. Calculate its plate ac resistance. [Ans. 800 kΩ]

EXPERIMENTAL EXERCISE 6.1

TITLE : Vacuum diode characteristics.

OBJECTIVES: To,

1. trace the given circuit;
2. measure the values of plate current for different values of plate voltage;
3. plot the plate characteristics (graph between v_P and i_P) of the given diode for different values of filament current I_F;

4. determine the plate resistance (dc and ac) of the given diode in space-charge limited region of the characteristics, at a given point.

APPARATUS REQUIRED: Experimental board, dc milliammeter (0-50 mA), ac ammeter (0-1 A), electronic dc voltmeter, power supply.

CIRCUIT DIAGRAM: The circuit diagram is shown in Fig. E.6.1.1. Vacuum diode EZ80 is used for which pin numbers are also shown.

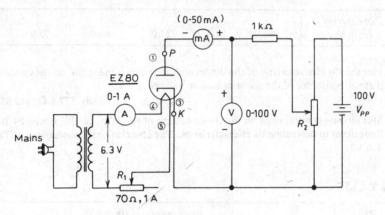

Fig. E.6.1.1

BRIEF THEORY: When the filament of a vacuum diode is supplied with ac voltage, it gets heated up. This in turn heats the cathode and electrons are emitted from the cathode. These emitted electrons move a short distance to form a negative space charge. When a positive voltage is applied to the plate with respect to the cathode, electrons from the space-charge region move towards the plate. This movement of electrons constitute the plate current. When the positive voltage of the plate is increased, the plate current increases. If the plate voltage is further increased, a stage comes when the plate current becomes constant. This current is known as temperature-limited current. Under this condition, all the electrons which are emitted from cathode are collected by the plate and no space charge is present. Reducing the filament current (or cathode temperature) does not affect the space charge current. But the temperature-limited current decreases. When the rated value of filament current (600 mA for EZ80) is supplied, the temperature-limited current may be very high. It may be much more than the maximum rated value of the plate current. So we should not try to get the temperature-limited region of characteristics, for this filament current. Reduce the filament current to observe the temperature limited region of characteristics.

PROCEDURE:

1. With the help of BEL data book find the important specifications and pin configuration for the tube used in the experiment. Trace the circuit.

2. Make the connections as shown in Fig. E.6.1.1. Use milliammeters and voltmeters of proper ranges.

3. Fix a certain value of filament current (say, 400 mA) with the help of a potentiometer in the filament circuit. Increase plate voltage in steps. For each step, note the plate current.

4. Increase the filament current I_F to a different value (say, 425 mA). Repeat the above. Similarly, take $I_F = 450$ mA and 475 mA and repeat observation. Be careful not to exceed the maximum ratings.

5. Plot the graph between the plate voltage and plate current for different filament currents. Now take a suitable point in the space-charge limited region of the characteristic and calculate the plate resistance (dc and ac) at a certain point on this curve.

OBSERVATIONS:

1. *Type number of the diode* = _____
2. *Information from the data book*:
   (a) Maximum plate current rating   = _____ mA
   (b) Maximum plate power dissipation rating = _____ W
3. *Plate characteristics*:

S. No.	Plate voltage, v_P (in V)	Plate current, i_P (in mA)		
		$I_F = 400$ mA	$I_F = 425$ mA	$I_F = 450$ mA
1.				
2.				
3.				

CALCULATIONS:

1. Plate dc resistance, $R_P = \dfrac{V_P}{I_P} = $ _____ = _____ Ω

2. Plate ac resistance, $r_p = \dfrac{\Delta v_P}{\Delta i_P} = $ _____ = _____ Ω

RESULTS:

1. The plate characteristics of the given vacuum diode is plotted in the graph.
2. (a) The plate dc resistance   = _____ Ω
   (b) The plate dynamic resistance = _____ Ω

EXPERIMENTAL EXERCISE 6.2

TITLE: Triode characteristics.

OBJECTIVES: To,

1. trace the given circuit;
2. measure plate current i_P for different plate voltages v_P when grid voltage V_G remains fixed;

3. plot the plate characteristics for different values of grid voltage V_G;
4. measure i_P for different values of v_G when plate voltage v_P remains fixed;
5. plot the transfer characteristics for different values of plate voltage V_P;
6. calculate triode parameters (μ, g_m and r_p) from the triode characteristics at a certain operating point.

APPARATUS REQUIRED: Experimental board, valve-type power supply, two electronic multimeters, milliammeter (0 to 25 mA).

CIRCUIT DIAGRAM: This is shown in Fig. E.6.2.1. Triode ECC82 (manufactured by BEL) is used. The pin numbers of the electrodes of the triode are also shown.

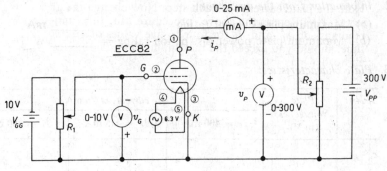

Fig. E.6.2.1

BRIEF THEORY: In a triode tube there are three electrodes. They are plate, control grid and cathode. When the cathode is heated, electrons are emitted from it. A negative space-charge region is formed around the cathode. A positive voltage at plate (with respect to the cathode) will attract these electrons. This movement of electrons constitute the plate current. Between the plate and the cathode is a fine wire-mesh, called control grid. When a negative voltage is given to the grid, electrons are repelled, and the space charge increases. As a result, the plate current decreases. If the grid is made more negative, the plate current is further decreased. In this way, the control grid is able to control the plate current.

There are two important triode characteristics: The plate characteristics and the transfer (or mutual) characteristics. The plate characteristic is a graph between plate voltage v_P and plate current i_P for a fixed value of grid voltage V_G. From this characteristic, the plate dynamic resistance can be calculated by using the following definition:

$$r_p = \frac{\Delta v_P}{\Delta i_P}\bigg|_{V_G = \text{const.}}$$

The variation of plate current i_P with grid voltage v_G, for fixed value of V_P is plotted in mutual (or transfer) characteristics. The transconductance g_m can be determined by making use of the following definition

$$g_m = \frac{\Delta i_P}{\Delta v_G}\bigg|_{V_P = \text{const.}}$$

The amplification factor (μ) can be determined with the help of the formula

$$\mu = g_m r_p$$

PROCEDURE:

1. Look into the valve data book and find all the important information (pin connection, maximum current ratings, etc.) about the triode ECC82 (or 6J5, or 6C5) used in the circuit. Trace the given circuit.
2. Make connections according to Fig. E.6.2.1. Use appropriate ranges of milliammeter and voltmeter.
3. For plate characteristics, first fix V_G at 0 V (i.e. short-circuit control grid with the cathode). Increase plate voltage in steps (of, say, 10 V). Note the plate current for each step. Now change the grid voltage to, say, -1 V. Repeat the above procedure. In this manner take the readings of v_p and i_p for four to five values of grid voltages. At no time, should you exceed the maximum plate dissipation ($i_p v_p = P_D$) and other maximum ratings.
4. For transfer characteristics, fix the plate voltage at, say, 100 V. Increase the grid voltage in steps, negatively, from zero onwards. Note the plate current for each step. In this manner repeat this process for two more values of V_P. For a given plate voltage V_P, the value of grid voltage V_G at which the plate current just reduces to zero is called cut-off voltage. Note this voltage.
5. Plot the plate and mutual characteristics.
6. Calculate triode parameters from the characteristics using their definitions.

OBSERVATIONS:

1. *Type number of the triode* = _____
2. *Information from the data book*:
   (a) Pin connections:

Connections for	*Pin number*
Cathode	_____
Plate	_____
Control grid	_____
Heater filament	_____

   (b) Maximum plate current rating = _____ mA
   (c) Maximum plate dissipation rating = _____ W

3. *Plate characteristics*:

S. No.	Plate voltage v_P (in V)	Plate current, i_P (in mA)				
		$V_G = 0$ V	$V_G = -1$ V	$V_G = -2$ V	$V_G = -3$ V	$V_G = -4$ V
1.						
2.						
3.						

4. *Mutual characteristics:*

S. No.	Grid voltage v_G (in V)	Plate current, i_P (in mA)		
		$V_P = 100$ V	$V_P = 120$ V	$V_P = 140$ V
1.	0			
2.	−1			
3.	−2			

CALCULATIONS:

1. From plate characteristics,

$$r_p = \frac{\Delta v_P}{\Delta i_P}\bigg|_{V_G = \underline{\quad} \text{V}} = \underline{\quad\quad} = \underline{\quad\quad} \text{ k}\Omega$$

2. From mutual characteristics,

$$g_m = \frac{\Delta i_P}{\Delta v_G}\bigg|_{V_P = \underline{\quad} \text{V}} = \underline{\quad\quad} = \underline{\quad\quad} \text{ mS}$$

3. $\quad \mu = r_p g_m = \underline{\quad\quad} = \underline{\quad\quad}$

RESULTS:

1. The plate characteristics and transfer characteristics of the given triode (_____) are drawn on the graph.

2. The triode parameters are as below:

Parameter	Value determined
1. r_p	_____ kΩ
2. g_m	_____ mS
3. μ	_____

at an operating point given by $V_P =$ _____ V and $I_P =$ _____ mA.

EXPERIMENTAL EXERCISE 6.3

TITLE: Pentode characteristics.

OBJECTIVES: To,

1. trace the given circuit;
2. plot the plate characteristics (graph between v_P and i_P for fixed values of V_G);
3. plot the transfer characteristics (graph between v_G and i_P for fixed values of V_P);
4. calculate the pentode parameters from the characteristics at a suitable operating point.

APPARATUS REQUIRED: Experimental board, dual power supply, two ammeters (0 to 25 mA), two electronic voltmeters.

CIRCUIT DIAGRAM: Circuit diagram using pentode EL84 is shown in Fig. E.6.3.1.

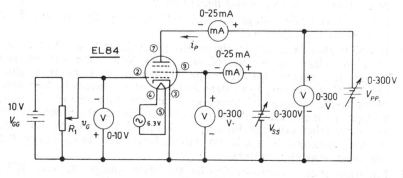

Fig. E.6.3.1

BRIEF THEORY: Besides the cathode and the plate, a pentode has three grids—control grid (or simply grid), screen grid and suppressor grid. The control grid has the same function as it has in a triode. It controls the number of electrons flowing from the cathode to the plate. It is kept at low negative potential. The screen grid is introduced to reduce the inter-electrode capacitance C_{pg} between the plate and the control grid. It is connected to the earth through a capacitor, so that it is effectively grounded for ac. It is given a positive potential (say, 100 V). The suppressor grid is usually connected (either internally or from outside the tube) to the cathode. It suppresses the effect of secondary emission. Because of this, the negative resistance region (of a tetrode) is avoided. In a pentode, there are two grids between the plate and the control grid. Because of this, a change in plate voltage is not much effective in changing the plate current. In other words, it has high plate dynamic resistance (of the order of 500 kΩ). The plate dynamic resistance is given as

$$r_p = \frac{\Delta v_P}{\Delta i_P}\bigg|_{V_G = \text{const.}}$$

It is the reciprocal of the slope of plate characteristic curve. The transconductance (g_m) is the slope of the transfer characteristics.

$$g_m = \frac{\Delta i_P}{\Delta v_G}\bigg|_{V_P = \text{const.}}$$

The amplification factor can either be obtained by its definition, i.e.,

$$\mu = \frac{\Delta v_P}{\Delta v_G}\bigg|_{I_P = \text{const.}}$$

or by using the formula

$$\mu = r_p g_m$$

PROCEDURE:

1. Find the type number of the pentode connected in the circuit. Find important information (pin configuration, maximum ratings, etc.) regarding this valve, from the data book.
2. Make the connections as shown in Fig. E.6.3.1. Choose proper ranges of electronic multimeters and milliammeters.
3. Fix the screen grid voltage at, say, 100 V. Keep this voltage fixed throughout the experiment.
4. For plate characteristics, fix the control grid voltage V_G at 0 V. Increase the plate voltage v_P in steps (of, say, 10 V). Note plate current and screen grid current for each step. Now change V_G to another value, say -1 V. Again vary v_P in steps and note plate current i_P. You need not note the variation of screen-grid current with change in plate voltage v_P for this (and other) values of V_G. Repeat the procedure for two-three other values of V_G (say, -2 V, -3 V, etc.). At no time should you exceed the maximum plate dissipation rating ($i_P \times v_P = P_D$) and other maximum ratings.
5. For transfer characteristics, first fix the plate voltage at, say, 80 V. Vary the grid voltage v_G, in steps, negatively from zero onwards. Note plate current for each step. Repeat this process for other values of V_P (say, 120 V and 150 V).
6. Plot the plate and transfer characteristics.
7. Select a suitable operating point in the linear portion of the characteristics. Calculate the tube parameters using the definitions given under Brief Theory.

OBSERVATIONS:

1. *Type number of the pentode* = _____
2. *Information from the data book:*
   (a) Pin connections:

Connections for	*Pin number*
Cathode	_____
Plate	_____
Control grid	_____
Screen grid	_____
Suppressor grid	_____
Heater filaments	_____

   (b) Maximum plate current rating = _____ mA
   (c) Maximum plate dissipation rating = _____ W

3. *Plate characteristics:*

S. No.	Plate voltage v_P (in V)	$V_G = 0$ V i_P (mA)	i_{SG} (mA)	$V_G = -1$ V i_P (mA)	$V_G = -2$ V i_P (mA)	$V_G = -3$ V i_P (mA)	$V_G = -4$ V i_P (mA)
1.							
2.							
3.							

4. *Mutual characteristics*:

S. No.	Grid voltage v_G (*in* V)	Plate current i_P (*in* mA)		
		$V_P = 80$ V	$V_P = 120$ V	$V_P = 150$ V
1.				
2.				
3.				

CALCULATIONS:

1. From plate characteristics:

$$r_p = \left.\frac{\Delta v_P}{\Delta i_P}\right|_{V_G = \underline{\quad} V} = \frac{\underline{\quad}}{\underline{\quad}} = \underline{\quad\quad} k\Omega$$

2. From mutual characteristics:

$$g_m = \left.\frac{\Delta i_P}{\Delta v_G}\right|_{V_P = \underline{\quad} V} = \frac{\underline{\quad}}{\underline{\quad}} = \underline{\quad\quad} mS$$

3. $\mu = r_p g_m = \underline{\quad\quad}$

RESULTS:

1. The plate and mutual characteristics of the given pentode type (_____) are plotted on the graph.
2. The tube parameters determined at $V_P = $_____V and $I_P = $_____ mA, are given below:

	Parameter	Value determined
1.	r_p	_____ kΩ
2.	g_m	_____ mS
3.	μ	_____

Transistor Biasing and Stabilization of Operating Point

OBJECTIVES: After completing this unit, you will be able to : ○ Draw different biasing arrangements in transistor circuits. ○ Explain with the help of simple equations as to how the operating point is obtained in different biasing circuits. ○ Calculate the operating point current and voltage in different biasing circuits. ○ Explain the effect of change in temperature on the operating point in different biasing circuits. ○ Explain the effect of change in transistor parameters on the operating point in different biasing circuits. ○ Explain with the help of simple equations as to why the potential divider biasing circuit is the most widely used circuit.

7.1 INTRODUCTION

Transistors are used in different kinds of circuits. These circuits are meant to serve a specific purpose. For example, a circuit may be used to increase the voltage or power level of an electrical signal; such a circuit is called an *amplifier*. There is another class of circuits which generates sine or square wave; such circuits are called *oscillators*. It is very difficult to study all the circuits in which transistors are used. However, the study of some fundamental aspects of transistor circuits may be helpful, because the knowledge gained during such a study can help us to understand other difficult circuits too. In this chapter, some basic concepts dealing with the dc biasing of transistors will be studied.

7.2 WHY BIAS A TRANSISTOR

The purpose of dc biasing of a transistor is to obtain a certain dc collector current at a certain dc collector voltage. These values of current and voltage are expressed by the term *operating point* (or *quiescent point*). To obtain the operating point, we make use of some circuits; and these circuits are called "biasing circuits". Of course, while fixing the operating point, it has to be seen that it provides proper dc conditions so that the specific function of the circuit is achieved. The suitability of an operating point for the specific application of the circuit should be seen on the transistor characteristics. In this chapter, we shall discuss the suitability of the operating point with a view that the specific use of the circuit is in amplifiers.

7.3 SELECTION OF OPERATING POINT

In order that the circuit amplifies the signal properly, a judicious selection of the operating point is very necessary. The biasing arrangement should be such as to make the emitter-base junction forward-biased and the collector-base junction reverse-biased. Under such biasing, the transistor is said to operate in the *active region* of its characteristics. Various transistor ratings are to be kept in view while designing the biasing circuit. These ratings—specified by the manufacturer—limit the range of useful operation of the transistor. $I_{C(max)}$ is the maximum current that can flow through the device and $V_{CE(max)}$ is the maximum voltage that can be applied across it safely. In no case should these current and voltage limits be crossed.

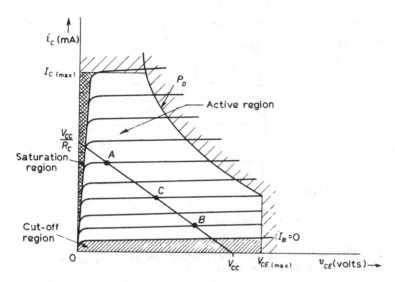

Fig. 7.1 Output characteristics of a transistor in common-emitter configuration. Maximum current, voltage and power ratings are indicated

If a transistor is to work as an amplifier, a load resistance R_C must be connected in the collector circuit. Only then the output ac signal voltage can develop across it. The dc load line corresponding to this resistance R_C and a given collector supply V_{CC} is shown in Fig. 7.1. The operating point will necessarily lie somewhere on this load line. Depending upon the base current, the operating point could be either at point *A*, *B*, or *C*. Let us now consider which one of these is the most suitable operating point.

After the dc (or static) conditions are established in the circuit, an ac signal voltage is applied to the input. Due to this voltage, the base current varies from instant to instant. As a result of this, the collector current and the collector voltage also vary with time. That is how an amplified ac signal is available at the output. The variations in collector current and collector voltage corresponding to a given variation (which may be assumed sinusoidal) of base current can be seen on the output characteristics of the transistor.

These variations are shown in Figs. 7.2, 7.3 and 7.4 for the operating points *A*, *B* and *C*, respectively. In Fig. 7.2 point *A* is very near to the saturation region. Even though the base current is varying sinusoidally, the output current (and also output voltage) is seen to be clipped at the positive peaks. This results in *distortion* of the signal. At the positive peaks, the base current varies, but collector current remains constant at saturation value. Thus we see that point *A* is not a suitable operating point.

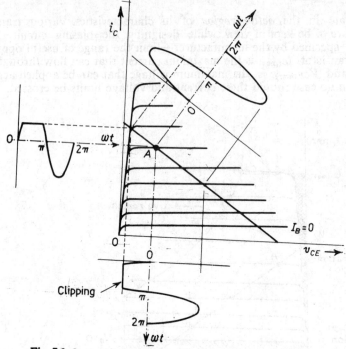

Fig. 7.2 Operating point near saturation region gives clipping at the positive peaks

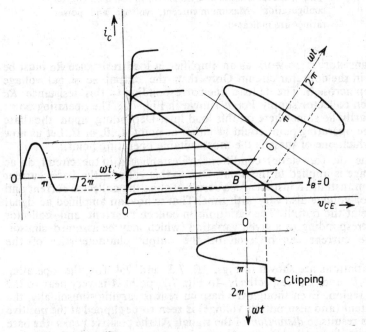

Fig. 7.3 Operating point near cut-off region gives clipping at the negative peaks

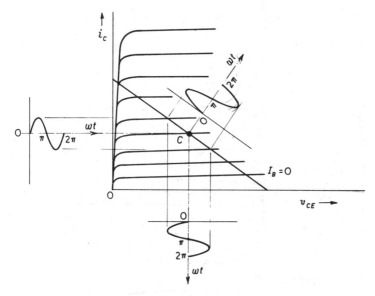

Fig. 7.4 Operating point at the centre of active region is
most suitable

In Fig. 7.3, the point B is very near to the cut-off region. The output
signal is now clipped at the negative peaks. Hence, this too is not a suitable
operating point.

It is clear from Fig. 7.4 that the output signal is not at all distorted if
point C is chosen as the operating point. A good amplifier amplifies signals
without introducing distortion, as much as possible. Thus, point C is the
most suitable operating point.

Even for the operating point C, distortion can occur in the amplifier if
the input signal is large. As shown in Fig. 7.5, the output current and output
voltage is clipped at both the positive and the negative peaks. Thus, the
maximum signal that can be handled by an amplifier is decided by the choice
of the operating point.

7.4 NEED FOR BIAS STABILIZATION

Only the fixing of a suitable operating point is not sufficient. It must also be
ensured that it remains where it was fixed. It is unfortunate that in the
transistor circuits the operating point shifts with the use of the circuit. Such
a shift of operating point may drive the transistor into an undesirable
region. The amplifier then becomes useless.

There are two reasons for the operating point to shift. Firstly, the transistor parameters are temperature dependent. Secondly, the parameters
(such as β) change from unit to unit. In spite of tremendous advancement
in semiconductor technology, the transistor parameters vary between wide
limits even among different units of the same type. Thus, when a transistor
is replaced by another of the same type, the operating point may shift. Such
problems do not arise in case of vacuum tube circuits.

Flow of current in the collector circuit produces heat at the collector
junction. This increases the temperature. More minority carriers are genera-

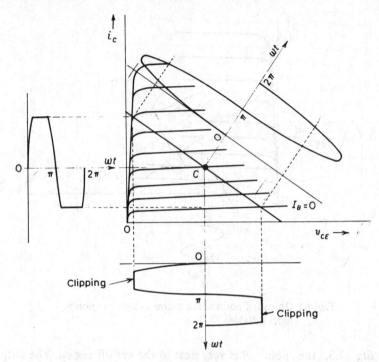

Fig. 7.5 Distortion may result because of too large an input signal

ted in base-collector region (since more bonds are broken). The leakage current I_{CBO} increases. Since

$$I_{CEO} = (1+\beta)I_{CBO} \qquad (7.1)$$

and
$$I_C = \beta I_B + I_{CEO} \qquad (7.2)$$

the increase in I_{CBO} will cause I_{CEO} to increase, which in turn increases the collector current I_C. This further raises the temperature of the collector-base junction, and the whole cycle repeats again. Such a cumulative increase in I_C will ultimately shift the operating point into the saturation region. This situation may prove to be very dangerous. The excess heat produced at the junction may even burn the transistor. Such a situation is described by the term *thermal runaway*.

The sequence of events resulting in thermal runaway of the transistor can be summarized as shown in Fig. 7.6. Here an upward arrow indicates an increase in the quantity written with it. Thus, the above statement means: "As temperature T increases, the leakage current I_{CBO} also increases: as I_{CBO} increases, the leakage current in common-emitter configuration I_{CEO} also increases: as I_{CEO} increases, ... and so on".

7.5 REQUIREMENTS OF A BIASING CIRCUIT

The discussion in the above sections may be summarized by stating that the biasing circuit should:

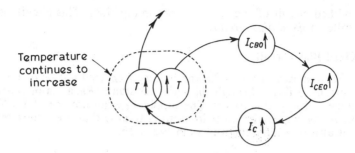

Fig. 7.6 Increase in I_{CEO} with temperature leads to thermal runaway

(*i*) Establish the operating point in the centre of the active region of the characteristics, so that on applying the input signal the instantaneous operating point does not move either to the saturation region or to the cut-off region, even at the extreme values of the input signal.

(*ii*) Stabilize the collector current against temperature variations.

(*iii*) Make the operating point independent of the transistor parameters so that it does not shift when the transistor is replaced by another of the same type in the circuit.

7.6 DIFFERENT BIASING CIRCUITS

The simplest biasing circuit could be the one drawn in Fig. 7.7. The emitter-base junction is forward-biased by the battery V_{BB} and the collector-base junction is reverse-biased by the battery V_{CC}. The voltage V_{BE} across the forward-biased junction is very low (for a germanium transistor, $V_{BE} = 0.3$ V; and for silicon transistor, $V_{BE} = 0.7$ V). This requires that the battery voltage V_{BB} must also be of the same order. The voltage V_{CC} should be of a much larger value than the voltage V_{BB}; only then is the collector-base junction reverse-biased.

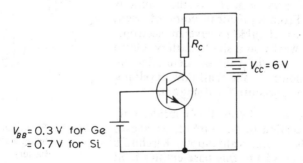

Fig. 7.7 Simplest biasing circuit

Though the circuit of Fig. 7.7 achieves forward-biasing of the emitter-base junction and reverse-biasing of the collector-base junction, it is not a practical circuit. It is extremely difficult to have a battery V_{BB} of either 0.3 V or 0.7 V. This circuit, therefore, is never used.

A modified circuit of Fig. 7.7 is shown in Fig. 7.8a. This circuit is commonly called a *fixed-bias circuit*.

7.6.1 Fixed-Bias Circuit

In the circuit shown in Fig. 7.8a, the battery V_{BB} need not be of low value. When current I_B flows through the series resistance R_B, a major portion of the voltage is dropped across it. The supply V_{BB} can now be of 1.5 V. Such a supply is easily available. It is interesting to note that the same circuit can also be drawn in a different way as in Fig. 7.8b.

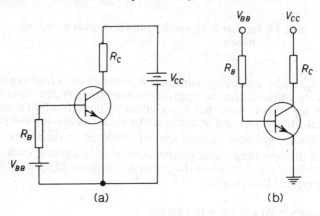

(a)   (b)

Fig. 7.8 Fixed-bias circuit

The circuit in Fig. 7.8 uses two batteries, V_{CC} and V_{BB}. The positive terminals of both the batteries are connected to the collector and base resistors. We can use a single battery instead of the two, as shown in Fig. 7.9. The value of R_B is then suitably modified. From a practical point of view, it is always preferable to have an electronic circuit that works on a single battery. Given a fixed-bias circuit, can we determine its operating point? We shall now develop a step-by-step procedure for doing this.

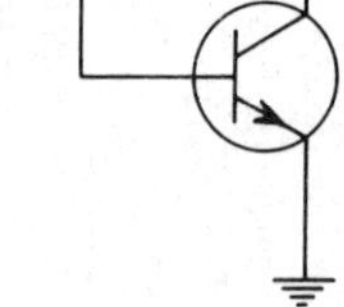

Fig. 7.9 Fixed-bias circuit using a single battery

Input Section: Let us first consider only the input section of the circuit, as shown in Fig. 7.10a. We can apply Kirchoff's voltage law (KVL) to this base-emitter loop and get

$$V_{CC} = I_B R_B + V_{BE}$$

from which the base current is given as

$$I_B = \frac{V_{CC} - V_{BE}}{R_B} \tag{7.3}$$

Since V_{BE} is very small compared to V_{CC}, not much error will be committed if it is neglected. The base current is then given by the simple expression

$$I_B \simeq \frac{V_{CC}}{R_B} \qquad (7.4)$$

The supply voltage V_{CC} being of fixed value—once the resistance R_B is selected—the base current I_B is also fixed. Hence the name *fixed-bias circuit*.

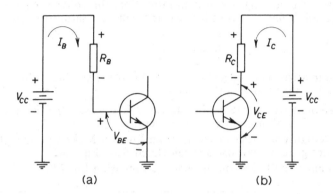

(a)   (b)

Fig. 7.10 (a) Input section of the fixed-bias circuit
(b) Output section of the fixed-bias circuit

Output Section: We now consider the output section of the circuit, as shown in Fig. 7.10b. The collector current I_C that flows through the resistor R_C is given as

$$I_C = \beta I_B + I_{CEO} \qquad (7.5)$$

In this equation, βI_B is the portion of current transferred from the input side. The current I_{CEO} is the leakage current in the CE configuration. Though the current I_{CEO} is not as small as the leakage current in the CB configuration I_{CBO}, yet it is very small compared to the usual values of I_C. A very small error will be introduced if we neglect the current I_{CEO} in our calculations. Therefore, to a good approximation, the collector current I_C is given as

$$I_C = \beta I_B \qquad (7.6)$$

Applying KVL to the output section of Fig. 7.10b, we get

$$V_{CC} = I_C R_C + V_{CE} \qquad (7.7)$$

One word of caution: It is clear from the above equation that the supply voltage V_{CC} provides the voltages across the resistor R_C and also across the collector-emitter terminals. Obviously, the voltage drop $I_C R_C$ can never be more than the supply voltage, V_{CC}, or

$$I_C \leqslant \frac{V_{CC}}{R_C} \qquad (7.8)$$

If the value of I_C turns out to be greater than the maximum value given by Eq. 7.8, it is certainly wrong. It is so, because the operating point is lying in the saturation region of the characteristics. Here, the collector current I_C is limited due to saturation, and its value remains at its maximum (given by Eq. 7.8) whatever the value of base current I_B is. Equation 7.6 then becomes invalid.

Having taken care of the above caution, we are to find V_{CE} now, to determine the operating point. Equation 7.7 can be written as

$$V_{CE} = V_{CC} - I_C R_C \tag{7.9}$$

Of course, when the transistor is in saturation, the voltage V_{CE} is almost zero (actually a few tenths of a volt), and collector saturation current

$$I_{C(\text{sat})} = \frac{V_{CC}}{R_C}$$

To summarize: The operating point in the fixed-bias circuit can be calculated in the following three steps:

1. Calculate base current I_B using Eq. 7.4. In case V_{BE} is known, use Eq. 7.3 to obtain more accurate results.
2. Calculate collector current I_C from Eq. 7.6. Make sure that this value is not greater than the one calculated from Eq. 7.8.
3. Calculate collector-emitter voltage V_{CE} using Eq. 7.9.

Example 7.1 Calculate the collector current and the collector-to-emitter voltage for the circuit given in Fig. 7.11.

Solution:

(a) The base current I_B is given as

$$I_B = \frac{(V_{CC} - V_{BE})}{R_B} \simeq \frac{V_{CC}}{R_B} = \frac{9}{300 \times 10^3}$$

$$= 3 \times 10^{-5}\text{A} = \mathbf{30\ \mu A}$$

(b) The collector current I_C is given as

$$I_C = \beta I_B = 50 \times 30 \times 10^{-6}\text{A} = \mathbf{1.5\ mA}$$

Let us check if this current is less than the collector saturation current.

$$I_{C(\text{sat})} = \frac{V_{CC}}{R_C} = \frac{9}{2 \times 10^3}$$

$$= 4.5 \times 10^{-3}\text{A} = 4.5\ \text{mA}$$

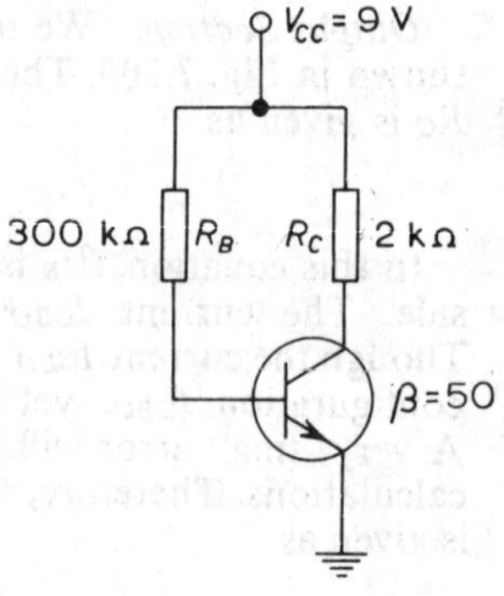

Fig. 7.11

Thus, the transistor is not in saturation.

(c) The collector-to-emitter voltage

$$V_{CE} = V_{CC} - I_C R_C = 9 - 1.5 \times 10^{-3} \times 2 \times 10^3 = \mathbf{6\ V}$$

Example 7.2 Calculate the coordinates of the operating point as fixed in the circuit in Fig. 7.12a. Given: $R_C = 1$ kΩ, $R_B = 100$ kΩ.

Solution:

(a) The base current is

$$I_B = \frac{(V_{CC} - V_{BE})}{R_B} \simeq \frac{V_{CC}}{R_B} = \frac{10}{100 \times 10^3}\ \text{A} = \mathbf{100\ \mu A}$$

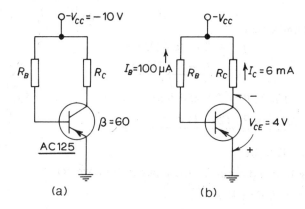

Fig. 7.12

(b) The collector current is

$$I_C = \beta I_B = 60 \times 100 \times 10^{-6} A = 6\ \text{mA}$$

We shall now check if this current is less than the collector saturation current

$$I_{C(\text{sat})} = \frac{V_{CC}}{R_C} = \frac{10}{1 \times 10^3} A = 10\ \text{mA}$$

Therefore, the transistor is not in saturation.

(c) The voltage between the collector and emitter terminals is

$$V_{CE} = V_{CC} - I_C R_C = 10 - 6 \times 10^{-3} \times 10^3 = 4\ \text{V}$$

Figure 7.12b shows the value and the direction of base current I_B, collector current I_C and collector-emitter voltage V_{CE}.

Example 7.3 In the circuit in Fig. 7.12a, the transistor is replaced by another unit of AC125. This new transistor has $\beta = 150$ instead of 60. Determine the quiescent operating point.

Solution:

(a) The base current remains the same, i.e. 100 μA.
(b) The collector current is

$$I_C = \beta I_B = 150 \times 100 \times 10^{-6}\ A = 15\ \text{mA}$$

The collector saturation current was 10 mA in the last example. Here also, this current remains the same. But the calculated current I_C is seen to be greater than $I_{C(\text{sat})}$. Hence, the transistor is now in saturation. In this case, the operating point is specified as

$$I_C = I_{C(\text{sat})} = 10\ \text{mA}$$
$$V_{CE} = 0\ \text{V}$$

Example 7.4 In the biasing circuit in Fig. 7.13, a supply of 6 V and a load resistance of 1 kΩ is used. (a) Find the value of resistance R_B so that a germanium transistor with $\beta = 20$ and $I_{CBO} = 2\ \mu$A draws an I_C of 1 mA. (b) What I_C is drawn if the transistor parameters change to $\beta = 25$ and $I_{CBO} = 10\ \mu$A due to rise in temperature ?

Solution:

(a) We know that

$$I_C = \beta I_B + (\beta+1)I_{CBO}$$

or
$$I_B = \frac{I_C - (\beta+1)I_{CBO}}{\beta}$$

Here, $I_C = 1$ mA; $\beta = 20$; and $I_{CBO} = 2\ \mu$A

$$\therefore \quad I_B = \frac{1 \times 10^{-3} - (20+1) \times 2 \times 10^{-6}}{20}\ \text{A}$$

$$= 47.9\ \mu\text{A}$$

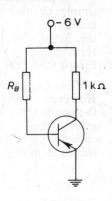

Fig. 7.13

Writing loop equation for the input section, we get

$$6 = I_B R_B + V_{BE}$$

Since it is stated that the transistor used in the circuit is a germanium transistor, V_{BE} can be assumed to be 0.3 V. Thus, from the above equation we have

$$R_B = \frac{6 - 0.3}{I_B} = \frac{5.7}{47.9 \times 10^{-6}}\ \Omega = 118.998\ \text{k}\Omega$$

It is worthwhile to see how much error is committed if we neglect I_{CBO} and V_{BE} from the above calculations. Since,

$$I_C = \beta I_B$$

$$I_B = \frac{I_C}{\beta} = \frac{1 \times 10^{-3}}{20}\ \text{A} = 50\ \mu\text{A}$$

The input loop equation now becomes

$$6 = I_B R_B$$

Therefore, the resistance R_B is given as

$$R_B = \frac{6}{50 \times 10^{-6}}\ \Omega = 120\ \text{k}\Omega$$

The percentage error $= \dfrac{120 - 118.998}{118.998} \times 100 = 0.842\ \%$

Comment: This error is too small to bother about. Moreover, the resistors available in the market ordinarily have a tolerance of $\pm 10\ \%$. It is, therefore, not very *incorrect* to neglect V_{BE} and I_{CBO} while making calculations.

(b) Here, due to rise in temperature, the transistor parameters have changed to $\beta = 25$ and $I_{CBO} = 10\ \mu$A. The collector current is now given as

$$I_C = \beta I_B + (\beta+1)I_{CBO}$$

$$= 25 \times 47.9 \times 10^{-6} + (25+1) \times 10 \times 10^{-6}\ \text{A}$$

$$= 1.46\ \text{mA}$$

Comment: It may be noted that the collector current has increased by almost 50 % due to rise in temperature.

Why fixed-bias circuit is seldom used The fixed-bias circuit in Fig. 7.9 is a simple circuit. It uses very few components (only two resistors and one battery). It is very easy to fix the quiescent operating point any where in the active region of the characteristics by simply changing the value of resistor R_B. It provides maximum flexibility in the design. In spite of all this, it is seldom used in practice.

This circuit meets the first requirement stated in Sec. 7.5 very well. However, it miserably fails to meet the second and third requirements. With the rise in temperature, a cumulative action takes place, and the collector current goes on increasing. The circuit provides no check on the increase in collector current. The operating point is not stable. This situation can be shown as in Fig. 7.14.

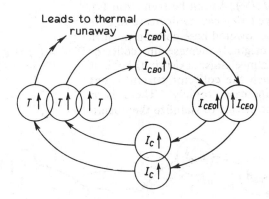

Fig. 7.14 Fixed-bias circuit leads to thermal
runaway of the transistor

As regards the third requirement—i.e. the Q point should not shift on replacing the transistor with another of same type—the circuit utterly fails. Since, $I_C = \beta I_B$, and the base current I_B is already fixed, the current I_C is solely dependent on β. When the transistor is replaced by another with different value of β, the operating point will shift. The stabilization of the operating point is very poor. Therefore, the biasing circuit needs some modification.

7.6.2 Collector-to-base Bias Circuit

Figure 7.15 shows a modified biasing circuit. Here, the base resistor R_B is connected to the collector instead of connecting it to the battery V_{CC}. Writing the loop equation for the input circuit, we get

$$V_{CC} = R_C(I_C + I_B) + I_B R_B + V_{BE}$$

or
$$V_{CC} = R_C I_C + (R_C + R_B) I_B + V_{BE} \tag{7.10}$$

or
$$I_B = \frac{(V_{CC} - I_C R_C) - V_{BE}}{R_C + R_B} \tag{7.11}$$

From the output section of the circuit, we have

$$V_{CE} = V_{CC} - (I_C + I_B) R_C$$

or
$$V_{CE} \simeq V_{CC} - I_C R_C \qquad \text{(since } I_B \ll I_C) \tag{7.12}$$

Substituting Eq. 7.12 in Eq. 7.11, we get

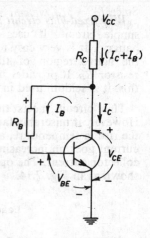

$$I_B = \frac{V_{CE} - V_{BE}}{R_C + R_B} \qquad (7.13)$$

Let us now see what happens when the temperature rises.

Suppose the temperature increases, causing the leakage current (and also β) to increase. This increases the collector current (since $I_C = \beta I_B + I_{CEO}$). As the collector current increases, the voltage V_{CE} decreases (since $V_{CE} = V_{CC} - I_C R_C$). As can be seen from Eq. 7.13, the reduced V_{CE} causes decrease in base current I_B. The lowered base current in turn reduces the original increase in collector current. Thus, a mechanism exists in the circuit because of which the collector current is not allowed to increase rapidly. There is a tendency in the circuit to stabilize the operating point.

Fig. 7.15 Collector-to-base bias circuit

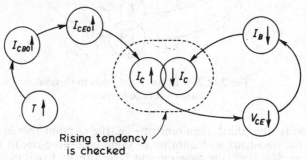

Rising tendency is checked

Fig. 7.16 Collector-to-base bias circuit checks the rising tendency of collector current

Note that the resistor R_B connects the collector (the output) with the base (the input). This means that a feedback exists in the circuit. The base current is dependent on the collector voltage. And this dependence is such as to nullify the changes in base current. That is why this circuit is also called a *voltage feedback* bias circuit.

Suppose the transistor in this circuit is replaced by another having different value of β. The shift in the operating point will not be as much as it occurs in case of fixed-bias circuit. This can be seen as follows.

For determining the operating point, we substitute βI_B for I_C in Eq. 7.10 to get

$$V_{CC} = R_C \beta I_B + (R_C + R_B) I_B + V_{BE}$$

or

$$V_{CC} = V_{BE} + [R_B + (\beta + 1) R_C] I_B$$

or

$$I_B = \frac{V_{CC} - V_{BE}}{R_B + (\beta + 1) R_C} \simeq \frac{V_{CC}}{R_B + \beta R_C} \qquad (7.14)$$

Since $I_C = \beta I_B$, we can determine the collector current. To determine the collector voltage, we write the loop equation for the output section of the circuit (Fig. 7.15).

$$V_{CC} - (I_B+I_C)R_C - V_{CE} = 0$$

or
$$V_{CE} = V_{CC}-(I_C+I_B)R_C \cong V_{CC}-I_CR_C \qquad (7.15)$$

Why collector-to-base bias circuits is seldom used This circuit has a tendency to stabilize the operating point against temperature and β variations. But the circuit is not used very much. The resistor R_B not only provides a dc feedback for the stabilization of operating point, but it also causes an ac feedback. This reduces the voltage gain of the amplifier. It is not desirable. After all, the biasing of a transistor was needed so that it could amplify the ac signals properly. Because of this drawback, the circuit is not very commonly used.

Example 7.5 How much is the emitter current in the circuit in Fig. 7.17 ? Also calculate V_C.

Solution: From Eq. 7.14, the base current is given as

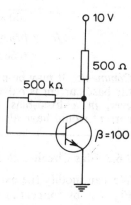

Fig. 7.17

$$I_B = \frac{V_{CC}}{R_B+\beta R_C}$$

Here, $V_{CC} = 10$ V; $R_B = 500\times10^3$ Ω;

$R_C = 500$ Ω; $\beta = 100$.

Therefore,

$$I_B = \frac{10}{500\times10^3+100\times500}$$

$$= 18\times10^{-6}\ A = 18\ \mu A$$

The emitter current is then given as

$$I_E \cong I_C = \beta I_B = 100\times18\times10^{-6} = 1.8\times10^{-3}\ A = \textbf{1.8 mA}$$

The collector voltage

$$V_C = V_{CE} = V_{CC}-I_CR_C = 10-1.8\times10^{-3}\times500 = \textbf{9.1 V}$$

Comment: Since the collector voltage $V_C = V_{CE}$ is only slightly less than V_{CC}, the quiescent operating point is near the cut-off region.

Example 7.6 Calculate the minimum and maximum collector current in Fig. 7.18, if the β of the transistor varies within the limits indicated.

Solution: From Eq. 7.14, the base current is given as

$$I_B = \frac{V_{CC}}{R_B+\beta R_C}$$

(*a*) Let us first take the minimum value of β, so that $\beta = 50$;

$$V_{CC} = 20 \text{ V}; \quad R_B = 200 \text{ k}\Omega = 200 \times 10^3 \ \Omega$$
$$R_C = 2 \text{ k}\Omega = 2 \times 10^3 \ \Omega$$

Therefore,

$$I_B = \frac{20}{200 \times 10^3 + 50 \times 2 \times 10^3}$$
$$= 66.5 \times 10^{-6} \text{ A}$$

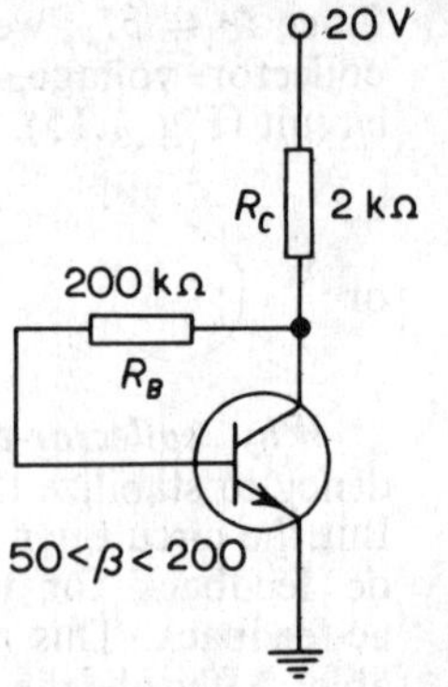

The collector current is given as

$$I_C = \beta I_B = 50 \times 66.6 \times 10^{-6}$$
$$= 3.33 \times 10^{-3} \text{ A} = 3.33 \text{ mA}$$

(*b*) Now we take the maximum value of β, i.e. $\beta, = 200$, so that,

$$I_B = \frac{20}{200 \times 10^3 + 200 \times 2 \times 10^3} = 33.33 \times 10^{-6} \text{ A}$$
$$\therefore \qquad I_C = \beta I_B = 200 \times 33.33 \times 10^{-6} \text{ A}$$
$$= 6.66 \times 10^{-3} \text{ A} = 6.66 \text{ mA}$$

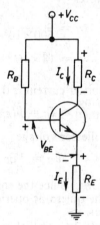

Fig. 7.18

Comment: It may be noted that in this circuit when β increases four times, the base current is halved and the collector current becomes double. However, in a fixed-bias circuit, if β had increased four times, the collector current would have also increased four times.

7.6.3 Bias Circuit with Emitter Resistor

We can modify the fixed-bias circuit of Fig. 7.9 by connecting a resistor to the emitter terminal. The modified circuit is shown in Fig. 7.19. In this circuit we have three resistors R_C, R_B and R_E and a battery V_{CC}.

We shall now see what happens to the Q point when the temperature increases. For this, we write the loop equation for the input section of the circuit.

$$V_{CC} = R_B I_B + V_{BE} + I_E R_E \qquad (7.16)$$

or $\qquad I_B = \dfrac{(V_{CC} - I_E R_E - V_{BE})}{R_B}$

Fig. 7.19 Bias circuit with emitter resistor

or $\qquad I_B \cong \dfrac{(V_{CC} - I_E R_E)}{R_B}$ (since V_{BE} is very small) $\qquad (7.17)$

As the temperature tends to increase, the following sequence of events occur (Fig. 7.20).

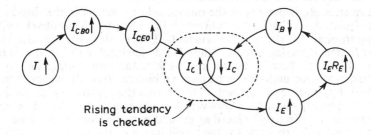

Fig. 7.20

Because of the temperature rise, the leakage current increases. This increases the collector current as well as the emitter current. As a result, the voltage drop across resistor R_E also increases. This reduces the numerator of Eq. 7.17 and hence the current I_B also reduces. This results in reduction of the collector current. Thus we see that the collector current is not allowed to increase to the extent it would have been in the absence of the resistor R_E.

In case the transistor is replaced by another of the same type (which may have different value of β), then also this circuit provides stabilization of the Q point, as is shown in Fig. 7.21.

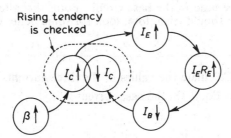

Fig. 7.21

Having seen that the operating point is stable in this circuit, let us determine the Q point. To do this, let us rewrite Eq. 7.16 as

$$V_{CC} = I_B R_B + V_{BE} + (\beta+1)I_B R_E$$

$$\text{(since} \quad I_E = (\beta+1)I_B)$$

or $$I_B = \frac{V_{CC}-V_{BE}}{R_B+(\beta+1)R_E} \simeq \frac{V_{CC}}{R_B+\beta R_E} \qquad (7.18)$$

We can calculate the collector current easily, since

$$I_C = \beta I_B = \frac{\beta V_{CC}}{R_B+\beta R_E} = \frac{V_{CC}}{R_E+(R_B/\beta)} \qquad (7.19)$$

To find V_{CE}, we write the loop equation for the output section,

$$V_{CC} = I_C R_C + V_{CE} + I_E R_E$$

$$V_{CE} = V_{CC}-(R_C+R_E)I_C \qquad \text{(since} \ I_C \simeq I_E) \qquad (7.20)$$

Operating point is thus determined by Eqs. 7.19 and 7.20.

The resistor R_E is present in the output side as well as in the input side of the circuit. A feedback occurs through this resistor. The feedback voltage is proportional to the emitter current. Hence, this circuit is also called *current feedback* biasing circuit. While the dc feedback helps in the stabilization of the Q point, the ac feedback reduces the voltage gain of the amplifier; again, an undesirable feature. Of course, this drawback can be remedied by putting a capacitor C_E across the resistor R_E, as shown in Fig. 7.22. The capacitor C_E offers very low impedance to the ac current. The emitter is effectively placed at ground potential for the ac signal. The circuit provides dc feedback for the stabilization of the Q point, but does not give any ac feedback. The process of amplification of the ac signal remains unaffected.

Why this circuit is not used   The circuit in Fig. 7.19 does provide some stabilization of the Q point. But as you can see from Eq. 7.19, the denominator can be independent of β only if

$$R_E \gg \frac{R_B}{\beta}$$

This means we should either have a very high value of R_E or a very low value of R_B. A high value of R_E will cause a large dc drop across it. To obtain a particular operating point under this condition, it will require a high dc source V_{CC}. On the other hand if R_B is low, a separate low voltage supply has to be used in the base circuit. Both the alternatives are quite impractical. We should, therefore, look for some better circuit.

Example 7.7   Calculate the values of the three currents in Fig. 7.22.

Solution: From Eq. 7.18, the base current is given as

$$I_B = \frac{V_{CC}}{R_B + (\beta + 1)R_E}$$

Here, $V_{CC} = 10$ V; $R_B = 1$ MΩ $= 1 \times 10^6$ Ω
   $R_E = 1$ kΩ $= 1 \times 10^3$ Ω; $\beta = 100$

Therefore, $I_B = \dfrac{10}{1 \times 10^6 + (100+1) \times 1 \times 10^3}$

   $= 9.09 \times 10^{-6}$ A

   $= 9.09$ μA

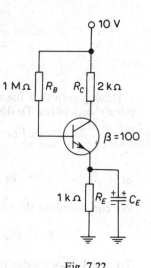

Fig. 7.22

Now, the collector current
   $I_C = \beta I_B = 100 \times 9.09 \times 10^{-6}$

   $= 0.909 \times 10^{-3}$ A

   $= 0.909$ mA

The emitter current
   $I_E = I_C + I_B \simeq I_C = 0.909$ mA

Example 7.8 Calculate the minimum and maximum values of emitter current for the biasing circuit of Fig. 7.23. Also calculate the corresponding values of collector-to-emitter voltage. The transistor used in the circuit is a germanium transistor.

Solution:

Since a germanium transistor is used in the circuit, $V_{BE} = 0.3$ V. The base current is given by Eq. 7.18 as

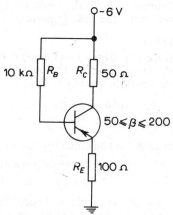

Fig. 7.23

$$I_B = \frac{V_{CC} - V_{BE}}{R_B + (\beta+1)R_E}$$

Multiplying both sides of the above equation by $(\beta+1)$, we get

$$(\beta+1)I_B = \frac{(V_{CC} - V_{BE})(\beta+1)}{R_B + (\beta+1)R_E}$$

Since $I_E = (\beta+1) I_B$, the emitter current is given as

$$I_E = \frac{(V_{CC} - V_{BE})(\beta+1)}{R_B + (\beta+1)R_E} \tag{7.21}$$

(a) Let us first take the minimum value of β, so that $\beta = 50$; $V_{CC} = 6$ V. $R_B = 10$ k$\Omega = 10 \times 10^3$ Ω; $R_E = 100$ Ω. Therefore,

$$I_E = \frac{(6-0.3)(50+1)}{10 \times 10^3 + (50+1) \times 100} = \frac{5.7 \times 51}{15100}$$

$$= 19.25 \times 10^{-3} \text{ A} = \textbf{19.25 mA}$$

The collector-to-emitter voltage is given by

$$V_{CE} = V_{CC} - (R_C + R_B)I_E \quad \text{(since } I_C \simeq I_E\text{)}$$
$$= 6 - (50 + 100) \times 19.25 \times 10^{-3} = \textbf{3.1 V}$$

(b) Let us now consider the maximum value of β, i.e. $\beta = 200$. The emitter current becomes

$$I_E = \frac{(6-0.3)(200+1)}{10 \times 10^3 + (200+1) \times 100} = \frac{5.7 \times 201}{(10+20) \times 10^3}$$

$$= 38.2 \times 10^{-3} \text{ A} = \textbf{38.2 mA}$$

The collector-to-emitter voltage is

$$V_{CE} = V_{CC} - (R_C + R_E)I_E = 6 - (50 + 100) \times 38.2 \times 10^{-3}$$
$$= 6 - 5.7 = \textbf{0.3 V}$$

Comment: (i) Note that when β becomes four times, the emitter current becomes almost double. Had it been a fixed-bias circuit, the emitter current would have increased four times. Thus, the circuit does provide some stability of the Q point.

(ii) When β changes from 50 to 200, the Q point shifts from the active region to very near the saturation. With $\beta = 50$, $V_{CE} = 3.1$ V (almost half of V_{CC}). But, with $\beta = 200$, $V_{CE} = 0.3$ V (nearing zero volts).

Example 7.9 If the collector resistance R_C in Fig. 7.23 is changed to 1 kΩ, determine the new Q points for the minimum and maximum values of β.

Solution:
Since the value of the emitter current does not depend upon the value of R_C (see Eq. 7.21), the emitter current I_E remains the same as calculated in the previous example. That is

(i) For $\beta = 50$, $I_E = 19.25$ mA

(ii) For $\beta = 200$, $I_E = 38.2$ mA

In case (i), the collector-to-emitter voltage is given by
$$V_{CE} = V_{CC} - (R_C + R_E)I_E = 6 - (1000 + 100) \times 19.25 \times 10^{-3}$$
$$= 6 - 21.17 = -15.17 \text{ V}$$

The above result is absurd! Sum of the voltage drops across R_C and R_E cannot be greater than the supply voltage V_{CC}. Is our calculation wrong? Certainly not. We face such difficulties when the transistor is in saturation. The maximum possible current that can be supplied by the battery V_{CC} to the output section is
$$I_{C(sat)} = \frac{V_{CC}}{R_C + R_E} = \frac{6}{1000 + 100}$$
$$= 5.45 \times 10^{-3} \text{ A} = 5.45 \text{ mA}$$

Under saturation, the collector-to-emitter voltage is
$$V_{CE(sat)} = 0 \text{ V}$$

(ii) We have seen that the transistor is in saturation when its $\beta = 50$. In case $\beta = 200$, there is all the more reason for the transistor to be in saturation. So, the Q point will be the same as calculated earlier, i.e.
$$I_{C(sat)} = 5.45 \text{ mA}, V_{C(sat)} = 0 \text{ V}$$

Example 7.10 Calculate the value of R_B in the biasing circuit of Fig. 7.24 so that the Q point is fixed at $I_C = 8$ mA and $V_{CE} = 3$ V.

Solution: The current I_B is given as
$$I_B = \frac{I_C}{\beta}$$

Here, $I_C = 8$ mA $= 8 \times 10^{-3}$ A and $\beta = 80$.

Therefore, $\quad I_B = \dfrac{8 \times 10^{-3}}{80} = 1 \times 10^{-4}$ A $= 100 \ \mu$A

From Eq. 7.18, we have
$$I_B R_B + (\beta + 1)I_B R_E = V_{CC} - V_{BE} \simeq V_{CC}$$
or $\qquad R_B = \dfrac{V_{CC} - (\beta + 1)I_B R_E}{I_B}$

Here, $V_{CC} = 9$ V; $\beta = 80$; $I_B = 1 \times 10^{-4}$ A, $R_E = 500 \ \Omega$

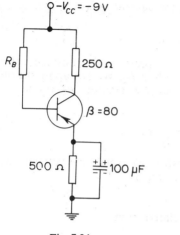

Fig. 7.24

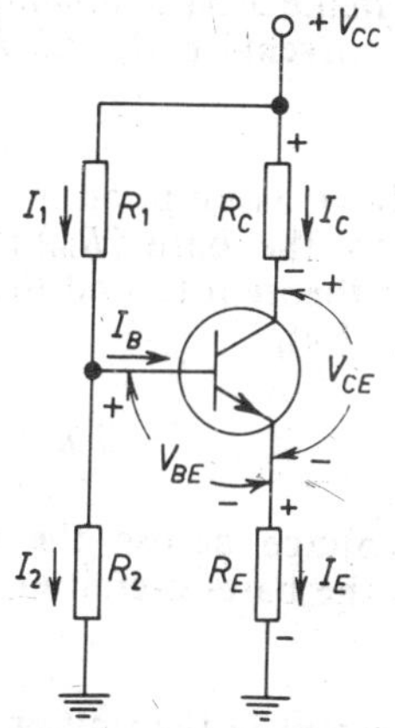

Fig. 7.25 Voltage divider
biasing circuit

Therefore, $R_B = \dfrac{9-(80+1)\times 1 \times 10^{-4} \times 500}{1 \times 10^{-4}} = \dfrac{4.95}{1\times 10^{-4}}\Omega$

$= 49.5\,\text{k}\Omega$

7.6.4 Voltage Divider Biasing Circuit

This is the most widely used biasing circuit. It is shown in Fig. 7.25. Compare this circuit with the one shown in Fig. 7.19. Here, an additional resistor R_2 is connected between base and ground. The name "voltage divider" comes from the voltage divider formed by the resistors R_1 and R_2. By suitably selecting this voltage divider network, the operating point of the transistor can be made almost independent of beta (β). This is why this circuit is also called "biasing circuit independent of beta".

Approximate analysis To determine the operating point, we first consider the input section of the circuit, redrawn in Fig. 7.26.

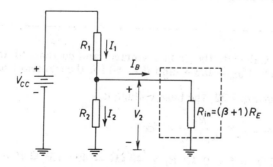

Fig. 7.26 Input section of the voltage divider
biasing circuit

We make a basic assumption: The base current I_B is very small compared to the currents in R_1 and R_2. That is

$$I_1 \simeq I_2 \gg I_B$$

The above assumption is valid because, in practice, the resistance seen looking into the base (R_{in}) is much larger than R_2. We can apply the voltage-divider theorem to find the voltage across the resistor R_2 (same as base voltage V_B),

$$V_B = V_2 = \frac{R_2}{R_1 + R_2} \times V_{CC} \qquad (7.22)$$

The voltage across the emitter resistor R_E equals the voltage across R_2 minus the base-to-emitter voltage V_{BE}. That is

$$V_E = V_2 - V_{BE}$$

The current in the emitter is then calculated from

$$I_E = \frac{V_E}{R_E} = \frac{V_2 - V_{BE}}{R_E} \qquad (7.23)$$

The voltage at the collector (measured with respect to ground) V_C equals the supply voltage V_{CC} minus the voltage drop across R_C,

$$V_C = V_{CC} - I_C R_C$$

The collector-to-emitter voltage is then given as

$$V_{CE} = V_C - V_E = (V_{CC} - I_C R_C) - I_E R_E$$

or $$V_{CE} \simeq V_{CC} - (R_C + R_E)I_C \qquad (7.24)$$

since I_C and I_E are approximately equal.

Note that in the above analysis, nowhere does β appear in any equation. It means that the operating point does not depend upon the value of β of the transistor. This is why the voltage divider circuit is most widely used. In the mass production of transistors, one of the main problems is the wide variation in β. It varies from transistor to transistor of the same type. For example, the transistor AC127 has a minimum β of 50 and maximum β of 150 for an I_C of 10 mA and a temperature of 25 °C. If this biasing circuit is used, no problem is faced on replacement of the transistor in the circuit. The operating point remains where it was fixed in the original design.

Example 7.11   Calculate the dc bias voltages and currents for the circuit of Fig. 7.27. Assume $V_{BE} = 0.3$ V and $\beta = 60$ for the transistor used.

Solution: From Eq. 7.22, the base voltage is

$$V_B = V_2 = \frac{R_2}{R_1 + R_2} \times V_{CC}$$

Here, $R_2 = 5$ k$\Omega = 5 \times 10^3\ \Omega$; $R_1 = 40$ k$\Omega = 40 \times 10^3$; Ω $V_{CC} = 12$ V

Therefore, $V_2 = \dfrac{5 \times 10^3}{(40 + 5) \times 10^3} \times 12 = 1.3$ V

The emitter voltage
$$V_E = V_2 - V_{BE} = 1.3 - 0.3 = 1.0 \text{ V}$$

Therefore, the emitter current
$$I_E = \frac{V_E}{R_E} = \frac{1.0}{1 \times 10^3} = 1.0 \times 10^{-3} \text{ A}$$
$$= 1.0 \text{ mA}$$

The collector current,
$$I_C \simeq I_E = 1.0 \text{ mA}$$

The collector voltage
$$V_C = V_{CC} - I_C R_C$$
$$= 12 - 1 \times 10^{-3} \times 5 \times 10^3 = 7 \text{ V}$$

Finally, the collector-to-emitter voltage
$$V_{CE} = V_C - V_E = 7 - 1 = 6 \text{ V}$$

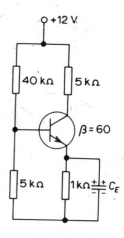

Fig. 7.27

Example 7.12 To set up 100 mA of emitter current in the power amplifier circuit of Fig. 7.28, calculate the value of the resistor R_E. Also calculate V_{CE}. The dc resistance of the primary of the output transformer is 20 Ω.

Solution: Given: $R_1 = 200 \ \Omega$; $R_2 = 100 \ \Omega$; $R_C = 20 \ \Omega$

$$V_{CC} = 15 \text{ V};$$
$$I_C \simeq I_E = 100 \text{ mA} = 0.1 \text{ A}$$

From Eq. 7.22, the base voltage is
$$V_B = \frac{R_2}{R_1 + R_2} \times V_{CC}$$
$$= \frac{100}{200 + 100} \times 15 = 5 \text{ V}$$

Neglecting V_{BE}, $V_E = V_B = 5 \text{ V}$
From Eq. 7.23, the emitter resistance is given by

$$R_E = \frac{V_E}{I_E} = \frac{5}{0.1} = 50 \ \Omega.$$

The collector-to-emitter voltage is then calculated using Eq. 7.24,

$$V_{CE} = V_{CC} - (R_C + R_E)I_C$$
$$= 15 - (20 + 50) \times 0.1 = 8 \text{ V}$$

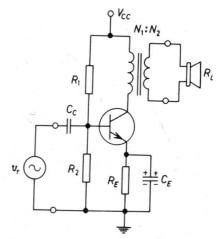

Fig. 7.28

Accurate analysis You may be wondering why the operating point does not change when the transistor is replaced by another, in the circuit of Fig. 7.25. Is it really so ? To verify this, let us try to analyse the circuit more accurately. For such an analysis, Thevenin's theorem is of great help. A brief review of this theorem for dc circuits is given below.

Thevenin's Theorem: Suppose we have a complicated network containing resistors and voltage sources (see Fig. 7.29a). A and B are two terminals in this network. Thevenin's theorem simply states that this circuit acts as if a voltage V_{TH} in series with a resistor R_{TH} is connected between this pair of terminals, as shown in Fig. 7.29b. Now, when we connect the resistor R_L across the terminals A and B, only one loop is seen. It becomes very easy to calculate the current in this resistor. The power of Thevenin's theorem lies in converting the complicated network into a single loop circuit.

Thevenin's equivalent voltage source V_{TH} is the open circuit voltage across terminals AB. Thevenin's resistor R_{TH} is the resistance from A to B when all the sources in the network are reduced to zero. After *Thevenizing* the circuit at AB terminals (Thevenize means "find the Thevenin's equivalent of") we may connect any resistor across AB and calculate the current flowing in it (see Fig. 7.29c).

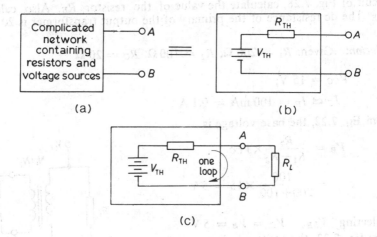

(a) (b)

(c)

Fig. 7.29 Thevenin's theorem

The voltage-divider biasing circuit is drawn in Fig. 7.30a. Let us Thevenize the circuit on the left of the terminals AB. The result is shown in Fig. 7.30b. Here, V_{TH} is the Thevenin's voltage given as

$$V_{TH} = \frac{R_2}{R_1 + R_2} \times V_{CC} \qquad (7.25)$$

The resistor R_{TH} is found by reducing the battery V_{CC} to zero and calculating the equivalent resistance between terminals A and B. When V_{CC} is shorted, the two resistors R_1 and R_2 are in parallel as shown in Fig. 7.31. Thus

$$R_{TH} = \frac{R_1 R_2}{R_1 + R_2} \qquad (7.26)$$

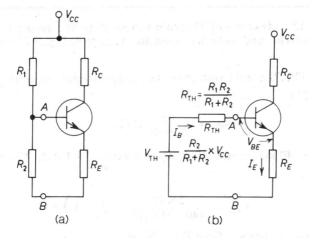

Fig. 7.30 Voltage divider biasing circuit

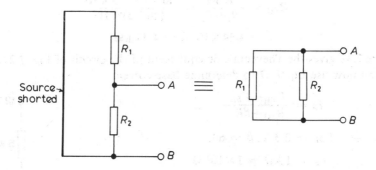

Fig. 7.31 R_{TH} in Thevenin's equivalent

We shall now analyse the circuit of Fig. 7.30b for calculation of th‸ operating point. The loop equation for the input section can be written as

$$V_{TH} = I_B R_{TH} + V_{BE} + R_E I_E$$

or $\qquad V_{TH} = I_B R_{TH} + V_{BE} + (\beta+1) I_B R_E \qquad$ (since $I_E = (\beta+1)I_B$)

or $\qquad I_B[R_{TH} + (\beta+1)R_E] = V_{TH} - V_{BE}$

or $$I_B = \frac{V_{TH} - V_{BE}}{R_{TH} + (\beta+1)R_E} \simeq \frac{V_{TH}}{R_{TH} + \beta R_E} \qquad (7.27)$$

Once the base current is fixed, the collector current can be calculated as

$$I_C = \beta I_B$$

The collector-to-base emitter voltage is then found by the familiar equation,

$$V_{CE} = V_{CC} - (R_C + R_E)I_C$$

It will be worthwhile to see to what extent the approximate analysis is valid. This is done in the next example.

Example 7.13 Make use of Thevenin's theorem to find accurate values of collector current and collector-to-emitter voltage in Fig. 7.27 (of Example 7.11).

Solution: The Thevenin voltage of the voltage-divider circuit is (see Eq. 7.25)

$$V_{TH} = \frac{R_2}{R_1 + R_2} \times V_{CC}$$

Here, $R_1 = 40$ kΩ $= 40 \times 10^3$ Ω; $R_2 = 5$ kΩ $= 5 \times 10^3$ Ω; $V_{CC} = 12$ V

Therefore,

$$V_{TH} = \frac{5 \times 10^3}{(40+5) \times 10^3} \times 12 = \mathbf{1.3 \ V}$$

The Thevenin resistance, from Eq. 7.26, is

$$R_{TH} = \frac{R_1 R_2}{R_1 + R_2} = \frac{40 \times 10^3 \times 5 \times 10^3}{(40+5) \times 10^3}$$

$$= 4.44 \times 10^3 \ \Omega = \mathbf{4.44 \ k\Omega}$$

Figure 7.32 gives the Thevenin's dc equivalent of the circuit of Fig. 7.27. We can now use Eq. 7.27 to determine base current

$$I_B = \frac{V_{TH} - V_{BE}}{R_{TH} + \beta R_E}$$

Here, $V_{BE} = 0.3$ V; $\beta = 60$,

$$R_E = 1 \text{ k}\Omega = 1 \times 10^3 \ \Omega$$

Therefore, $$I_B = \frac{1.3 - 0.3}{4.44 \times 10^3 + 60 \times 1 \times 10^3}$$

$$= 15.52 \times 10^{-6} \text{ A}$$

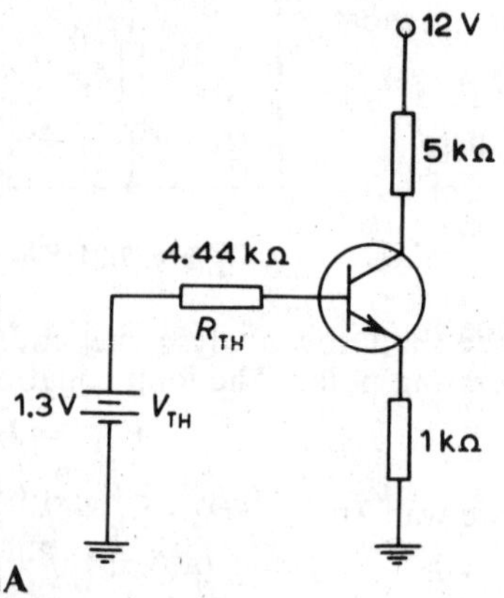

The collector current is then

$$I_C = \beta I_B = 60 \times 15.52 \times 10^{-6}$$

$$= 0.93 \times 10^{-3} \text{ A} = \mathbf{0.93 \ mA}$$

The collector-to-emitter voltage is

$$V_{CE} = V_{CC} - (R_C + R_E)I_C$$

$$= 12 - (5 \times 10^3 + 1 \times 10^3) \times 0.93 \times 10^{-3}$$

$$= 12 - 5.58 = \mathbf{6.42 \ V}$$

Fig. 7.32

Comment: The values of I_C and V_{CE}, obtained above, may be compared with those obtained in Example 7.11. Note that the error committed is within 7 % only. Thus the approximations made are quite reasonable.

7.6.5 Emitter-Bias Circuit

Figure 7.33a shows an emitter-bias circuit. The circuit gets this name because the negative supply V_{EE} is used to forward-bias the emitter junction through resistor R_E. As usual, the V_{CC} supply reverse-biases the collector junction. This circuit uses only three resistors and it provides almost as much stability of operating point as a voltage divider circuit does. However, the emitter biasing can be used only when two supplies— one positive and the other negative—are available. Figure 7.33b shows a simple way to draw this circuit using split supply.

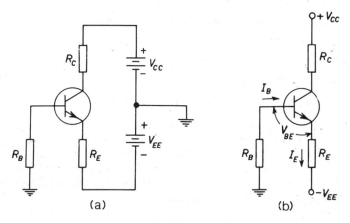

(a)   (b)

Fig. 7.33 Emitter-bias circuit

To determine the operating point, we apply Kirchhoff's voltage law to the emitter-base loop.

$$I_B R_B + V_{BE} + I_E R_E - V_{EE} = 0$$

Since $I_C \simeq I_E$ and $I_B = I_E/\beta$, we can rearrange the above equation to get

$$I_E = \frac{V_{EE} - V_{BE}}{R_E + R_B/\beta} \qquad (7.28)$$

If we want the operating point to be independent of β, we should have

$$R_E \gg \frac{R_B}{\beta}$$

This condition can easily be met in practice. Since the supply V_{EE} is much greater than V_{BE}, we may approximate Eq. 7.28 to give

$$I_E \simeq \frac{V_{EE}}{R_E} \qquad (7.29)$$

The above equation shows that the emitter is virtually at ground potential. All the V_{EE} supply voltage appears across R_E. If the emitter is at ground point, the collector-to-emitter voltage V_{CE} is simply given as

$$V_{CE} = V_{CC} - I_C R_C \qquad (7.30)$$

Example 7.14 Calculate I_C and V_{CE} for the emitter-bias circuit of Fig. 7.33, where $V_{CC} = 12$ V; $V_{EE} = 15$ V, $R_C = 5$ kΩ, $R_E = 10$ kΩ, $R_B = 10$ kΩ, $\beta = 100$.

Solution: From Eq. 7.29, the emitter current is

$$I_E = \frac{V_{EE}}{R_E} = \frac{15}{10 \times 10^3} = 1.5 \times 10^{-3} \text{ A} = 1.5 \text{ mA}$$

The collector current is

$$I_C \cong I_E = \textbf{1.5 mA}$$

Using Eq. 7.30, the collector-to-emitter voltage V_{CE} is

$$V_{CE} = V_{CC} - I_C R_C = 12 - 1.5 \times 10^{-3} \times 5 \times 10^3$$
$$= 12 - 7.5 = \textbf{4.5 V}$$

7.7 *PNP* TRANSISTOR BIASING CIRCUITS

You may be wondering why we have been using only *NPN* transistors ? How will a biasing circuit change if a *PNP* transistor is used in place of an *NPN* transistor ? To forward-bias the emitter diode of a *PNP* transistor, V_{BE} must have the polarity as shown in Fig. 7.34. The collector junction is to be reverse-biased and the polarity of V_{CE} is also to be reversed, as shown.

In fact, the *PNP* transistor is the *complement* of the *NPN* transistor. Here, the word complement signifies that all voltages and currents are opposite to those of the *NPN* transistor. Therefore, to find the complementary *PNP* circuit of a given *NPN* circuit, all you have to do is:

(i) to replace the *NPN* transistor by a *PNP* transistor, and

(ii) to complement or reverse all voltages and currents.

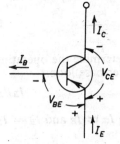

Fig. 7.34 Notations for a *PNP* transistor

Note that if you use *magnitudes* of voltages and currents, all formulae derived for *NPN* circuits apply to *PNP* circuits as well. For instance, in Example 7.2, use is made of the same formulae as in Example 7.1. After calculating the magnitudes of I_C and V_{CE}, the direction of I_C and the polarity of V_{CE} are properly marked in Fig. 7.12b, remembering that it is a *PNP* transistor.

REVIEW QUESTIONS

7.1 Explain the term "biasing".

7.2 Explain why a transistor should be biased.

7.3 Connect two external batteries between the two junctions of a transistor in its three configurations so that it works in the active region.

7.4 In case of the CE configuration, what are the approximate voltages of the dc batteries connected between base-emitter, and collector-emitter terminals?

7.5 Explain why the battery connected between the emitter and base terminals requires a high resistance in series with it.

7.6 Draw transistor-biasing circuits using a 9 V battery and two resistors (1 kΩ and 150 kΩ) in two different ways. Point out the circuit in which bias stabilization exists.

7.7 Draw a simple circuit in which only one battery is used and biasing is achieved by fixing the base current.

7.8 In the circuit given in Fig. R.7.1, derive the expressions for I_C and V_{CE}.

7.9 Draw a biasing circuit using the following components:

    (i) two resistors of 1 kΩ each;   (iii) one dc source of 6 V, and

    (ii) one resistor of 100 kΩ;   (iv) one *PNP* transistor (say AC126)

7.10 Draw a potential-divider biasing circuit making use of a 9 V battery. Mark the direction of current flowing through each resistor of the circuit.

7.11 For the circuit given in Fig. R.7.2, derive the following expression

$$I_E = \frac{V_{CC} - V_{BE}}{R_B/(\beta+1) + R_E}$$

7.12 Prove mathematically that the operating point in a potential-divider biasing circuit is independent of β. Make relevant assumptions.

7.13 Explain why the fixed-bias circuit, in spite of its simplicity, is not much used in amplifiers.

7.14 Explain how stabilization of operating point is achieved when one end of the base resistor R_B is connected to the collector terminal instead of the dc supply.

7.15 Explain the function of the emitter resistor R_E in the potential-divider biasing circuit.

7.16 Explain why operating point is fixed in the centre of the active region of transistor characteristics in a good voltage amplifier.

7.17 Explain why the operating point is not selected near the saturation region of the transistor characteristics in a voltage amplifier.

7.18 State the factors to be considered while designing a biasing circuit for a good transistor voltage amplifier.

7.19 Explain how the circuits given in Fig. R.7.3 to R.7.6 respond to temperature and beta variations.

7.20 Derive the expressions for Q point in the circuit given in Fig. R.7.6, using Thevenin's theorem.

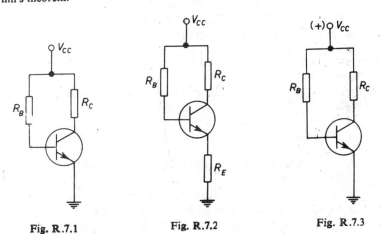

Fig. R.7.1   Fig. R.7.2   Fig. R.7.3

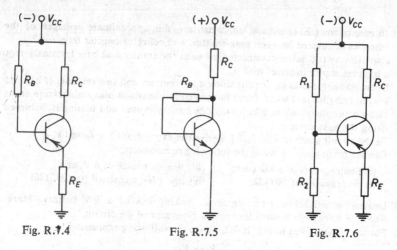

Fig. R.7.4   Fig. R.7.5   Fig. R.7.6

OBJECTIVE-TYPE QUESTIONS

I. Here are some incomplete statements. Four alternatives are provided below each. Tick the alternative that completes the statement correctly.

1. The emitter resistor R_E bypassed by a capacitor

   (a) reduces the voltage gain   (c) causes thermal runaway
   (b) increases the voltage gain   (d) stabilizes the Q point

2. The Q point in a voltage amplifier is selected in the middle of the active region because

   (a) it gives a distortionless output
   (b) the operating point then becomes very stable
   (c) the circuit then requires less number of resistors
   (d) it then requires a small dc voltage

3. The operating point of an *NPN* transistor amplifier should not be selected in the saturation region as

   (a) it may drive the transistor to thermal runaway
   (b) it may cause output to be clipped in the negative half of the input signal
   (c) it may cause output to be clipped in the positive half of the input signal
   (d) it may require high dc collector supply

4. The potential-divider method of biasing is used in amplifiers to

   (a) limit the input ac signal going to the base
   (b) make the operating point almost independent of β
   (c) reduce the dc base current
   (d) reduce the cost of the circuit

5. A transistor is operating in the active region. Under this condition

   (a) both the junctions are forward-biased
   (b) both the junctions are reverse-biased
   (c) emitter-base junction is reverse-biased, and collector-base junction is forward-biased
   (d) emitter-base junction is forward-biased and collector-base junction is reverse-biased

6. The signal handling capacity of an amplifier is high if

   (a) the operating point is selected near the cut-off region
   (b) the operating point is selected near the saturation region
   (c) the operating point is selected in the middle of the active region
   (d) an *NPN* transistor of similar characteristics is used instead of 'PNP one

II. Some statements are written below. Write whether they are **TRUE** or **FALSE** in the space provided against each.

1. The purpose of biasing a transistor is to obtain a certain dc collector current at a certain dc collector voltage. _____

2. In a certain biasing circuit, V_{CC} and V_{CE} are equal. This is because the transistor is heavily conducting. _____

3. A good biasing circuit should stabilize the collector current against temperature variations. _____

4. The emitter resistor R_E is bypassed by a capacitor so as to improve the stabilization of Q point. _____

5. The dc collector current in a transistor circuit is limited by the junction capacitance. _____

6. Negative dc feedback through R_E is responsible for the stabilization of the operating point in a potential-divider bias circuit. _____

III. Amplifier circuits shown in Fig. O.7.1 may be either incomplete or wrongly drawn or both. If so, detect the same and redraw them correctly.

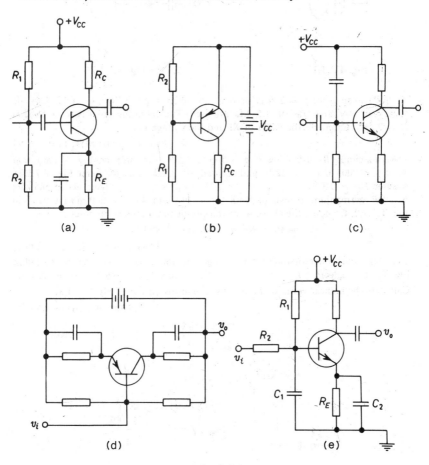

Fig. O.7.1

Ans. I. 1. *d*; 2. *a*; 3. *c*; 4. *b*; 5. *d*; 6. *c*.
   II. 1. T; 2. F; 3. T; 4. F; 5. F; 6. T.

TUTORIAL SHEET 7.1

1. Calculate the value of V_{CE} in a fixed-bias circuit given in Fig. T.7.1.1. Assume $\beta_{dc} = 100$, $R_B = 200$ kΩ, $R_C = 1$ kΩ and $V_{CC} = 10$ V. [**Ans.** $V_{CE} = 5$ V]

2. Calculate the value of R_C and R_B if the dc operating point is to be fixed at $V_{CE} = 7$ V, $I_C = 5$ mA. Following data are given (refer Fig. T.7.1.1): $V_{CC} = 12$ V, $\beta_{dc} = 100$. [**Ans.** $R_B = 240$ kΩ, $R_C = 1$ kΩ]

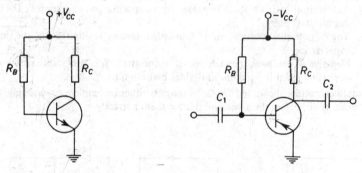

Fig. T.7.1.1   Fig. T.7.1.2

3. A *PNP* transistor of $\beta = 200$ is used in the circuit given in Fig. T.7.1.2. A dc supply of 9 V and R_C of 1.5 kΩ are used. The operating point is to be fixed at $I_C = 2$ mA. Calculate the value of R_B and the voltage V_{CE}.
[**Ans.** $R_B = 0.9$ MΩ; $V_{CE} = 6$ V]

4. Design a simple fixed-biasing circuit for a *PNP* transistor having β such that $50 < \beta < 200$, if $V_{CC} = 12$ V and a load of 3 kΩ is used. (Refer Fig. T.7.1.2). Assume $V_{BE} = 0.3$ V. [**Ans.** $R_B < 585$ kΩ]

5. A *PNP* germanium transistor with $\beta = 100$ and $V_{BE} = 200$ mV is used in Fig. T.7.1.2. Compute the Q point for the circuit conditions given below:
$$V_{CC} = 16 \text{ V}; \quad R_C = 5 \text{ k}\Omega; \quad R_B = 790 \text{ k}\Omega$$
[**Ans.** $V_{CE} = 6$ V; $I_C = 2$ mA]

6. Calculate the collector-to-emitter voltage for the *PNP* transistor connected in Fig. T.7.1.3 neglecting V_{BE}. [**Ans.** $V_{CE} = -5.4$ V; $I_C = 3.6$ mA]

7. Calculate the highest value of R_C permissible in the circuit of Fig. T.7.1.3.
[**Ans.** $R_{C(max)} = 2.5$ kΩ]

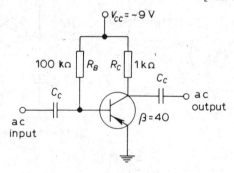

Fig. T.7.1.3

TUTORIAL SHEET 7.2

1. Calculate the Q point for the dc-bias circuit in Fig. T.7.2.1, given the following:
   $R_C = 3\,k\Omega$; $R_1 = 60\,k\Omega$; $V_{CC} = 12\,V$; $\beta = 60$;
   assume V_{BE} negligible. [**Ans.** $V_{CE} = 3\,V$; $I_C = 3\,mA$]

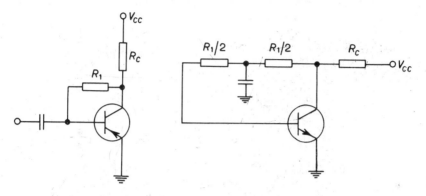

Fig. T.7.2.1   Fig. T.7.2.2

2. Calculate all the dc currents and voltage V_{CE} in the transistor of Fig. T.7.2.2 for the following given data:
   $V_{CC} = 10\,V$; $R_C = 3\,k\Omega$, $R_1 = 250\,k\Omega$; $\beta = 50$; neglect V_{BE}.
   [**Ans.** $I_E = I_C = 1.25\,mA$;
   $I_B = 25\,\mu A$; $V_{CE} = 6.25\,V$]

3. Select the value of R_1 to set up the biasing condition such that $V_{CE} = 0.5\,V_{CC}$, for the following circuit components (refer Fig. T.7.2.1).
   $V_{CC} = 30\,V$; $R_C = 5\,k\Omega$; $\beta = 40$. [**Ans.** 200 kΩ]

4. Calculate the biasing point of the transistor (refer Fig. T.7.2.1) for the following data:
   $R_C = 5\,k\Omega$; $V_{CC} = 15\,V$; $\beta = 100$, $R_1 = 215\,k\Omega$; $V_{BE} = 0.7\,V$.
   [**Ans.** $I_C = 2\,mA$; $V_{CE} = 5\,V$]

5. Calculate the new operating point if the transistor of Problem 4 is replaced by the other silicon PNP having $\beta = 300$. [**Ans.** $I_C = 2.5\,mA$; $V_{CE} = 2.5\,V$]

TUTORIAL SHEET 7.3

1. Calculate V_{CE} and I_C in the circuit of Fig. T.7.3.1 if $V_{CC} = 9\,V$; $R_B = 50\,k\Omega$; $R_C = 250\,\Omega$; $R_E = 500\,\Omega$ and $\beta = 80$. [**Ans.** $V_{CE} = 3\,V$, $I_C = 8\,mA$]

2. Compute the Q point of the transistor (refer Fig. T.7.3.1) if $R_B = 400\,k\Omega$; $R_C = 2\,k\Omega$; $R_E = 1\,k\Omega$; $\beta = 100$ and $V_{CC} = 20\,V$, neglecting V_{BE}. Mark the direction of I_C. [**Ans.** $I_C = 4\,mA$, $V_{CE} = 8\,V$]

3. The NPN transistor in the circuit given in Fig. T.7.3.2 has a $\beta = 56$. Calculate the Q point if the following circuit components are used:
   $V_{CC} = 18\,V$; $R_B = 50\,k\Omega$; $R_E = 0.75\,k\Omega$ and $R_C = 500\,\Omega$.
   Assume $V_{BE} = 0.7\,V$. [**Ans.** $I_C = 10.53\,mA$, $V_{CE} = 4.83\,V$]

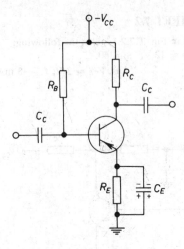

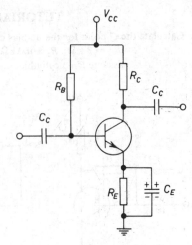

Fig. T.7.3.1   Fig. T.7.3.2

4. Calculate the Q point for the transistor given in Fig. T.7.3.2 for the given circuit parameters. $V_{CC} = 10$ V; $V_{BE} = 0.25$ V; $\beta = 80$; $R_B = 75$ kΩ; $R_C = 0.5$ kΩ and $R_E = 470$ Ω.   [**Ans.** $I_C = 6.92$ mA, $V_{CE} = 3.29$ V]

5. A *PNP* transistor having a dc current gain in CE equal to 100 is to be biased at $I_C = 5$ mA and $V_{CE} = 3.8$ V. The collector load has a resistance of 500 Ω. If $V_{CC} = 10$ V and $V_{BE} = 0.3$ V, calculate the value of R_B and R_E. (Refer Fig. T.7.3.1).   [**Ans.** $R_E = 740$ Ω, $R_B = 120$ kΩ]

TUTORIAL SHEET 7.4

Note : Use approximate method of solving biasing circuit unless specifically asked otherwise.

1. Calculate the collector current and collector-to-emitter voltage of the circuit given in Fig. T.7.4.1 assuming the following circuit components and transistor specifications:

$R_1 = 40$ kΩ	$V_{BE} = 0.5$ V
$R_2 = 4$ kΩ	$\beta = 40$
$R_C = 10$ kΩ	$V_{CC} = 22$ V
$R_E = 1.5$ kΩ	[**Ans.** $I_C \simeq I_E = 1$ mA, $V_{CE} = 10.5$ V]

2. Calculate the bias voltage, and currents for the *PNP* silicon transistor used in Fig. T.7.4.2 assuming the following data:

$R_1 = 100$ kΩ	$V_{CC} = 12$ V
$R_2 = 27$ kΩ	$V_{BE} = 0.751$ V
$R_C = 2$ kΩ	$\beta = 75$
$R_E = 1$ kΩ	[**Ans.** $I_C \simeq I_E = 1.8$ mA; $V_{CE} = 6.6$ V]

3. Solve Problems 1 and 2, using an accurate method.

4. Calculate the value of resistors R_1 and R_L to place the Q point at $I_E = 2$ mA and $V_{CE} = 6$ V, in the two circuits of Fig. T.7.4.3a and T.7.4.3b. In both circuits, a transistor of $\alpha = 0.985$, $I_{CBO} = 4$ μA and $V_{BE} = 200$ mV, is used. The V_{CC} supply used is 16 V.   [**Ans.** (a) $R_1 = 5.54$ kΩ, $R_l = 3$ kΩ; (b) $R_1 = 56.3$ kΩ, $R_l = 4$ kΩ]

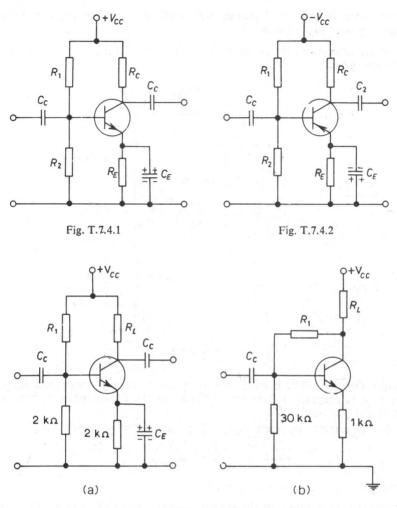

Fig. T.7.4.1

Fig. T.7.4.2

(a)

(b)

Fig. T.7.4.3

EXPERIMENTAL EXERCISE 7.1

TITLE: Fixed-bias circuit with and without emitter resistor.

OBJECTIVES: To,

1. trace the given biasing circuit;
2. measure the Q point collector current and collector-to-emitter voltage with and without emitter resistor R_E;
3. note the variation of the Q point by increasing the temperature of the transistor in fixed-bias circuit with and without emitter resistor R_E;
4. note the variation in Q point by changing the base resistor in bias circuit, when emitter resistor is present and not present;

APPARATUS REQUIRED: Experimental board, electronic multimeter, milliam-meter, power supply unit.

CIRCUIT DIAGRAM: As given in Fig. E.7.1.1. (typical values of components are also given).

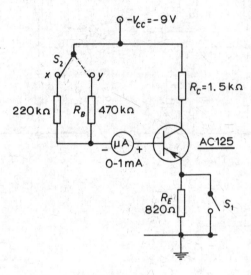

Fig. E.7.1.1

BRIEF THEORY: The biasing circuit when the emitter resistance R_E is not present, is generally referred to as fixed-bias circuit (i.e. when switch S_1 is in closed position).

In this circuit, the operating point current $I_C \simeq I_E$ is given by

$$I_B = \frac{(V_{CC} - V_{BE})}{R_B} \cong \frac{V_{CC}}{R_B} \tag{1}$$

and $$V_{CE} = V_{CC} - I_E R_C \tag{2}$$

When the temperature of the collector-to-base junction changes, the leakage current I_{CEO} increases. Since

$$I_C = \beta I_B + I_{CEO}$$

the operating point may go into the saturation region. Sometimes, thermal runaway may also take place. When the emitter resistor R_E is added in the circuit (i.e. when switch S_1 is in the open position) the operating point is given by

$$I_E = \frac{V_{CC} - V_{BE}}{R_E + R_B/\beta} \cong \frac{V_{CC}}{R_E + R_B/\beta}$$

and $$V_{CE} = V_{CC} - I_E R_C$$

If the temperature increases, the following sequence of events takes place.

$$T\uparrow I_{CEO}\uparrow I_C\uparrow I_E\uparrow R_E I_E\uparrow V_E\uparrow V_{BE}\downarrow I_B\downarrow I_C\downarrow$$

This shows that there is a tendency to make the operating point stable.

PROCEDURE:

1. Take the experimental board and identify the resistors R_B, R_C, and R_E. Also find out the values of these resistors.
2. Apply $V_{CC} = 9$ V and close the switch S_1. Connect milliammeter and voltmeter. Note the values of I_C and V_{CE}.
3. Increase the temperature of the transistor (by rubbing your hands together and touching the transistor with one of the fingers; or by putting a lamp near it) and note the effect on collector current.
4. Now put off the switch S_1 so that the emitter resistor R_E comes in the circuit. Note the new operating emitter voltage.
5. Now increase the temperature and note the effect on the operating point.
6. Now change the switch S_2 in such a way that the base resistor R_B changes. With the new value of base resistor, repeat the above experiment.

OBSERVATIONS:

1. When the switch S_1 is in closed position, i.e. R_E is not in the circuit. Assume $\beta = 100$ for transistor.

S. No.	V_{CC}	R_B	I_C		V_{CE}	
			Theor.	Pract.	Theor.	Pract.
1.						
2.						

2. When the switch S_1 is in the open condition, i.e. when R_E is present in the circuit.

S. No.	V_{CC}	R_B	I_C		V_{CE}	
			Theor.	Pract.	Theor.	Pract.
1.						
2.						

RESULTS: It is found that the effect of increasing temperature on the collector current, is much reduced when R_E is brought into the circuit.

EXPERIMENTAL EXERCISE 7.2

TITLE: Collector-to-base feedback bias circuit.

OBJECTIVES: To,

1. trace the given circuit of collector-to-base feedback bias circuit;
2. measure the Q point collector current and collector-to-emitter voltage;
3. note the change in Q point when the temperature of the transistor changes;

4. compare the stability of the Q point against temperature variations in the two circuits, i.e. collector-to-base feedback circuit and fixed-bias circuit.

APPARATUS REQUIRED: Experimental board, electronic multimeter, milliam-meter, power supply unit.

CIRCUIT DIAGRAM: As given in Fig. E.7.2.1 (typical values of components are also given).

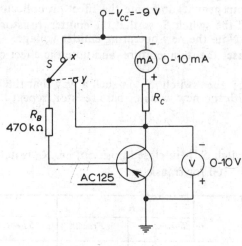

Fig. E.7.2.1

BRIEF THEORY: In collector-to-base feedback circuit, the operating point is given as

$$Ic \simeq I_E = \frac{V_{CC} - V_{BE}}{R_C + R_B/\beta} \simeq \frac{V_{CC}}{R_C + R_B/\beta}$$

and $$V_{CE} = V_{CC} - I_E R_C$$

In this circuit, the operating point has a tendency to be stable against temperature variations, as seen by the following sequence of events:

$$T \uparrow I_{CEO} \uparrow Ic \uparrow Vc \downarrow I_B \downarrow Ic \downarrow$$

PROCEDURE:

1. Trace the given bias circuit. Note the values of various resistors in the circuit.
2. Apply Vcc (say -9 V). Put the switch S in position x so that R_B is connected between the supply terminal and base terminal. Note the Q point collector current and collector-to-emitter voltage.
3. Increase the temperature of the transistor (either with the help of a lamp or by rubbing hands together and touching it with fingers). Note the variation in Q point.
4. Now change the switch S to position Q so that the base resistor gets connected between collector and base. Note the Q point.
5. Repeat step 3 for this circuit.

OBSERVATIONS:

1. *For fixed-bias circuit:*
   $V_{CC} =$ _____ V; $V_{CE} =$ _____ V; $I_C =$ _____ mA;
   At higher temperature, $I_C =$ _____ mA

2. *For collector-to-base feedback circuit:*
   $V_{CC} =$ _____ V; $V_{CE} =$ _____ V; $I_C =$ _____ mA;
   At higher temperature, $I_C =$ _____ mA.

RESULTS: It is observed that the Q point shifts much slowly in collector-to-base feedback circuit as compared with the fixed-bias circuit, when the temperature is changed.

EXPERIMENTAL EXERCISE 7.3

TITLE: Potential-divider biasing circuit.

OBJECTIVES: To,

1. trace the circuit diagram of the potential-divider bias circuit;
2. measure the operating-point collector current and collector-to-emitter voltage;
3. measure the operating point when one of the bias resistors changes;
4. note the effect of change in temperature on the operating point.

APPARATUS REQUIRED: Experimental board, transistorized regulated power supply, electronic multimeter, milliammeter.

CIRCUIT DIAGRAM: As shown in Fig. E.7.3.1 (typical values of the components are also shown.)

BRIEF THEORY: Potential-divider bias circuit is a widely-used biasing circuit in amplifiers. The most significant advantage of this circuit is that the operating point in this circuit is almost independent of β. The expression for emitter current (which is also equal to collector current) is given by

$$I_E = \frac{\dfrac{V_{CC} \times R_2}{R_1 + R_2} - V_{BE}}{R_E}$$

The collector-to-emitter voltage can be given by

$$V_{CE} = V_{CC} - (R_E + R_C)I_E$$

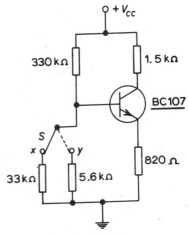

Fig. E.7.3.1

The operating point can be changed by changing one of the resistors of the potential-divider network. In the experiment, two values of the resistor R_2 are provided.

PROCEDURE:

1. From the given circuit, find out whether the transistor is *PNP* type or *NPN* type. Trace the circuit and note down the values of different resistors.
2. Connect the collector supply dc voltage in the circuit. Adjust the dc voltage to, say, 9 V.
3. Measure the collector supply voltage V_{CC}, the collector voltage V_C and collector-to-emitter voltage V_{CE}. Calculate the collector current by finding the voltage drop across the collector resistor R_C. This drop is $(V_{CC} - V_C)$ V.
4. Now put the switch *S* in the second position. Note the new *Q* point by measuring the collector current and collector-to-emitter voltage.

OBSERVATIONS:

D.C. supply voltage $V_{CC} = 9$ V.

S. No.	(Base to ground resistor) R_2	V_C	I_C	V_{CE}
1.				
2.				
3.				

RESULT: When resistor R_2 connected between base and ground decreases, the collector current also decreases.

Small-Signal Amplifiers

OBJECTIVES: After completing this unit, you will be able to: ○ Draw a single stage amplifier circuit (CE configuration). ○ Calculate the voltage gain of a single-stage amplifier when supplied with; (a) collector characteristics of the transistor; (b) values of different resistors used in the circuit; (c) value of the dc supply voltage; and (d) dynamic input resistance of transistor. ○ Explain the phase relationship between the input and the output signal in a single-stage amplifier circuit. ○ Calculate the voltage gain, input impedance and output impedance of a single-stage amplifier circuit when circuit parameters and transistor parameters like β and r_{in} (dynamic input resistance), or the h-parameters, are given. ○ Explain the working of a single-stage triode amplifier. ○ Calculate the voltage gain of a single-stage triode amplifier using the plate characteristics of triode. ○ Calculate the voltage gain of a single-stage triode amplifier with the help of ac equivalent circuit. ○ Explain the working of a single-stage pentode amplifier. ○ Calculate the voltage gain of pentode amplifier graphically as well as with the help of equivalent circuit. ○ Draw the circuit diagram of a single-stage field-effect transistor amplifier. ○ Calculate the voltage gain of a single-stage field effect transistor amplifier by graphical as well as equivalent circuit method. ○ Compare the performance of transistor amplifier and vacuum-tube amplifier.

8.1 INTRODUCTION

Almost no electronic system can work without an amplifier. Could the voice of a singer reach everybody in the audience, in a hall, if the PA system (Public Address system) fails? It is only because of the *enlargement* or the *amplification* of the signal picked up by microphone that we can enjoy a music orchestra. We are able to hear the news or the cricket commentary on our radio, simply because the amplifier in the radio *amplifies* the weak signals received by its antenna. The up and down motion of the needle of a record player produces a signal with the help of a piezoelectric crystal. But this signal cannot drive the loudspeakers unless it is amplified to a sufficient level. The signal can only be of any use if it is amplified to give a suitable output (such as sound in radio, picture in TV etc.).

In the previous chapter, the transistor circuits were analysed purely from the dc point of view. After a transistor is biased in the active region, it can work as an amplifier. We apply an ac voltage between the base and emitter terminals to produce fluctuations in the collector current. An amplified output signal is obtained when this fluctuating collector current flows through a collector resistor R_C. When the input signal is so weak as to produce *small* fluctuations in the collector current compared to its quiescent value, the amplifier is called *small-signal amplifier* (also "voltage amplifier"). Such an amplifier is used as the first stage of the amplifier used in receivers (radio and TV), tape recorders, stereos and measuring instruments.

A vacuum triode (or a pentode) is also capable of amplifying signals.

8.2 SINGLE-STAGE TRANSISTOR AMPLIFIER

We have seen in the previous chapter that the voltage divider method of biasing is the best. The circuit is shown in Fig. 8.1a. Almost all amplifiers use this biasing circuit, because the design of the circuit is simple and it provides good stabilization of the operating point. If this circuit is to amplify ac voltages, some more components must be added to it. The result is shown in Fig. 8.1b. We have added three capacitors.

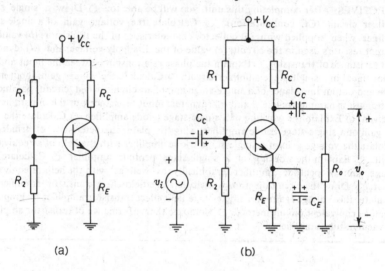

(a)   (b)

Fig. 8.1 (a) Voltage-divider biasing circuit; (b) Same circuit connected into an amplifier

The capacitors C_C are called the *coupling capacitors*. A coupling capacitor passes an ac signal from one side to the other. At the same time, it does not allow the dc voltage to pass through. Hence, it is also called a *blocking capacitor*. For instance, it is due to the capacitor C_C (connected between collector and output) in Fig. 8.1b that the output across the resistor R_o is free from the collector's dc voltage.

The capacitor C_E works as a *bypass capacitor*. It bypasses all the ac currents from the emitter to the ground. If the capacitor C_E is not put in the circuit, the ac voltage developed across R_E will affect the input ac voltage. Such a feedback of ac signal is reduced by putting the capacitor C_E. If the capacitor C_E is good enough to provide an effective bypass to the lowest frequency of the signal, it will do so better to the higher frequencies. We, therefore, select such a value of capacitor C_E that gives quite a low impedance compared to R_E at the lowest frequency present in the input signal. As a practical guide, we make the reactance of the capacitor C_E at the lowest frequency, not more than one-tenth the value of R_E. That is

$$X_{CE} \leqslant \frac{R_E}{10} \qquad (8.1)$$

The resistor R_O represents the resistance of whatever is connected at the output. Quite often, the amplification of the signal given by one amplifier may not suffice. More stages of amplifiers are then needed. The resistor R_O in Fig. 8.1b will then represent the input resistance of the next stage.

To what extent an amplifier enlarges signals is expressed in terms of its *voltage gain*. The voltage gain of an amplifier is given as

$$A_v = \frac{\text{output ac voltage}}{\text{input ac voltage}} = \frac{V_o}{V_i} \qquad (8.2)$$

The other quantities of interest for a voltage amplifier are current gain (A_i), input impedance (Z_i), and output impedance (Z_o). The amplifier can be analysed for its performance by the following two methods:

(i) Graphical method
(ii) Equivalent circuit method

8.3 GRAPHICAL METHOD

For analysing an amplifier by this method, we need the output characteristics of the transistor. These are supplied by the manufacturer. When the ac voltage is applied to the input, the base current varies. The corresponding variations in collector current and collector voltage can be seen on the characteristics. This method involves no approximating assumptions. Hence, the results obtained by this method are more accurate than the equivalent circuit method. One can also visualize the maximum ac voltage that can be properly handled by this amplifier. In fact, for *large-signal amplifiers* (power amplifiers) this is the only suitable method.

8.3.1 Is dc Load line Same as ac Load line

In the amplifier circuit of Fig. 8.1, the resistors R_1 and R_2 form a voltage divider arrangement for fixing a certain dc base voltage. This base voltage and the resistor R_E fix the emitter current. The collector current is almost the same as the emitter current. The resistor R_C then decides the value of V_{CE}. Writing the KVL equation for the output section of the circuit, we get

$$V_{CC} = I_C R_C + V_{CE} + I_E R_E$$
$$= V_{CE} + I_C (R_C + R_E) \qquad [\text{since } I_C \simeq I_E]$$

or
$$I_C = \frac{-1}{(R_C + R_E)} V_{CE} + \frac{V_{CC}}{(R_C + R_E)} \qquad (8.3)$$

This is the equation of the dc load line. By plotting this line on the output characteristics, the dc collector voltage and current can be determined for the given value of base current. As regards the dc currents and voltages, the amplifier circuit of Fig. 8.1b behaves like the circuit shown in Fig. 8.2a. This is obtained by opening all the capacitors in the original circuit. The capacitors are as good as open circuits for dc.

Suppose we had changed the dc bias, giving a different value of base current. The collector current and collector voltage both will change. As a result, the Q point will shift on the dc load line. This is what roughly happens when we apply an input ac signal. But, in the ac signals, the variations occur very fast. The capacitors can no longer be considered as an open circuit. In fact, the variations in the currents and voltages occur so fast that the capacitors in the circuit may be treated as short-circuits. Also while dealing with ac currents and voltages, we need not consider the dc supplies. If we do this, the original circuit of Fig. 8.1b reduces to the one shown in Fig. 8.2b. This circuit explains the behaviour of the amplifier from the ac point of view. You may now see that the resistor R_C comes in parallel with R_O and forms the ac load for the amplifier. The variation in the collector current and voltage are seen with the help of the ac load line corresponding to this ac load. How the ac load line is drawn is made clear in the next section.

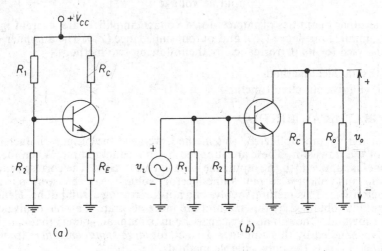

Fig. 8.2 Amplifier circuit of Fig. 8.1b for: (a) DC behaviour
(b) AC behaviour

8.3.2 Calculation of Gain

To understand how to calculate the current gain and voltage gain by the graphical method, we consider a typical amplifier circuit. One such circuit is shown in Fig. 8.3. The output characteristics of the transistor used in this circuit, are shown in Fig. 8.4.

We first plot the dc load line on the output characteristics. The equation of this dc load line is given by Eq. 8.3. This line is drawn by simply joining the points $(V_{CC}, 0)$ and $(0, V_{CC}/R_{dc})$. It may be seen that the slope of this line is $-1/R_{dc}$, where the dc load, $R_{dc} = R_C + R_E$. In this case, $V_{CC} = 9$ V; and $R_{dc} = R_C + R_E = 1$ kΩ $+ 0.1$ kΩ $= 1.1$ kΩ. Thus, the two points for plotting the dc load line are $(9$ V, $0)$ and $(0, 8.2$ mA$)$. Let us assume that the biasing arrangement is such that the dc base current is 30 μA. The quiescent operating point Q is given by the intersection of the dc load line and the output characteristic corresponding to $I_B = 30$ μA. The dc collector current and collector-to-emitter voltage of the Q point may be seen to be 4 mA and 4.5 V, respectively.

When we apply an ac input signal V_i, the circuit behaves like the one shown in Fig. 8.2b. It is clear that for ac, the load resistance is R_C in parallel with R_O. This is the ac load R_{ac} for which the load line should be plotted. In our case, $R_{ac} = 1$ kΩ ‖ 470 Ω $= 320$ Ω. The ac load line will have a slope of $-1/R_{ac}$. Since the Q point describes the zero-signal conditions of the circuit, the ac load line should also pass through Q point. To draw such

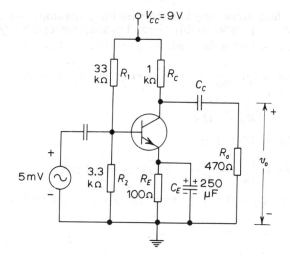

Fig. 8.3 Typical transistor amplifier circuit

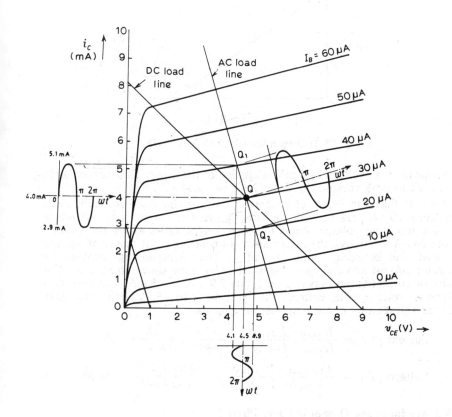

Fig. 8.4 Calculation of gain by graphical method

a line, we can first draw any line AB with the given slope. We can then draw the ac load line parallel to this line and passing through the Q point.

Figure 8.5 explains how a line with the given slope $-1/R_{ac}$ is drawn.

$$\text{Slope} = -\frac{1}{R_{ac}} = \tan \theta = \tan (180 - \alpha) = -\tan \alpha$$

$$\therefore \qquad \tan \alpha = \frac{1}{R_{ac}} = \frac{OB}{OA}$$

Let us take $OA = 1$ V. Then

$$OB = \frac{OA}{R_{ac}} = \frac{1}{320} = 0.0031 \text{ A} = 3.1 \text{ mA}$$

After locating the points B and A, line AB whose slope $= 1/R_{ac}$ can be drawn.

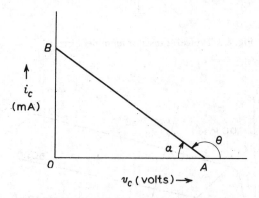

Fig. 8.5 To draw a line with a slope of $-1/R_{ac}$

Suppose an ac voltage of 5 mV is applied at the input. This corresponds to a peak-to-peak variation of $5 \times \sqrt{2} \times 2 = 14.14$ mV.

Assume that the input characteristics of the transistor are such, as to produce a 20 μA peak-to-peak variation in the base current corresponding to this input ac voltage. When the base current varies within these limits (from 20 μA to 40 μA), the instantaneous operating point moves along the ac load line between points Q_1 and Q_2. The corresponding variations in collector current and collector-to-emitter voltage are shown in Fig. 8.4. The collector current varies between the limits 2.9 mA to 5.1 mA. The collector-to-emitter voltage varies between the limits 4.1 V to 4.9 V. The current gain and the voltage gain of the amplifier are given as :

$$\text{Current gain} = \frac{I_{C(max)} - I_{C(min)}}{I_{B(max)} - I_{B(min)}} = \frac{(5.1 - 2.9) \text{ mA}}{(40 - 20) \text{ }\mu\text{A}} = 110$$

$$\text{Voltage gain} = \frac{V_{CE(max)} - V_{CE(min)}}{V_{i(max)} - V_{i(min)}} = \frac{(4.9 - 4.1) \text{ V}}{14.14 \text{ mV}} = 56.58$$

8.3.3 Are Input and Output in Same Phase ?

In Fig. 8.4 observe the waveforms of the base current, the collector current and the collector-to-emitter voltage. When the input voltage increases, the base current also increases. The instantaneous Q point moves towards Q_1; as a result, the current I_C increases, but the voltage V_{CE} reduces. For clear understanding of amplifier operation, the variations of input voltage, base current, collector current, collector-to-emitter (output) are again drawn in Fig. 8.6. From these diagrams, it is clear that *the input voltage and output voltage are out of phase by 180°*.

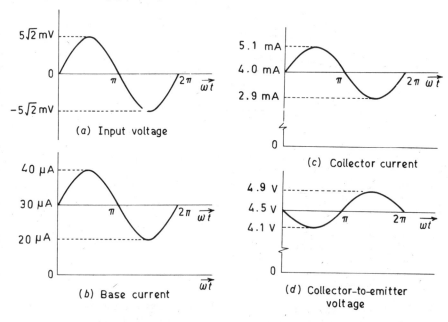

Fig. 8.6 Phase relationships between input and output

8.4 EQUIVALENT CIRCUIT METHOD

Our main concern in analysing an amplifier circuit is to determine its ac behaviour. We are interested in calculating the ac current gain, voltage gain, input impedance and output impedance. For this purpose, the given amplifier circuit is converted into its equivalent circuit from the ac point of view. All the capacitors and the dc supplies are replaced by short circuits. The CE amplifier circuit of Fig. 8.1*b* then reduces to the form of Fig. 8.2*b*. In the equivalent circuit method of analysis of the amplifier, the transistor is also replaced by its ac equivalent.

8.4.1 Development of Transistor AC Equivalent Circuit

Figure 8.7 shows the typical output characteristics of a transistor in the CE mode. The curves are almost horizontal. For a given value of base current, the collector current hardly depends upon the value of the collector-to-emitter voltage. Keeping I_B constant, the change in I_C corresponding to certain change in V_{CE} is very small. It means that the output section of the transistor offers very high dynamic resistance. The transistor, therefore, can be replaced by a current source between its output terminals. This is shown in Fig. 8.8.

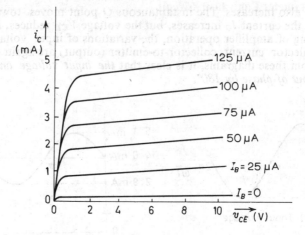

Fig. 8.7 Typical output characteristics of a transistor
in CE mode

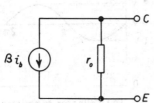

Fig. 8.8 Transistor equivalent circuit
between collector and emitter

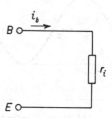

Fig. 8.9 Transistor equivalent circuit
between base and emitter

The current source βi_b depends, as it should, on the input ac current i_b and the current amplification factor β. The resistance r_o represents the dynamic output resistance of the transistor and its value is quite high (typically 40 kΩ).

In the input section, the emitter-base junction of the transistor is forward-biased. The input characteristic of the transistor is similar to that of a forward-biased diode. The junction, therefore, can be replaced by a resistance r_i. The value of this resistance is low (typically 800 Ω). The input section of the transistor therefore simply becomes the one shown in Fig. 8.9.

The complete ac equivalent circuit of the transistor is obtained by combining the input and output section. This is shown in Fig. 8.10.

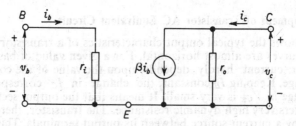

Fig. 8.10 Transistor equivalent circuit

8.4.2 *h*-Parameter Equivalent Circuit

Quite often, the manufacturers specify the characteristics of a transistor in terms of its *h* parameters (the letter *h* stands for hybrid). The word *hybrid* is used with these parameters because they are a mixture of constants having different units. The hybrid parameters have become popular because they can be measured easily.

A transistor is a three-terminal device. If one of the terminals is common between the input and the output, there are two ports (pairs of terminals). See Fig. 8.11. For our purpose, the pair of terminals at the left represents the input terminals and the pair of terminals at the right, the output terminals. For each pair of terminals, there are two variables (current and voltage). There are a number of ways in which these four variables can be related. One of the ways, which is most frequently employed in transistor circuit analysis is as follows :

$$v_1 = h_{11}i_1 + h_{12}v_2 \tag{8.4}$$

$$i_2 = h_{21}i_1 + h_{22}v_2 \tag{8.5}$$

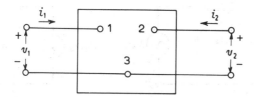

Fig. 8.11 A transistor as a two-port network

The parameters h_{11}, h_{12}, h_{21} and h_{22} which relate the four variables of the two-port system by the Eqs. 8.4 and 8.5 are called hybrid parameters. These parameters can be defined from the above equations by first putting $v_2 = 0$ (i.e. short-circuiting the output terminals) and then putting $i_1 = 0$ (i.e. opening the input terminals)

$$h_{11} = \left.\frac{v_1}{i_1}\right|_{v_2=0} = \text{Input impedance (with output shorted)} = h_i$$

$$h_{21} = \left.\frac{i_2}{i_1}\right|_{v_2=0} = \text{Forward current ratio (with output shorted)} = h_f$$

$$h_{12} = \left.\frac{v_1}{v_2}\right|_{i_1=0} = \text{Reverse voltage ratio (with input open)} = h_r$$

$$h_{22} = \left.\frac{i_2}{v_2}\right|_{i_1=0} = \text{Output admittance (with input open)} = h_o$$

It may be noted that h_i, being the ratio of voltage to current, has units of Ω. Similarly, h_o being the ratio of current to voltage has units of siemens (earlier known as mhos). However, h_f and h_r being the ratio of similar quantities are pure numbers having no units. Thus, the parameters are hybrid in nature.

An additional suffix e is added to the symbols of the h parameters to indicate that the transistor is used in the CE mode. In this mode, the terminal 1 is the base terminal and terminal 2 is the collector. Therefore, v_1 and i_1 become v_b and i_b, respectively, and at the output port, v_2 and i_2 become v_c and i_c, respectively. With this understanding, the Eqs. 8.4 and 8.5 can be written as

$$v_b = h_{ie}i_b + h_{re}v_c \tag{8.6}$$

$$i_c = h_{fe}i_b + h_{oe}v_c \tag{8.7}$$

Since each term of Eq. 8.6 has the units of volts, we can use Kirchhoff's voltage law to find a circuit that 'fits' this equation. The result is shown in Fig. 8.12. Similarly, we observe that each term of Eq. 8.7 has the units of current. Using Kirchhoff's current law, we get the circuit shown in Fig. 8.13 to fit this equation. Combining both of these figures, we get Fig. 8.14. This circuit satisfies both the Eqs. 8.6 and 8.7; and therefore this is the complete ac equivalent circuit of the transistor using h parameters.

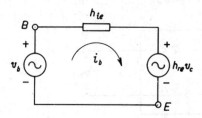
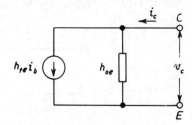

Fig. 8.12 Hybrid input equivalent circuit   Fig. 8.13 Hybrid output equivalent circuit

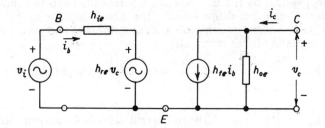

Fig. 8.14 Complete hybrid equivalent circuit of a transistor

Let us now compare the hybrid equivalent circuit of Fig. 8.14 with the one developed in Fig. 8.10. We find that

$h_{ie} = r_i$, the dynamic input resistance

$h_{fe} = \beta$, the current amplification factor

$1/h_{oe} = r_o$, the dynamic output resistance.

The only difference in the two circuits is the presence of a voltage source $h_{re}v_c$ in the input of the hybrid model. The magnitude of this voltage source depends upon the output voltage v_c. The parameter h_{re}, therefore, represents a "feedback" of the output voltage to the input circuit. In the normal operation of the transistor, this effect is very small. It will make practically no difference if we neglect the term $h_{re}v_c$ from the hybrid equivalent circuit. The typical values of h parameters are

$$h_{ie} = 1 \text{ k}\Omega$$
$$h_{re} = 2.5 \times 10^{-4}$$
$$h_{fe} = 50$$
$$h_{oe} = 25 \text{ }\mu\text{S} \quad (\text{or, } 1/h_{oe} = 40 \text{ k}\Omega)$$

The h parameters at a given operating point can be determined from the static characteristics of the transistor, as illustrated in Example 8.1.

Example 8.1 Determine the hybrid parameters from the given transistor characteristics (Figs. 8.15 and 8.16) at an operating point, $I_C = 2$ mA, and $V_{CE} = 8.5$ V.

Solution:

On the collector characteristics of Fig. 8.15, draw a vertical line corresponding to $V_{CE} = 8.5$ V. Draw a horizontal line corresponding to $I_C = 2$ mA. The intersection of these two lines fixes the operating point. This is marked as Q. Note that this Q point lies in the middle of the two characteristic curves corresponding to base currents $I_{B1} = 10 \text{ }\mu\text{A}$ and $I_{B2} = 20 \text{ }\mu\text{A}$. This indicates that the base current at the operating point is 15 μA. An additional characteristic curve for $I_B = 15 \text{ }\mu\text{A}$ is drawn.

Refer Fig. 8.15. At constant V_{CE} of 8.5 V, if I_B changes, say, by a small amount around the Q point from 10 μA to 20 μA, the collector current changes from 1.7 mA to 2.7 mA. Therefore,

$$h_{fe} = \left.\frac{\Delta i_C}{\Delta i_B}\right|_{v_{CE} = \text{const.}} = \left.\frac{(2.7-1.7)\times 10^{-3}}{(20-10)\times 10^{-6}}\right|_{V_{CE} = 8.5 \text{ V}}$$

$$= \frac{10^{-3}}{10\times 10^{-6}} = \textbf{100}$$

Refer Fig. 8.15b. At constant I_B of 15 μA, suppose the voltage V_{CE} changes around the Q point from 7 to 10 V. The corresponding change in collector current is from 2.1 mA to 2.2 mA. Therefore,

$$h_{oe} = \left.\frac{\Delta i_C}{\Delta v_{CE}}\right|_{i_B = \text{const.}} = \left.\frac{(2.2-2.1)\times 10^{-3}}{10-7}\right|_{I_B = 15 \text{ }\mu\text{A}}$$

$$= \frac{0.1\times 10^{-3}}{3} = 33 \text{ }\mu\text{A/V} = 33 \text{ }\mu\text{S}$$

To determine the parameters h_{ie} and h_{re}, we first fix the Q point on the input characteristics, as shown in Fig. 8.16. As shown in Fig. 8.16a, an additional curve corresponding to $V_{CE} = 8.5$ V is drawn. A small change in V_{BE} is then chosen, resulting in a corresponding change in I_B. We may then calculate h_{ie} as follows:

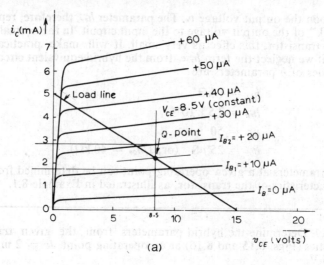

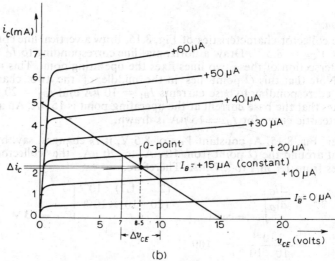

Fig. 8.15 Collector characteristics of a transistor for the calculation of h_{fe} and h_{oe}

$$h_{ie} = \frac{\Delta V_{BE}}{\Delta i_B}\bigg|_{V_{CE} = \text{const.}} = \frac{0.730 - 0.715}{(20 - 10)\,\mu A}\bigg|_{V_{CE} = 8.5\,V}$$

$$= \frac{0.015}{10 \times 10^{-6}} = 1.5\ k\Omega$$

The last parameter h_{re} can be found by first drawing a horizontal line through the Q point of $I_B = 15\ \mu A$. As shown in Fig. 8.16b, when V_{CE} changes from 0 V to 20 V, the corresponding change in V_{BE} is from 0.73 V to 0.72 V. Therefore,

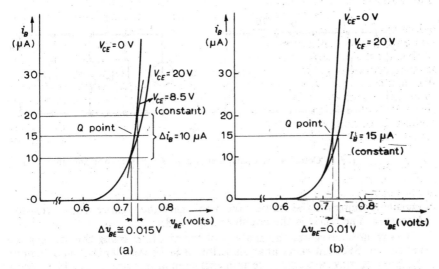

Fig. 8.16 Input characteristics of a transistor for the calculation of h_{ie} and h_{re}.

$$h_{re} = \frac{\Delta V_{BE}}{\Delta V_{CE}}\bigg|_{i_B\,=\,const.} = \frac{(0.73-0.72)\text{ V}}{(20-0)\text{ V}}\bigg|_{I_B\,=\,15\,\mu A}$$

$$= \frac{0.01}{20} = 5\times 10^{-4}$$

The value of the parameter h_{re} is very small. Change in V_{BE} corresponding to a large change in V_{CE} is quite small. Such a small change in the voltage V_{BE} is difficult to determine from the graph. The value of h_{re} obtained from the graphical method may not be very accurate.

8.4.3 Amplifier Analysis

The CE transistor amplifier circuit of Fig. 8.1b was redrawn in Fig. 8.2b from the point of view of its ac behaviour. In equivalent circuit method of analysing the amplifier, the transistor is also replaced by its ac equivalent. Once the complete ac equivalent circuit is available to us, we can determine the current gain, voltage gain, input impedance and output impedance of the amplifier.

Figure 8.17 shows the complete ac equivalent circuit of the transistor amplifier of Fig. 8.1b. On the output side, the two resistors R_C and R_o can be replaced by a single resistor R_{ac} such that

$$R_{ac} = R_C \parallel R_o = \frac{R_C R_o}{R_C + R_o}$$

On the input side, if the input voltage source v_i is assumed to be ideal (with zero internal resistance), the presence or the absence of the resistors R_1 and R_2 is immaterial. Whatever may be the values of R_1 and R_2 the

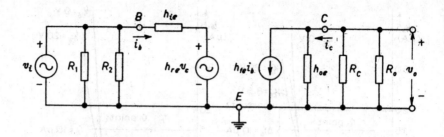

Fig. 8.17 Complete ac equivalent circuit of the transistor amplifier

current i_b remains the same. We can therefore ignore these resistors altogether. The result is the circuit in Fig. 8.18.

We can now carry out the analysis of the amplifier using this ac equivalent circuit. Such an exact analysis is found to be very lengthy and tedious. Much effort may be saved if certain approximations are made. The results obtained from such an *approximate analysis* will not be much different from those obtained by the exact analysis.

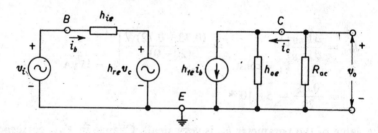

Fig. 8.18 AC equivalent circuit where R_C and R_b are replaced by R_{ac} and the biasing resistors R_1 and R_2 are omitted

For a typical transistor amplifier circuit R_{ac} is of the order of 1 kΩ. Whereas, $h_{oe} = 25 \ \mu S$, so that $1/h_{oe} = 40 \ k\Omega$. As $1/h_{oe}$ is in parallel with R_{ac}, the equivalent resistance is $(1/h_{oe}) \parallel R_{ac} \simeq R_{ac}$, because $(1/h_{oe})$ is about 40 times greater than R_{ac}. Therefore, for the approximate analysis, we may omit h_{oe} from the equivalent circuit. The value of h_{re} is typically 1×10^{-4}. This means that the feedback voltage $h_{re} v_c$ is very small and therefore can be omitted from the equivalent circuit. With these approximations, the ac equivalent circuit of the amplifier becomes the one shown in Fig. 8.19.

Current gain From Fig. 8.19, the output current i_c is seen to be the same as $h_{fe}i_b$. The current in the input is i_b. Therefore,

$$\text{Current gain, } A_i = \frac{\text{Output current}}{\text{Input current}} = \frac{i_c}{i_b}$$

$$= \frac{h_{fe}i_b}{i_b} = h_{fe} \tag{8.8}$$

or

$$A_i = \beta$$

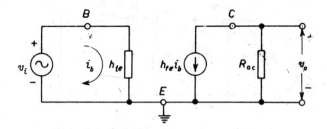

Fig. 8.19 AC equivalent circuit for approximate analysis of
the amplifier

Voltage gain The output voltage $v_o = -h_{fe}i_b R_{ac}$. Note the negative sign.
The flow of the output current is such that it makes the collector negative
with respect to the ground. The input voltage $v_i = i_b h_{ie}$. Therefore,

$$\text{Voltage gain, } A_v = \frac{\text{Output voltage}}{\text{Input voltage}} = \frac{-h_{fe}\, i_b\, R_{ac}}{i_b h_{ie}}$$

$$= \frac{-h_{fe} R_{ac}}{h_{ie}} \tag{8.9}$$

or, simply, $\qquad A_v = \frac{\beta R_{ac}}{r_i} \angle 180°$

The angle 180° indicates that output and input voltages have a phase difference of 180°.

Power gain The power gain of the amplifier is simply the product of
current gain and voltage gain. Thus, power gain,

$$A_p = A_i A_v \tag{8.10}$$

Input impedance The closed loop equation for input side is

$$v_i = i_b h_{ie}$$

Therefore,

$$\text{input impedance, } Z_{in} = \frac{\text{input voltage}}{\text{input current}} = \frac{v_i}{i_b}$$

$$= \frac{i_b h_{ie}}{i_b} = h_{ie} \tag{8.11}$$

or, simply, $\qquad Z_{in} = r_i$

In case, the biasing resistors R_1 and R_2 are to be considered, the input
section of the equivalent circuit is as shown in Fig. 8.20. Then, the input
impedance becomes

$$Z'_{in} = R_1 \parallel R_2 \parallel h_{ie} \simeq h_{ie} \quad \text{(since } h_{ie} \text{ is much smaller than } R_1 \text{ or } R_2)$$

Output impedance The output impedance of an amplifier is defined as the
ratio of the output voltage to the output current with the input v_i set at
zero. When we set $v_i = 0$, the current i_b is also zero, and consequently the
output current i_c also becomes zero even if we connect an external voltage

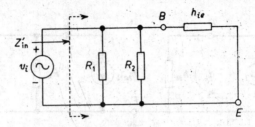

Fig. 8.20 Input impedance when biasing resistors
are considered

source at the output terminals. In other words, the output impedance Z_o will be infinite. If we take into account R_{ac}, the output impedance (Fig. 8.21) Z_o' is simply R_{ac}. Had we not neglected h_{oe}, the output impedance would have been $Z_o = 1/h_{oe}$ and the impedance Z_o' would have been

$$Z_o' = (1/h_{oe}) \parallel R_{ac} \simeq R_{ac} \tag{8.12}$$

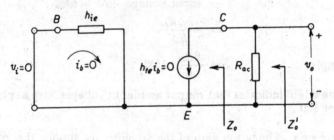

Fig. 8.21 Output impedance of an amplifier

Example 8.2 Figure 8.22 shows a common-emitter amplifier using fixed bias. Draw its ac equivalent circuit. Calculate (a) input impedance; (b) voltage gain; and (c) current gain. Assume $h_{ie} = r_i = 2$ kΩ, and $h_{fe} = \beta = 100$.

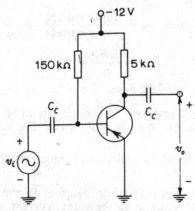

Fig. 8.22 Common-emitter amplifier

Solution:

The ac equivalent circuit of the given amplifier circuit is shown in Fig. 8.23. It is assumed that the capacitors offer a short-circuit at the signal frequency. As is clear in the equivalent circuit, the base resistor (150 kΩ) is much greater than the input resistance of the transistor (2 kΩ). Under this situation, the input impedance of the circuit may be taken as 2 kΩ. (If exact calculations are made, this impedance is equal to 150 kΩ ∥ 2 kΩ=1.973 kΩ.)

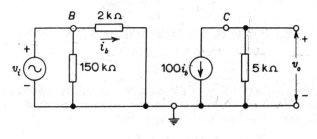

Fig. 8.23

For calculation of voltage gain we make use of the formula

$$Av = \frac{\beta R_{ac}}{r_{in}} \angle 180°$$

Here β is 100, r_{in} is 2 kΩ and R_{ac} is 5 kΩ. Substituting these values, we get

$$Av = \frac{100 \times 5 \times 10^3}{2 \times 10^3} \angle 180° = \mathbf{250} \angle \mathbf{180°}$$

Output current is equal to 100 i_b, whereas input current is i_b. Therefore, current gain is **100**.

Example 8.3 In the single stage amplifier circuit shown in Fig. 8.24, an *NPN* transistor is used. The parameters of this transistor are $\beta_{ac} = 150$ and $r_{in} = 2$ kΩ. Calculate (*a*) voltage gain; (*b*) input impedance; and (*c*) Q point. (neglect V_{BE})

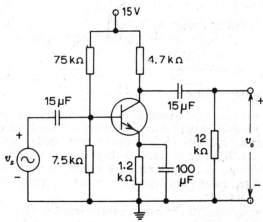

Fig. 8.24

Solution:

In the amplifier circuit, the ac load resistance is

$$R_{ac} = 4.7 \text{ k}\Omega \parallel 12 \text{ k}\Omega = \frac{4.7 \times 10^3 \times 12 \times 10^3}{(4.7 + 12) \times 10^3}$$

$$= 3.38 \text{ k}\Omega$$

(a) The voltage gain is given as

$$A_v = \frac{\beta_{ac} R_{ac}}{r_{in}} \angle 180°$$

Here, $\beta_{ac} = 150$; $r_{in} = 2 \text{ k}\Omega$ and $R_{ac} = 3.38 \text{ k}\Omega$

$$\therefore A_v = \frac{150 \times 3.38 \times 10^3}{2 \times 10^3} = 253.5$$

(b) As far as ac operation of the amplifier circuit is concerned, the resistors 75 kΩ and 7.5 kΩ are both connected between the base and ground. Hence, the input impedance of the amplifier is

$$Z_{in} = 75 \text{ k}\Omega \parallel 7.5 \text{ k}\Omega \parallel r_{in}$$

$$= 75 \text{ k}\Omega \parallel 7.5 \text{ k}\Omega \parallel 2 \text{ k}\Omega \simeq 7.5 \text{ k}\Omega \parallel 2 \text{ k}\Omega$$

$$= 1.5 \text{ k}\Omega$$

(c) The resistors 75 kΩ and 7.5 kΩ make a potential divider. The voltage at the base is given as

$$V_B = V_{CC} \frac{R_2}{R_1 + R_2} = \frac{15 \times 7.5 \times 10^3}{75 \times 10^3 + 7.5 \times 10^3} = \frac{15}{11}$$

$$= 1.36 \text{ V}$$

Since, $V_{BE} = 0$, the voltage at the emitter is same as that at the base, i.e.

$$V_E = V_B = 1.36 \text{ V}$$

Therefore, emitter current is

$$I_E = \frac{V_E}{R_E} = \frac{1.36}{1.2 \times 10^3} = 1.13 \text{ mA}$$

The collector-to-emitter voltage is

$$V_{CE} = V_{CC} - (R_C + R_E)I_E \qquad \text{[since } I_C \simeq I_E]$$

$$= 15 - (4.7 + 1.2) \times 10^3 \times 1.13 \times 10^{-3}$$

$$= 8.33 \text{ V}$$

8.5 TRIODE AMPLIFIER

Before the transistor was invented in 1948, the triodes with their sister devices were extensively used for amplification of weak electrical signals. Although most of the electronic circuits now make use of solid-state devices, valves are still used in very high-power amplifiers and in some microwave amplifiers. Though valves have a shrinking future, they will still be used till a proper solid state device comes up to replace them from their limited region of usage.

You have already studied the triode parameters and their characteristics. Now, we shall discuss the basic amplifier circuit using a triode valve.

Figure 8.25 represents a simple triode amplifier circuit. In this circuit the dc supply V_{PP} provides a high dc voltage for the plate, through the load resistor R_L. The low voltage supply V_{GG} provides the required negative bias to the grid through the grid-leak resistor R_G. The grid-leak resistor R_G has another function also. It provides a leakage path for those electrons which got stuck to the grid wire while going to the plate from the cathode. In the absence of this resistor, the negative charge of the electrons collected at the grid would have adversely affected the operation of the circuit. The load resistor R_L has two functions. Firstly, it helps in selecting the proper plate voltage (dc). Secondly, the amplifier output voltage v_o is made available only because of the presence of R_L. The voltage v_s is a weak ac signal which is to be amplified. This voltage gets applied to the grid through the coupling capacitor C_C. The output voltage v_o is available at the output points through the coupling capacitor connected to the output side. These coupling capacitors serve the same purpose as that in a transistor amplifier circuit.

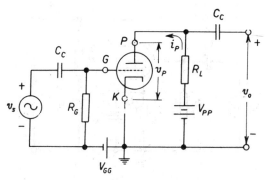

Fig. 8.25 Triode amplifier

8.5.1 Amplification of Signal by Triode Amplifier

Initially, when the ac signal is not applied to the circuit, and the dc sources are connected, some dc plate current flows in the circuit. As a result, there is a particular plate voltage. The grid voltage is decided by the value of supply V_{GG}. The plate voltage (v_P), plate current (i_P) and HT voltage (V_{PP}) are related by the equation

$$v_P = V_{PP} - i_P R_L \tag{8.13}$$

When the ac signal is applied at the grid, its voltage changes. During the positive half of the input signal, the grid voltage becomes less negative, while during the negative half, the grid voltage becomes more negative. Accordingly, the plate current changes. This changing plate current passes through the load resistance R_L. Because of this, the voltage drop across R_L changes. Therefore, an ac voltage is available at the output. The ac voltage at the output is much larger in magnitude than the ac signal at the input. This is how the triode circuit works as an amplifier.

8.5.2 Self-Biasing Arrangement

In the amplifier circuit of Fig. 8.25, two dc supplies V_{PP} and V_{GG} are used. The sole purpose of the supply V_{GG} is to provide a negative voltage to the grid with respect to cathode. In practical situations, it is always desirable to have a single dc source only. The purpose served by V_{GG} can be achieved by the biasing arrangement shown in Fig. 8.26. In this diagram, a resistor R_K and the capacitor C_K are connected between the cathode and ground. When the dc plate current I_P passes through this resistor, a dc voltage drop $(I_P R_K)$ develops across it. The polarity of this voltage drop is such as to make the cathode positive with respect to the ground. Since resistor R_G is connected between the ground and grid and no current flows through it, the potential of the grid and the ground is the same. This means that the voltage drop across R_K makes the grid negative with respect to the cathode. The capacitor C_K offers very low impedance to the ac signal, thus preventing the ac current to pass through R_K. This is why the capacitor C_K is called a *bypass capacitor*. In this way, the negative feedback of the signal is avoided, and the gain of the amplifier does not decrease.

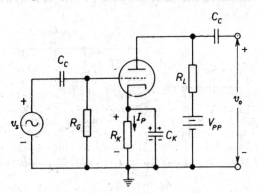

Fig. 8.26 Triode amplifier circuit with self-biasing arrangement

The voltage gain of the amplifier can be determined by the following two methods:

(i) Graphical method
(ii) Equivalent circuit method

8.5.3 Calculation of Gain by Graphical Method

In the graphical method of calculating the voltage gain of a given triode amplifier, the plate characteristics of the triode is used. Assume that the triode ECC83 is used in the circuit in Fig. 8.27. The static plate characteristics of this tube are shown in Fig. 8.28. As in case of a transistor amplifier, to locate the Q point, it is necessary to draw the dc load line on the plate characteristics. Writing the KVL equation for the output section, we get

$$V_{PP} - i_P R_L - v_P - i_P R_K = 0$$

or
$$i_P = -\frac{v_P}{(R_L + R_K)} + \frac{V_{PP}}{(R_L + R_K)} \qquad (8.14)$$

In practical circuits, the value of R_K is much smaller than that of R_L. We can neglect R_K and write Eq. 8.14 as

$$i_P = -\frac{v_P}{R_L} + \frac{V_{PP}}{R_L} \qquad (8.15)$$

This is the *equation of the dc load line.* It represents a straight line having a slope equal to $-1/R_L$ and an intercept on the vertical axis equal to V_{PP}/R_L. The dc load line can be drawn on the plate characteristics by joining any two points which satisfy Eq. 8.15. Two such convenient points are:

When $i_P = 0$; $v_P = V_{PP}$

When $v_P = 0$; $i_P = V_{PP}/R_L$

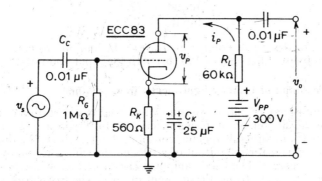

Fig. 8.27 A practical triode amplifier circuit

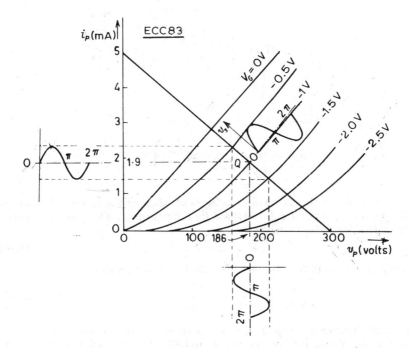

Fig. 8.28 Determination of voltage gain of a triode amplifier by graphical method

In the circuit of Fig. 8.27, V_{PP} is 300 V and R_L is 60 kΩ. Therefore, the two points are (300 V, 0) and (0, 5 mA). These points are located and joined by a straight line on the plate characteristics. If the grid-bias voltage in the circuit is −1 V, the Q point is the point of intersection of the dc load line, and the characteristic curve for $V_G = -1$ V. At this Q point, I_P is 1.9 mA and V_P is 186 V. Suppose a sinusoidal voltage of 0.5 V (peak) is applied to the input of the amplifier, the corresponding variation in plate current and in plate voltage are shown in Fig. 8.28. The peak value of the plate-voltage swing can be seen to be $(210 - 155)/2 = 27.5$ V. Therefore, the voltage gain of the amplifier is

$$A_v = \frac{\text{Output}}{\text{Input}} = \frac{27.5}{0.5} = 55$$

Note that, as the input voltage increases, the output voltage decreases. There is a phase difference of 180° between the input and output.

8.5.4 Calculation of Gain by Equivalent Circuit Method

For a small-signal amplifier, we can assume that the operation of the triode is in the linear region of its characteristics. In the linear region, the tube parameters (μ, r_p and g_m) remain constant. Under these conditions, the equivalent circuit for the output of a triode tube can be found by applying Thevenin's theorem. The triode amplifier circuit of Fig. 8.25 looks like the circuit in Fig. 8.29 from the ac point of view. The capacitors and the dc supply are shown shorted.

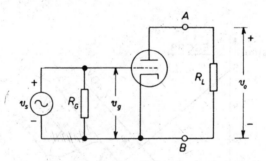

Fig. 8.29 A basic triode amplifier circuit from the ac point of view

We now obtain the Thevenin's equivalent between the points A and B. The open-circuit voltage is obtained by removing R_L. The voltage between terminals A and B is then

$$V_{TH} = \mu v_g \qquad (8.16)$$

where μ is the amplification factor of the triode and is defined as

$$\mu = \frac{\Delta v_P}{\Delta v_G}\bigg|_{i_P\ =\ \text{const.}} = \frac{v_p}{v_g}\bigg|_{\text{output open}} \qquad (8.17)$$

To obtain Thevenin's impedance, we short-circuit the voltage source v_s and find the impedance between points A and B. This impedance is r_p, the dynamic plate resistance of the triode. Thus

$$R_{TH} = r_p = \frac{\Delta v_p}{\Delta i_p}\bigg|_{v_G = \text{const.}} = \frac{v_p}{i_p}\bigg|_{\text{input shorted}} \tag{8.18}$$

The Thevenin's equivalent of the circuit in Fig. 8.29 can be drawn as in Fig. 8.30. Note the polarity of the voltage source μv_g with reference to voltage v_g. At an instant when the grid voltage v_g is positive, the plate voltage is negative.

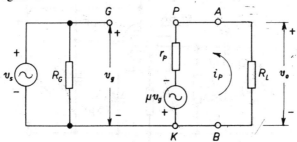

Fig. 8.30 AC equivalent circuit of a triode amplifier

We can now find the voltage gain of the amplifier. From Fig. 8.30, the KVL equation for the output loop is

$$\mu v_g = i_p(r_p + R_L)$$

or

$$i_p = \frac{\mu v_g}{r_p + R_L} \tag{8.19}$$

Here, $v_g = v_s$. The output voltage v_o is given as

$$v_o = -i_p R_L = -\frac{\mu v_s}{r_p + R_L} R_L$$

Therefore, the gain of the amplifier is

$$A = \frac{v_o}{v_s} = -\frac{\mu R_L}{r_p + R_L} = \frac{\mu R_L}{r_p + R_L} \angle 180° \tag{8.20}$$

Example 8.4 A small signal triode amplifier uses a load resistor of 10 kΩ and a triode tube having $\mu = 20$ and $r_p = 10$ kΩ. Find its voltage gain.

Solution: Voltage gain of the amplifier is

$$A = \frac{\mu R_L}{r_p + R_L} \angle 180°$$

Here, $\mu = 20$; $r_p = 10$ kΩ $= 10 \times 10^3$ Ω; $R_L = 10$ kΩ $= 10 \times 10^3$ Ω.

$$\therefore \quad A = \frac{20 \times 10 \times 10^3}{10 \times 10^3 + 10 \times 10^3} \angle 180° = 10 \angle 180°$$

8.6 A PRACTICAL PENTODE AMPLIFIER

Because of inter-electrode capacitances, a triode is not suitable for operation at high frequencies. We then replace the triode by a pentode tube. A typical pentode amplifier circuit is shown in Fig. 8.31.

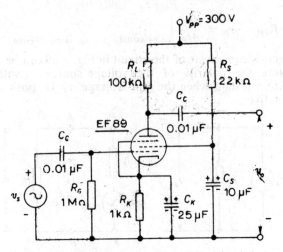

Fig. 8.31 Typical pentode amplifier circuit

The resistor R_S is known as the screen resistor and it helps in fixing particular dc voltage at the screen. The capacitor C_S bypasses the resistor R_S for ac. The screen grid remains effectively at ground potential from the ac point of view. The $R_K C_K$ combination provides self-bias as was the case in the triode amplifier circuit. The resistor R_G is the grid leak resistor. The ac signal to be amplified (v_s in this case) is fed to the grid through the coupling capacitor C_C. The amplified voltage is available at the output. The voltage gain of this amplifier can be calculated by either the graphical method or by the equivalent circuit method. The ac equivalent circuit of a pentode is exactly same as that of a triode. The screen grid and the suppressor grid do not affect the ac equivalent circuit. The plate resistance (r_p) of a pentode being very high (100 to 500 kΩ), it is generally represented by a current-source equivalent circuit. Such an equivalent circuit is shown in Fig. 8.32.

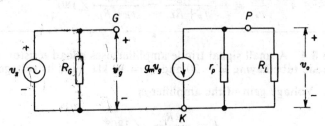

Fig. 8.32 Current-source ac equivalent circuit of a pentode amplifier

When this circuit is compared with voltage-source equivalent circuit of Fig. 8.30, it is observed that the resistance r_p is connected in parallel with the current source whose value is

$$\frac{\mu v_g}{r_p} = g_m v_g$$

The current $g_m v_g$ is divided into two parallel branches. The current i_L flowing through the load resistance R_L is

$$i_L = g_m v_g \frac{r_p}{R_L + r_p} = \frac{\mu v_g}{R_L + r_p}$$

Therefore, the output voltage is

$$v_o = -i_L R_L = -\frac{\mu v_g}{R_L + r_p} R_L$$

Here, v_g being same as v_s, the voltage gain of the amplifier is

$$A = \frac{v_o}{v_s} = -\frac{\mu R_L}{R_L + r_p} = \frac{\mu R_L}{R_L + r_p} \angle 180° \qquad (8.21)$$

This expression is exactly the same as that of the gain of a triode amplifier given in Eq. 8.20. We can also write Eq. 8.21 in a different way as

$$A = -\frac{\mu/r_p}{1 + R_L/r_p} R_L = -\frac{g_m R_L}{1 + R_L/r_p}$$

Usually r_p is very high compared to R_L so that

$$A \simeq -g_m R_L = g_m R_L \angle 180° \qquad (8.22)$$

8.7 AMPLIFIER USING FIELD-EFFECT TRANSISTOR (FET)

Figure 8.33 shows an amplifier using an N-channel field-effect transistor. Unlike a bipolar transistor, the input impedance of an FET is very high and it works like a vacuum pentode. The combination $R_S C_S$ provides self bias. R_G provides dc path for reverse-biasing of gate-source junction. R_D is the load resistance of the amplifier. Typical values of the components are shown in the figure.

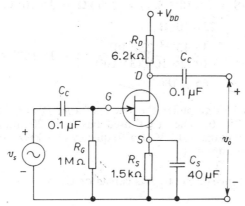

Fig. 8.33 Amplifier using FET

8.7.1 FET AC Equivalent Circuit

The ac equivalent circuit of an FET is given in Fig. 8.34. Like a vacuum pentode, an FET also has three parameters: (i) dynamic drain resistance r_d; (ii) amplification factor μ; and (iii) transconductance g_m. The value of r_d is very high ($\simeq 100$ kΩ), but the g_m has a very low value ($\simeq 200$ μS). The

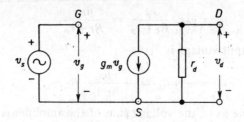

Fig. 8.34 AC equivalent circuit of FET

amplification factor is also low ($\simeq$ 20). Therefore, the FET amplifier has very low voltage gain. Since dynamic drain resistance r_d of an FET is very high, it is better represented by an equivalent current source rather than an equivalent voltage source. In this respect, the FET works more like a pentode tube. We can find the expression for the voltage gain of an FET amplifier as

$$A_v = g_m R_L \angle 180° \qquad (8.23)$$

Example 8.5 A single-stage amplifier uses a vacuum pentode whose $g_m = 3000 \; \mu S$ and $r_p = 300$ kΩ. Find the gain of the amplifier if the load resistance is 22 kΩ.

Solution: The voltage gain of a pentode amplifier is

$$A = -\frac{g_m R_L}{1 + R_L/r_p}$$

Here, $g_m = 3000 \; \mu S = 3 \times 10^{-3}$ S; $r_p = 300$ kΩ $= 3 \times 10^5$ Ω;

$R_L = 22$ kΩ $= 22 \times 10^3$ Ω

$\therefore \qquad A = -\dfrac{3 \times 10^{-3} \times 22 \times 10^3}{1 + (22 \times 10^3)/(3 \times 10^5)} \simeq -3 \times 10^{-3} \times 22 \times 10^3$

$= 66 \angle 180°$

Example 8.6 In an FET amplifier, the load resistance $R_L = 12$ kΩ; $R_G = 1$ MΩ, $R_S = 1$ kΩ, $C_S = 25 \; \mu F$. The FET used has $\mu = 20$ and $r_d = 100$ kΩ. If the input signal voltage is 0.1 V at a frequency of 1 kHz, find the output signal voltage of the amplifier.

Solution: The reactance of the bypass capacitor C_S is

$$X_{CS} = \frac{1}{2\pi f C_S} = \frac{1}{2 \times 3.141 \times 1 \times 10^3 \times 25 \times 10^{-6}}$$

$= 6.3 \; \Omega$

This is much smaller than $R_S = 1$ kΩ. We can assume R_S to be completely bypassed. The magnitude of the voltage gain of the amplifier is

$$A = \frac{\mu R_L}{R_L + r_d} = \frac{20 \times 12 \times 10^3}{12 \times 10^3 + 100 \times 10^3} = 2.14$$

Therefore the output signal voltage is

$$v_o = Av_i = 2.14 \times 0.1 = 0.214 \text{ V}$$

8.8 COMPARISON BETWEEN TRANSISTOR AND TUBE AMPLIFIERS

Decidedly, as an active device, a transistor is much better than a vacuum tube. However, in some respects, a vacuum-tube amplifier is better than a transistor amplifier. For example, input impedance of a vacuum-tube amplifier is much higher ($\simeq 1$ MΩ) than that of a transistor amplifier ($\simeq 1$ kΩ). Because of the high input impedance of vacuum-tube amplifier, the input signal source is not loaded. This is certainly a great advantage. However, the recently developed field-effect transistor (FET) has removed this drawback.

The transistor amplifier has a much higher voltage gain than a vacuum-tube triode amplifier. But when a pentode is used, the voltage gain becomes high. On the other hand, when a field-effect transistor is used, although input impedance is increased, the voltage gain is much reduced. Of course, the voltage gain can then be increased by adding more stages to the amplifier.

REVIEW QUESTIONS

8.1 Draw the circuit diagram of a single-stage transistor amplifier. State the function(s) of each component used in this circuit.

8.2 Explain how amplified voltage becomes available at the output points of a single-stage amplifier.

8.3 Explain how phase reversal of the signal takes place when it is amplified by a single-stage voltage amplifier.

8.4 Draw an ac equivalent circuit of a common-emitter transistor amplifier. Derive the following expression using this equivalent circuit:

$$A_v = \frac{\beta R_{ac}}{r_{in}} \angle 180°$$

Explain with the help of equivalent circuit, the phase reversal of the signal.

8.5 State the name of the four h parameters for a transistor in CE configuration. Define them. Write down the typical values of these parameters.

8.6 A step-up transformer can increase the voltage level of an ac signal. This can also be achieved by a transistor voltage amplifier. Explain the difference in the two processes.

8.7 Explain the following terms in brief (say, within 5 lines), in connection with a transistor voltage amplifier:

(a) Input impedance (Z)
(b) Output impedance (Z_o)
(c) Voltage gain
(d) Current gain
(e) Power gain

8.8 State what will happen to the voltage gain of an amplifier if the bypass capacitor (C_r) is open circuited.

8.9 Explain the difference between dc load line and ac load line. Why is it necessary to draw ac load line for calculating the voltage gain of an amplifier ?

8.10 Why have you to draw dc load line while you calculate the gain from an ac load line ?

8.11 Using the transistor characteristics and the load line, explain the phase reversal of the signal.

8.12 Draw a triode amplifier circuit. Explain how biasing is achieved in this circuit.

8.13 Explain the function of grid-leak resistor R_G in a triode amplifier circuit.

8.14 Make use of equivalent circuit of a triode amplifier and derive the following result:

$$A_v = \frac{\mu R_L}{r_p + R_L} \angle 180°$$

Explain the phase relationship between the output and input voltages.

8.15 Explain the limitation of the triode in amplifying very high frequency signals (30 MHz onwards).

8.16 Draw the circuit diagram of a pentode voltage amplifier. What additional components are added in this circuit when compared with a triode amplifier ? Explain the function of screen-grid resistor and screen-grid bypass capacitor.

8.17 Compare voltage gain of a pentode amplifier with that of a triode amplifier.

8.18 Compare the performance of a triode amplifier with that of a transistor amplifier with regards to (a) input impedance, (b) output impedance, and (c) voltage gain.

8.19 State the advantages of transistorized amplifiers over valve-type amplifiers.

8.20 State the main merit of valve amplifiers over transistorized amplifiers.

OBJECTIVE-TYPE QUESTIONS

I. A number of statements are given below. Tick ($\sqrt{}$) the statements you think to be correct.

1. The voltage gain of a transistor amplifier is a constant quantity and is independent of load resistance.

2. The voltage gain of a transistor amplifier increases as ac load resistance increases.

3. The voltage gain of a transistor amplifier in CE mode is always less than unity.

4. The input impedance of a transistor amplifier in CE configuration is very high (say 5 MΩ).

5. The input impedance of a transistor amplifier in CE mode is low (say 1.5 kΩ).

6. The input impedance of CE amplifier is extremely low (say 10 Ω).

7. The output impedance of a transistor amplifier is independent of the transistor configuration.

8. Valve-type amplifiers are lighter in weight than transistor amplifiers.

9. For same audio output, a transistor amplifier consumes more power than a valve-type amplifier.

10. A normal transistor amplifier gives higher voltage gain than a triode amplifier.

11. Life of a valve-type amplifier is longer compared to that of a transistor amplifier.

12. The voltage gain of a transistor amplifier decreases when the emitter-bypass capacitor C_E is present in the circuit.

13. The voltage gain of a transistor amplifier using potential-divider biasing arrangement with emitter-bypass capacitor depends upon the value of R_E.

14. The phase reversal between output and input takes place only for voltage waves and not for current waves, in a transistor amplifier in CE configuration.

II. Here are some incomplete statements. Some alternatives are suggested below each Tick ($\sqrt{}$) the alternative that completes the statement correctly.

1. In an amplifier, the coupling capacitors are used
   (a) to control the output
   (b) to limit the bandwidth
   (c) to match the impedances
   (d) to prevent dc mixing with input or output.

2. The main function of R_x (resistor connected between cathode and ground) in a triode amplifier is to
   (a) control plate current
   (b) increase bandwidth
   (c) prevent the tube from being directly grounded
   (d) provide predetermined bias voltage to the control grid.

3. If the power gain of an amplifier is X and its voltage gain is Y, then its current gain will be
   (a) X/Y
   (b) Y/X
   (c) $X \cdot Y$
   (d) $X + Y$

4. An amplifier circuit of voltage gain 100, gives 2 V output. The value of input voltage is
   (a) 200 V
   (b) 50 V
   (c) 20 mV
   (d) 2 mV

5. The input signal to an amplifier having a gain of 200 is given as $0.5 \cos (313\ t)$. The output signal may be represented by
   (a) $100 \cos (313\ t + 90°)$
   (b) $10 \cos (403\ t)$
   (c) $100 \cos (313\ t + 180°)$
   (d) $200 \cos (493\ t)$

6. The triode amplifier is not suitable for very high frequency signals because
   (a) it is costlier to use
   (b) tremendous distortion takes place
   (c) it is not suitable for musical sounds which are generally transmitted over a radio broadcasting system
   (d) inter-electrode capacitances reduce gain badly and practically amplifying action is stopped at high frequencies.

III. For a pentode valve, select the suitable group of words given in column B, matching with each of those given in column A.

Column A	Column B
1. The control grid_____	(a) reduces effect of inter-electrode capacitance
2. The suppressor grid_____	(b) helps cathode to remain at ground potential
3. The screen grid_____	(c) controls plate voltage in accordance with emission
	(d) controls plate current in accordance with input voltage
	(e) controls plate current in accordance with the load impedance
	(f) prevents secondary emitted electrons leaving plate.

V. Fill in the blanks in the following sentences using the most appropriate alternative from those given in bracket with each.

1. For a good voltage amplifier, its input impedance should be_____compared to the resistance of the source. (high/low/inductive/capacitive)

2. The coupling capacitors mainly affect_____cut-off frequency of an amplifier. (lower/upper/single/double)

3. In a *PNP* transistor, the emitter resistor (R_E) keeps the emitter at a _____ voltage compared to its ground potential. (positive/negative/zero)

4. The output current waveform in CE ampliffer is _____with input current wave. (in phase/out of phase by 180°/out of phase by 90°)

5. The output voltage waveform of CE amplifier is_____with its input voltage wave. (in phase/out of phase by 180°/ leading by 90°/lagging by 90°)

Aes. I. 2, 5, 10, 14.
II. 1. *d*, 2. *d*, 3. *a*, 4. *c*, 5. *c*, 6. *d*.
III. 1. *d*, 2. *f*, 3. *a*.
IV. 1. high, 2. lower, 3. negative, 4. in phase, 5. out of phase by 180°

TUTORIAL SHEET 8.1

1. The amplifier circuit given in Fig. T.8.1.1 uses a transistor AC125. The collector characteristics of this transistor are given in Fig. T.8.1.2. Locate the Q point if the quiescent base current is 15 μA. Draw the ac load line on the collector characteristics to determine the output voltage, if input base current swing is 5 μA (peak) sine wave. If the dynamic input resistance of the transistor is 800 Ω, calculate the voltage gain.

[**Ans.** $V_o = 0.95$ V (p–p); $A_V = 119$]

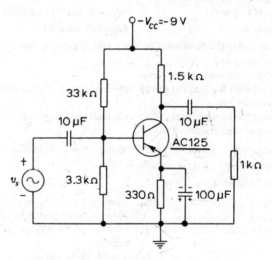

Fig. T.8.1.1

2. Figure T.8.1.3 shows the collector characteristics of a transistor used in an amplifier circuit. If the emitter resistor R_E is 200 Ω, determine the value of the collector resistance with the help of dc load line, drawn on the given characteristics. If the biasing resistors are such that the Q point base current is 40 μA, determine the collector current and collector-to-emitter voltage.

[**Ans.** $R_C = 2.5$ kΩ; $I_{CO} = 5.5$ mA; $V_{CE} = 10$ V]

3. If we connect a resistor across the output of the amplifier in Q. 2 such that the effective ac load resistance becomes 1 kΩ, draw the ac load line and calculate the voltage gain of the amplifier circuit assuming r_{in} to be 1 kΩ. Also calculate the value of the load resistor connected at the output points.

[**Ans.** $A_V = 150$; $R_L = 1.67$ kΩ]

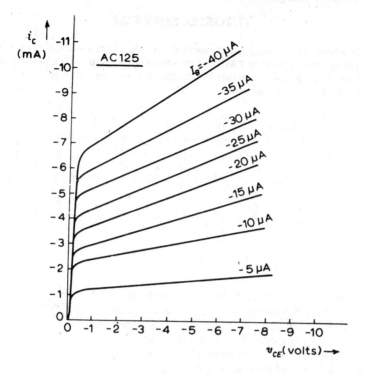

Fig. T.8.1.2

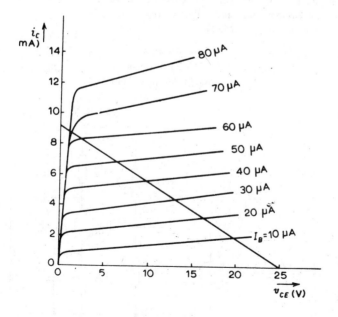

Fig. T.8.1.3

TUTORIAL SHEET 8.2

1. It is desired that the coupling capacitor C_C of Fig. T.8.2.1 should couple all frequencies from 500 Hz to 1 MHz of the source to the output. Calculate the value of the coupling capacitor such that its impedance is not more than 10 % of the load impedance, at any frequency. [Ans. 0.32 μF]

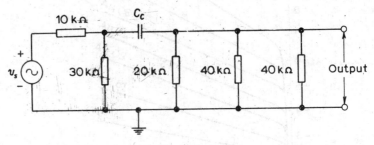

Fig. T.8.2.1

2. The point A in Fig. T.8.2.2 is desired to be effectively at the ground potential for the frequency range 40 Hz to 8 kHz. Calculate suitable value of bypass capacitor C_E. Assume that for effective bypassing, the impedance of the capacitor C_E should not be more than 10 % of the resistance R_E.

[Ans. 10 μF]

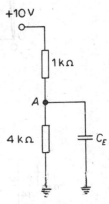

Fig. T.8.2.2

3. Work out the following quantities for the circuit given in Fig. T.8.2.3: (a) ac emitter current; (b) ac voltages at emitter, base and collector; (c) voltage gain. Assume h_{ie} or $r_{in} = 250$ Ω.

[Ans. (a) $i_{e(peak)} = 1.02$ mA $\simeq 1$ mA·
(b) $v_e = 0$ V, $v_b = 5$ mV (peak);
$v_c = 1$ V (peak); (c) $A_V \simeq 200$]

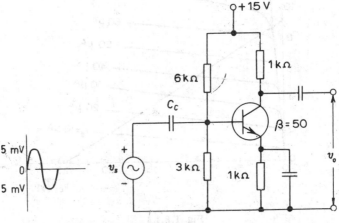

Fig. T.8.2.3

4. In the single-stage amplifier circuit in Fig. T.8.2.4, transistor AC126 is used. Draw its ac equivalent circuit and calculate the voltage gain (v_o/v_s) with and without R_L. Assume the following transistor parameters:

$$h_{fe} \text{ or } \beta_{ac} = 150; \quad h_{ie} \text{ or } r_{in} = 1.5 \text{ k}\Omega$$

[Ans. 50, 100]

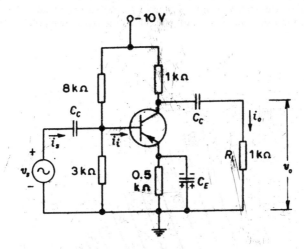

Fig. T.8.2.4

5. In the amplifier circuit of Fig. T.8.2.4, calculate
   (a) v_o/v_s, if the source resistance is 600 Ω.
   (b) i_o/i_i, where i_o is the current through the output resistance R_L and i_i is the input current as shown.
   (c) i_o/i_s, where i_s is the ac current supplied by the input source.

[Ans. (a) 29.85; (b) 75; (c) 44.44]

TUTORIAL SHEET 8.3

1. The Fig. T.8.3.1 shows a vacuum tube amplifier circuit. Fig. T.8.3.2 gives the characteristics of the tube used. The tube is biased at $V_{GK} = -12$ V. Draw the dc

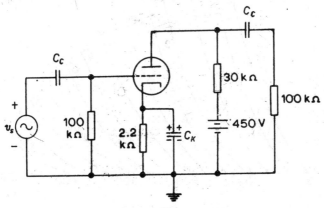

Fig. T.8.3.1

load line and locate Q point. Also draw the ac load line and determine the voltage gain. Assume an input swing of 4 V (peak).

[**Ans.** $I_P = 5$ mA; $V_P = 295$ V; $A_V = 12.5$]

2. An amplifier uses 6J5 triode tube whose characteristics are given in Fig. T.8.3.3. The dc load line intersects the axes at points (430 V, 0) and (0, 10 mA). Q point coordinates are (300 V, 3 mA). Value of R_{ac} (effective ac load) is 25 kΩ. Draw ac load line and calculate the voltage gain taking the input swing of 8 V (p−p).

[**Ans.** 13.1]

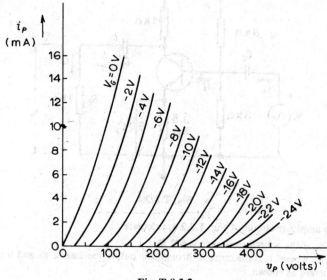

Fig. T.8.3.2

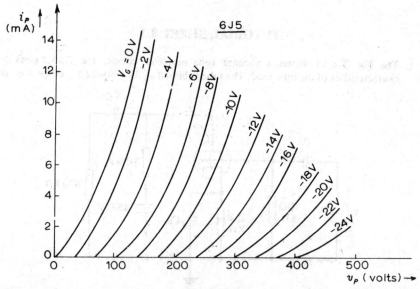

Fig. T.8.3.3

3. A pentode amplifier uses a tube whose characteristics are given in Fig. T.8.3.4. The two points on the dc load line are (200 V, 0) and (0, 10 mA). The control grid is biased at −1 V. Locate the Q point. If a signal of 2 V (p−p) is applied to the input, calculate voltage gain for an effective ac load resistance of 10 kΩ. [**Ans.** 45]

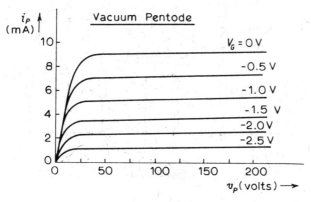

Fig. T.8.3.4

EXPERIMENTAL EXERCISE 8.1

TITLE: Single-stage transistor amplifier.

OBJECTIVES: To,

1. trace the circuit diagram of single-stage transistor amplifier;
2. measure the Q point collector current and collector-to-emitter voltage;
3. measure the maximum signal which can be amplified by the amplifier without having clipped output;
4. measure the voltage gain of the amplifier at 1 kHz;
5. measure the voltage gain of the amplifier for different values of load resistance.

APPARATUS REQUIRED: Amplifier circuit, electronic multimeter, ac milli-voltmeter, CRO.

CIRCUIT DIAGRAM: As given in Fig. E.8.1.1 Typical values of the components are also given.

BRIEF THEORY: In the amplifier circuit shown in the figure, the resistors R_1, R_2 and R_E fix a certain Q point. The resistor R_E stabilizes it against temperature variations. The capacitor C_E bypasses the resistor R_E for the ac signal. As it offers very low impedance path for ac, the emitter terminal is almost at ground potential. When the ac signal is applied to the base, the base-emitter voltage changes, because of which the base-current changes. Since collector current depends upon the base current, the collector current also changes. When this changing collector current passes through the load resistance R_C, an ac voltage is produced at the output. As the output voltage is much more than the input voltage, the circuit works as an amplifier circuit. The voltage gain of this amplifier is given by the formula

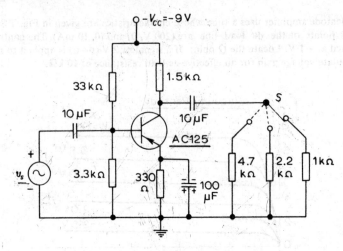

Fig. E.8.1.1

$$A_v = \frac{\beta R_{ac}}{r_{in}} \angle 180°$$

where r_{in} is the dynamic input resistance, β is the current amplification factor, and R_{ac} is the ac load resistance in the circuit.

PROCEDURE:

1. Look at the circuit and draw it accordingly in your notebook. With the help of the colour code, find the values of every resistor. Note the values of capacitors also.

2. Connect the dc supply V_{CC} (either from the regulated transistorized power supply or from IC power supply). Measure the dc voltage supplied.

3. For the measurement of quiescent collector current, measure the voltage of collector terminal with reference to ground (V_C). Calculate collector current from the formula

$$I_C = \frac{V_{CC} - V_C}{R_C}$$

Also measure V_{CE}, i.e. dc voltage between the collector and the emitter.

4. Make sure that the transistor is operating in the active region by noting that V_{CE} is about half of V_{CC}. Feed ac signal at 1 kHz at the input of the amplifier. Observe the amplified output on the CRO. Increase the input signal till the output waveshape starts getting distorted. Measure this input signal. This is the maximum signal that the amplifier can amplify without giving distorted output.

5. Now feed an ac signal that is less than the maximum signal handling capacity of the amplifier. Fix the frequency of the input signal at 1 kHz. Note the input and output voltages and calculate the voltage gain.

6. Connect different load resistors and find the voltage gain of the amplifier for each.

OBSERVATIONS:

1. *Q point of the amplifier*:

V_{cc}	V_c	$V_{cc} - V_c$	$I_c = \dfrac{V_{cc} - V_c}{R_c}$	V_{CE}

2. Maximum signal that can be handled by the amplifier without introducing distortion = _____ mV. Frequency of the input signal = 1 kHz.

3. *Voltage gain of the amplifier*:

S. no.	Load resistor	Input voltage	Output voltage	Gain $= \dfrac{v_o}{v_i}$
1.				
2.				
3.				

RESULT:

1. Q point of the transistor is
$$Ic = \text{_____ mA,} \quad V_{CE} = \text{_____ V}$$
Since $V_{CE} \simeq \frac{1}{2}V_{cc}$, the transistor is biased in active region.

2. Maximum signal handling capacity of the amplifier (at 1 kHz) = _____ mV.

3. The voltage gain reduces as the load resistance decreases.

EXPERIMENTAL EXERCISE 8.2

TITLE: Triode amplifier circuit.

OBJECTIVES: To,

1. trace the circuit of the given amplifier;
2. measure the operating point (i.e. the plate current and plate voltage);
3. measure the voltage gain of the amplifier in mid-frequency region (1 kHz);
4. plot the frequency response curve and thus determine the band-width;
5. verify the value of maximum voltage gain and bandwidth theoretically.

APPARATUS REQUIRED: HT supply, signal generator, electronic multimeter, CRO.

CIRCUIT DIAGRAM: As shown in Fig. E.8.2.1. Typical values of the components are also given.

BRIEF THEORY: In the triode amplifier, shown in Fig. E.8.2.1, the $R_K C_K$ combination provides grid bias. The grid-leak resistor R_G provides a dc path between the grid and ground. From ac point of view, the cathode is grounded through the capacitor C_K. The load resistor R_L has two functions. Firstly, it helps in fixing a particular dc voltage at the plate, and secondly it acts as load resistance for ac. The value of voltage gain is mainly decided

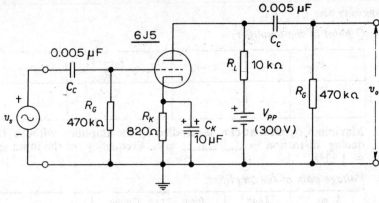

Fig. E.8.2.1

by this resistor. The coupling capacitor C_C in the output circuit allows ac voltage and blocks dc voltage from reaching the output. The coupling capacitor C_C in the input side helps in keeping grid bias independent of the dc content (if any) of the input signal.

It can be shown that the voltage gain of the amplifier is

$$Av = \frac{\mu R_L}{r_p + R_L} \angle 180° \qquad \text{(if } R_G \gg R_L\text{)}$$

The value of μ and r_p can be obtained from the data book. The measured value of voltage gain can be verified by calculating the gain, using the above formula.

PROCEDURE:

1. Find the details (parameters and pin configuration) of the triode from the data book. Trace the given circuit and identify the values of every component.

2. Connect the filament supply, i.e. 6.3 V (ac) and see that the filament glows. Now connect the high tension (dc) voltage and adjust its value to 300 V. Measure the value of plate voltage (V_P), plate current (I_P) and grid bias (V_G).

3. Now feed at the input a low (say, 0.5 V) ac signal so that grid never becomes positive. Adjust the frequency of the signal in mid-frequency range (say, 1 kHz). Measure the output voltage and find the voltage gain. Verify this value with the theoretically calculated value. You may observe some difference between the two values.

4. Slowly reduce the signal generator frequency to see at which frequency the voltage gain reduces to 70.7 % of the maximum gain. Similarly, find such a frequency on the higher frequency side. Keeping the signal voltage fixed, change the frequency of the signal. For each frequency, measure the output voltage. Take more readings around the cut-off frequencies, found approximately.

5. Plot the response curve on a semilog graph paper. Find the lower and upper cut-off frequencies from the response curve.

OBSERVATIONS:

1. *Operating point*:

    $V_{PP} =$ _____ V; $V_P =$ ____ ____ V

    $I_P = \dfrac{V_{PP} - V_P}{R_L} =$ _____ mA; $V_G =$ _____ V

2. *Mid-frequency voltage gain*:

    Frequency of the signal $= 1$ kHz

    Input voltage $V_i =$ _____ V

    Output voltage $V_o =$ _____ V

    Voltage gain $A = \dfrac{V_o}{V_i} =$ _____

3. *Approximate cut-off frequencies*:

    $f_1 =$ _____ Hz

    $f_2 =$ _____ kHz

4. *Information from data book*:

    $\mu =$ _____ ; $r_p =$ _____ kΩ

    Theoretical value of voltage gain $= \dfrac{\mu R_L}{r_p + R_L} =$ _____

5. *Frequency response*:

S. no.	Frequency, f	Output voltage, V_o	Voltage gain $A = \dfrac{V_o}{V_i}$
1.			
2.			
3.			

RESULTS:

1. Measured mid-frequency voltage gain = _____
2. Theoretical voltage gain = _____
3. Lower cut-off frequency $f_1 =$ _____ Hz
4. Upper cut-off frequency $f_2 =$ _____ kHz
5. Bandwidth $= f_2 - f_1 =$ _____ kHz

EXPERIMENTAL EXERCISE 8.3

TITLE: Pentode amplifier.

OBJECTIVES: To,

1. trace the circuit of the given amplifier;
2. identify the values of every component;
3. measure the operating point (i.e. the plate current, the plate voltage and the grid-bias voltage);
4. measure the mid-frequency gain of the amplifier;
5. compare the measured value of the voltage gain with the theoretical value;

6. plot the frequency response curve of the amplifier, and thus determine the bandwidth.

APPARATUS REQUIRED: Experimental board, dc power supply unit, electronic multimeter, signal generator.

CIRCUIT DIAGRAM: As given in Fig. E.8.3.1. Typical values of the components are also given.

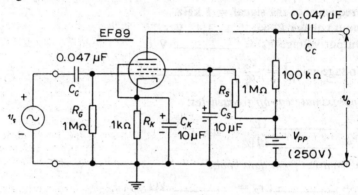

Fig. E.8.3.1

BRIEF THEORY: A pentode amplifier is not much different from a triode amplifier. It has an additional resistor R_S and a capacitor C_S. The resistor R_S fixes the required positive dc voltage at the screen grid. The screen-grid voltage is somewhat less than the supply voltage V_{PP}, since the screen-grid current flows through the resistor R_S causing a voltage drop. For the operation of the pentode as an amplifier, the screen grid should be effectively grounded from the ac point of view. This is achieved by connecting capacitor C_S between the screen grid and ground. The suppressor is connected to the cathode. It suppresses the secondary emitted electrons from the plate. This helps in removing the negative resistance portion of the tetrode characteristics. The dynamic plate resistance r_p and amplification factor for the pentode are much higher than those for the triode.

The voltage gain of the pentode amplifier is given as

$$A_v = g_m R_{eq} \angle 180°$$

where,

$$\frac{1}{R_{eq}} = \frac{1}{r_p} + \frac{1}{R_L} + \frac{1}{R_G}$$

The values of r_p and g_m can be found from the data book.

PROCEDURE:

1. Read the type number of the pentode. Find the details (r_p, g_m, pin configuration) of this pentode from the data book. Trace the given circuit.

2. Using the colour code, if required, find the values of all the resistors and capacitors in the circuit.

3. Connect the filament supply, i.e. 6.3 V (ac). After a little while, the filament should glow. Now connect the HT voltage and adjust its value at, say, 300 V. Measure plate voltage (V_P), plate current (I_P) and grid bias (V_G). This gives the operating point.

4. Feed at the input, an ac signal. Keep the voltage of the signal low (say 0.5 V) so that the grid never becomes positive. (Peak value of the ac signal should be less than the grid bias.) Adjust the frequency of the signal at 1 kHz. Measure the output voltage, V_o. Determine the mid-frequency voltage gain, by dividing V_o by V_i.

5. Using the formula, calculate the voltage gain. Compare this value with the one measured. Don't be surprised if the two values are a little different.

6. Now change the frequency of the input signal, keeping its voltage constant. At each frequency, measure the output voltage. Take more readings near about the cut-off frequencies.

7. Plot the frequency response curve on a semilog graph paper. Find the lower and upper cut-off frequencies from this curve.

OBSERVATIONS:

1. *Operating point*:

$V_{PP} =$ _____ V; $V_P =$ _____ V

$I_P = \dfrac{V_{PP} - V_P}{R_L} =$ _____ mA; $V_G =$ _____ V

2. *Mid-frequency voltage gain*:

Frequency of the signal = _____ kHz

Input voltage $V_i =$ _____ V; output voltage $V_o =$ _____ V

Voltage gain $A = \dfrac{V_o}{V_i} =$ _____

3. *Information from data book*:

$g_m =$ _____ μS; $r_p =$ _____ kΩ

4. *From the circuit*:

$R_L =$ _____ kΩ; $R_G =$ _____ MΩ

$R_{eq} = r_p \parallel R_L \parallel R_G =$ _____ kΩ

5. *Theoretical voltage gain*: $A = g_m R_L =$ _____

6. *Frequency response*: Input voltage $V_i =$ _____ V

S. no.	Frequency, f	Output voltage, V_o	Voltage gain $A = \dfrac{V_o}{V_i}$
1.			
2.			
3.			

RESULTS:

1. Measured mid-frequency voltage gain = _____
2. Theoretical voltage gain = _____
3. Lower cut-off frequency f_1 = _____ Hz
4. Upper cut-off frequency f_2 = _____ kHz
5. Bandwidth $= f_2 - f_1$ = _____ kHz

Multi-Stage Amplifiers

OBJECTIVES: After completing this unit, you will be able to: ○ Explain the need of multi-stage amplifiers in electronic systems. ○ Calculate the overall gain (as a ratio and also in dB) of a multi-stage amplifier, if the gain of each stage is known. ○ Explain the working of different types of multi-stage amplifiers (resistance-capacitance coupled, transformer-coupled and direct coupled) using transistors, triodes and FET. ○ State applications of *RC*-coupled, transformer-coupled and direct-coupled multi-stage amplifiers. ○ Explain the frequency response curve of an *RC*-coupled amplifier. ○ Compute the mid-frequency gain of a given two-stage *RC*-coupled amplifier. ○ Compute the lower and upper cutoff frequencies of a single-stage *RC*-coupled amplifier using a triode. ○ State the effect of cascading a number of stages on the bandwidth of an amplifier. ○ Explain different types of distortion that occur in the signal when it is amplified by an amplifier. ○ State the classification of amplifiers on the basis of frequency, coupling, purpose, and operating point.

9.1 DO WE REQUIRE MORE THAN ONE STAGE

An amplifier is the basic building block of most electronic systems. Just as one brick does not make a house, a single-stage amplifier is not sufficient to build a practical electronic system. In the last chapter, we had discussed the performance of a single-stage amplifier. Although the gain of an amplifier does depend on the device parameters and circuit components, there exists an upper theoretical limit for the gain obtainable from one stage. The gain of single stage is not sufficient for practical applications. The voltage level of a signal can be raised to the desired level if we use more than one stage. When a number of amplifier stages are used in succession (one after the other) it is called a *multi-stage amplifier* or a *cascaded amplifier*. Much higher gains can be obtained from the multi-stage amplifiers.

9.2 GAIN OF A MULTI-STAGE AMPLIFIER

A multi-stage amplifier (*n*-stages) can be represented by the block diagram as shown in Fig. 9.1. You may note that the output of the first stage makes the input of the second stage: the output of the second stage makes the input of the third stage, ..., and so on. The signal voltage v_s is applied to the input of the first stage. The final output v_o is then available at the output terminals of the last stage. The output of the first (or the input to the second stage) is

$$v_1 = A_1 v_s$$

where A_1 is the voltage gain of the first stage. Then the output of the second stage (or the input to the third stage) is

$$v_2 = A_2 v_1$$

Similarly, the final output v_o is given as

$$v_o = v_n = A_n v_{n-1}$$

where A_n is obviously the voltage gain of the last (nth) stage.

We may look upon this multi-stage amplifier as a single amplifier, whose input is v_s and output is v_o. The overall gain A of the amplifier is then given as

$$A = \frac{v_o}{v_s} = \frac{v_1}{v_s} \times \frac{v_2}{v_1} \times \dots \times \frac{v_{n-1}}{v_{n-2}} \times \frac{v_o}{v_{n-1}}$$

or
$$A = A_1 \times A_2 \times \dots \times A_{n-1} \times A_n \qquad (9.1)$$

The gain of an amplifier can also be expressed in another unit called *decibel*.

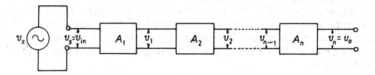

Fig. 9.1 Block diagram of a multistage amplifier having n stages

9.2.1 Decibel

In many problems it is found very convenient to compare two powers on a logarithmic scale rather than on a linear scale. The telephone industry proposed a logarithmic unit, named *bel* after Alexander Graham Bell. The number of bels by which a power P_2 exceeds a power P_1 is defined as

$$\text{Numbers of bels} = \log_{10} \frac{P_2}{P_1}$$

For practical purposes it has been found that the unit bel is quite large. Another unit, one-tenth as large, is more convenient. This smaller unit is called the *decibel* (abbreviated as dB), and since one decibel is one-tenth of a bel, we have

$$\text{Number of dB} = 10 \times \text{Number of bels} = 10 \log_{10} \frac{P_2}{P_1} \qquad (9.2)$$

Note that the unit dB denotes a power ratio. Therefore, the specification of a certain power in dB is meaningless unless a standard reference level is either implied or is stated explicitly. In communication applications, usually 6 mW or 1 mW is taken as standard reference level. When 1 mW is taken as reference, the unit dB is often referred to as dBm. A negative value of number of dB in Eq. 9.2 means that the power P_2 is less than the reference power P_1.

For an amplifier, P_1 may represent the input power and P_2 the output power. If V_1 and V_2 are the input and output voltages of the amplifier, then

$$P_1 = \frac{V_1^2}{R_i}$$

and
$$P_2 = \frac{V_2^2}{R_o}$$

where, R_i and R_o are the input and output impedances of the amplifier. Then, Eq. 9.2 can be written as

$$\text{Number of dB} = 10 \, \log_{10} \frac{V_2^2/R_o}{V_1^2/R_i} \tag{9.3}$$

In case the input and output impedances of the amplifier are equal, i.e. $R_i = R_o = R$, the Eq. 9.3 simplifies to

$$\text{Number of dB} = 10 \, \log_{10} \frac{V_2^2}{V_1^2} = 10 \, \log_{10} \left(\frac{V_2}{V_1}\right)^2$$

$$= 10 \times 2 \, \log_{10} \frac{V_2}{V_1} = 20 \, \log_{10} \frac{V_2}{V_1} \tag{9.4}$$

However, in general, the input and output impedances are not always equal. But the expression of Eq. 9.4 is adopted as a convenient definition of the decibel voltage gain of an amplifier, regardless of the magnitudes of the input and output impedances. Of course, this usage is technically improper.

As an example; if the voltage gain of an amplifier is 10, it can be denoted on the dB scale as

$$\text{Gain in dB} = 20 \, \log_{10} \frac{V_2}{V_1} = 20 \, \log_{10} 10$$

$$= 20 \times 1 = 20 \, \text{dB}$$

9.2.2 Gain of Multi-Stage Amplifier in dB

The gain of a multi-stage amplifier can be easily computed if the gains of the individual stages are known in dB. If we take logarithm (to the base 10) of Eq. 9.1 and then multiply each term by 20, we get

$$20 \, \log_{10} A = 20 \, \log_{10} A_1 + 20 \, \log_{10} A_2 + \cdots + 20 \, \log_{10} A_n$$

In the above equation, the term on the left is the overall gain of the multi-stage amplifier expressed in dB. The terms on the right denote the gains of the individual stages expressed in dB. Thus, the overall voltage gain in dB of a multi-stage amplifier is the sum of the decibel voltage gains of the individual stages. That is

$$A_{dB} = A_{dB1} + A_{dB2} + \cdots + A_{dB|n|} \tag{9.5}$$

9.2.3 Why dB is Used

You may wonder why we use a logarithmic scale to denote voltage or power gains, instead of using the simpler linear scale. The reasons for the popularity of dB scale are as follows:

(i) It permits gains to be directly added when a number of stages are cascaded. (Use of logarithms changes multiplication into an addition).

(ii) It permits us to denote, both very small as well as very large, quantities of linear scale by conveniently small figures. Thus a voltage gain of 0.000 001 (in fact, it represents a loss instead of gain) may be represented as a voltage gain of -120 dB, or a voltage loss of 120 dB. Similarly, a power gain of 456 000 is simply 56.59 ($\simeq$ 56.6) dB on the logarithmic scale.

(iii) The output of many amplifiers is ultimately converted into sound and this sound is received by the human ear. Experiments show that the ear responds to the sound intensities on a proportional or logarithmic scale rather than the linear scale. If the audio power increases from 4 W to 64 W, the hearing level does not increase by a factor of $64/4 = 16$. The response of the ear will increase by a factor of only 3, since $(4)^3 = 64$. Thus, the use of the dB unit is justified on a psychological basis too.

Example 9.1 A multistage amplifier consists of three stages. The voltage gains of the stages are 30, 50 and 80. Calculate the overall voltage gain in dB.

Solution: We know that the overall voltage gain in dB of the three-stage amplifier is given as

$$A_{dB} = A_{dB1} + A_{dB2} + A_{dB3}$$

But, we are given the voltage gains of the individual stages as ratios. So, we should first find the gains of the individual stages in decibels. Thus

$$A_{dB1} = 20 \log_{10} 30 = 29.54 \text{ dB}$$
$$A_{dB2} = 20 \log_{10} 50 = 33.98 \text{ dB}$$
$$A_{dB3} = 20 \log_{10} 80 = 38.06 \text{ dB}$$

Therefore

$$A_{dB} = 29.54 + 33.98 + 38.06 = \textbf{101.58 dB}$$

Alternatively, we could have determined A_{dB} as follows: The overall voltage gain is

$$A = A_1 \times A_2 \times A_3$$
$$= 30 \times 50 \times 80 = 120\,000$$

Therefore, the overall voltage gain in dB is

$$A_{dB} = 20 \log_{10} 120\,000 = \textbf{101.58 dB}$$

9.3 HOW TO COUPLE TWO STAGES

In a multi-stage amplifier, the output of one stage makes the input of the next stage (see Fig. 9.1). Can we connect the output terminals of one amplifier to the input terminals of the next amplifier directly? This may not always be possible due to practical difficulties. We must use a suitable coupling network between two stages so that a minimum loss of voltage occurs when the signal passes through this network to the next stage. Also, the dc voltage at the output of one stage should not be permitted to go to the input of the next. If it does, the biasing conditions of the next stage are disturbed.

The coupling network not only couples two stages; it also forms a part of the load impedance of the preceding stage. Thus, the performance of the amplifier will also depend upon the type of coupling network used. The three generally used coupling schemes are :

    (i) Resistance-capacitance coupling
    (ii) Transformer coupling
    (iii) Direct coupling

9.3.1 Resistance-Capacitance Coupling

Figure 9.2 shows how to couple two stages of amplifiers using resistance-capacitance (RC) coupling scheme. This is the most widely used method. In this scheme, the signal developed across the collector resistor R_C of the first stage is coupled to the base of the second stage through the capacitor C_C. The coupling capacitor C_C blocks the dc voltage of the first stage from reaching the base of the second stage. In this way, the dc biasing of the next stage is not interfered with. For this reason, the capacitor C_C is also called a *blocking capacitor*.

    Some loss of the signal voltage always occurs due to the drop across the coupling capacitor. This loss is more pronounced when the frequency of the input signal is low. (This point is discussed in more detail in Sec. 9.4.1). This is the main drawback of this coupling scheme. However, if we are interested in amplifying ac signals of frequencies greater than about 10 Hz, this coupling is the best solution. It is the most convenient and least expensive way to build a multi-stage amplifier.

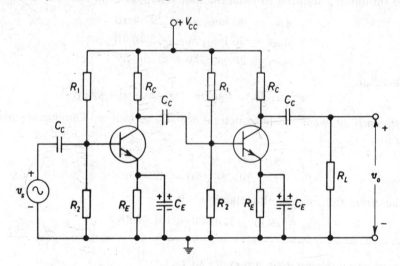

Fig. 9.2 Two-stage RC-coupled amplifier using transistors

    RC coupling scheme finds applications in almost all audio small-signal amplifiers used in record players, tape recorders, public-address systems, radio receivers, television receivers, etc.

    Triode (or pentode) amplifiers can also be cascaded by RC coupling. Figure 9.3 illustrates how RC coupling is used for two stages of triode amplifiers. Here, the cathode resistor R_K and capacitor C_K provide the self bias in the circuit. The operating point of a triode amplifier is independent of temperature. Therefore, the need of stabilization of the operating point does not arise. The circuit is much simpler than the one using transistors. But it requires high dc voltage (of the order of 300 V) supply.

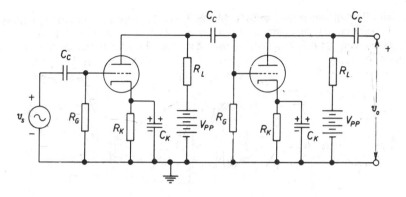

Fig. 9.3 Two-stage *RC*-coupled amplifier using triodes

The relatively new solid-state device—field-effect transistor (FET) has virtues of both the ordinary transistor (which is now called a bipolar-junction transistor, BJT, so as to distinguish it from the field-effect transistor, a unipolar device) and vacuum tubes. A FET is small in size, lighter in weight and it does not require any heater filament for its operation. At the same time, its input impedance is much higher as compared to that of an ordinary transistor. This difference arises because the gate-source junction in FET is reverse-biased, whereas the emitter-base junction in BJT is forward-biased. Figure 9.4 illustrates the use of *RC* coupling in the case of two stages of FET amplifiers. This circuit is similar to the *RC*-coupled amplifier in Fig. 9.3.

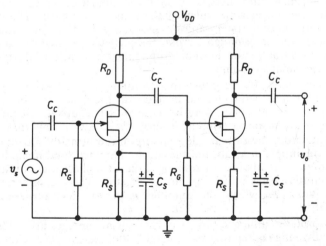

Fig. 9.4 Two-stage *RC*-coupled amplifier using FETs

9.3.2 Transformer Coupling

In this type of coupling, a transformer is used to transfer the ac output voltage of the first stage to the input of the second stage. The resistor R_C (see Fig. 9.2) is replaced by the primary winding of the transformer. The secondary winding of the transformer replaces the wire between the voltage divider (of the biasing network) and the base of the second stage. Figure 9.5 illustrates the transformer coupling between the two stages of amplifiers, using BJTs.

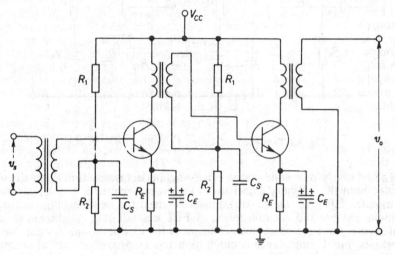

Fig. 9.5 Two stages, using transistors, are coupled by a transformer

Note that in this circuit there is no coupling capacitor. The dc isolation between the two stages is provided by the transformer itself. There exists no dc path between the primary and the secondary windings of a transformer. However, the ac voltage across the primary winding is transferred (with a multiplication factor depending upon the turns-ratio of the transformer) to the secondary winding.

The main advantage of the transformer coupling over RC coupling is that all the dc voltage supplied by V_{CC} is available at the collector. There is no voltage drop across the collector resistor R_C (of RC-coupled scheme). The dc resistance of the primary winding is very low (only a few ohms). The ac impedance across the primary depends upon the turns-ratio of the transformer and the input impedance of the second amplifier; and it can be made sufficiently high. The absence of resistor R_C in the collector circuit also eliminates the unnecessary power loss in the resistor. These considerations of power are important when the amplifier is to work as a power amplifier (see Unit 10).

The transformer coupling scheme has some disadvantages also. The most obvious disadvantage is the increased size of the system. The transformer is very bulky as compared to a resistor or a capacitor. It is also relatively costlier. Another disadvantage of this scheme arises from the fact that the transformer used differs in its working from an ideal one. In the transformer, there is some leakage inductance and interwinding capacitances. Because of these stray elements, the transformer-coupled amplifier does not amplify the signals of different frequencies equally well. The interwinding capacitance may give rise to a phenomenon of resonance at some frequency. This may make the gain of the amplifier very high at this frequency. At the same time, the gain may be quite low at another frequency.

Because of the above drawbacks, the transformer-coupling scheme is not used for amplifying low frequency (audio) signals. However, they are widely used for amplification of radio-frequency signals. Radio frequency means anything above 20 kHz. In radio receivers, the rf ranges from 550 kHz to 1600 kHz for the medium-wave band; and from 3 MHz to 30 MHz for the short wave band. In TV receivers, the rf signals have frequencies ranging from 54 MHz to 216 MHz. By putting suitable shunting capacitors across each winding of the transformer, we can get resonance at any desired rf frequency. Such amplifiers are called tuned-voltage amplifiers. These provide high gain at the desired rf frequency. For this reason, the transformer-coupled amplifiers are used in radio and TV receivers for amplifying rf signals. (Such amplifiers are discussed in Unit 11 of this book.)

The use of a transformer for coupling not only saves power loss in the collector resistor R_C, but also helps in proper impedance matching. By suitably selecting the turns ratio of the transformer, we can match any load with the output impedance of the amplifier. This helps in transferring maximum power from the amplifier to the load. This is discussed in more details in Unit 10 on power amplifiers.

A radio-frequency transformer-coupled amplifier using vacuum pentodes is shown in Fig. 9.6. In this circuit, shunt capacitors C_1 and C_2 across the primary and the secondary of the transformer are used for tuning purposes. For this reason, such amplifiers are called tuned-voltage amplifiers (for details see Unit 11). Pentodes are used in this circuit, because a pentode is more suitable than a triode for high-frequency amplification.

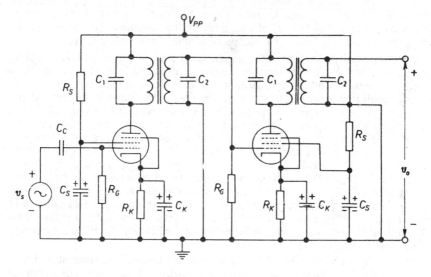

Fig. 9.6 Transformer-coupled, tuned voltage amplifier using pentodes

A tuned transformer-coupled amplifier using FETs is shown in Fig. 9.7.

9.3.3 Direct Coupling

In certain applications, the signal voltages are of very low frequency. For example, thermocouples are used for the measurement of temperature in furnaces. The voltage induced in the thermocouple is very small in magnitude

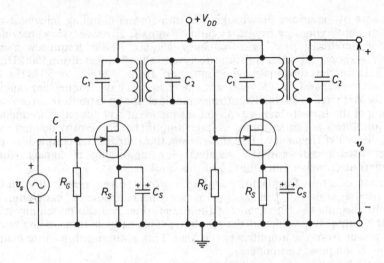

Fig. 9.7 Transformer-coupled, tuned voltage amplifier using FETs

(of the order of μV). This voltage needs to be amplified to a suitable level before it can be used to deflect the needle of a meter. The temperature of the furnace may change very slowly. The indicating meter should respond to such slowly varying changes. The amplifier used for the amplification of such slowly varying signals makes use of direct coupling. In this type of coupling scheme, the output of one stage of the amplifier is connected to the input of the next stage by means of a *simple connecting wire*.

For applications where the signal frequency is below 10 Hz, coupling capacitors and bypass capacitors cannot be used. At low frequencies, these capacitors can no longer be treated as short circuits, since they offer sufficiently high impedance. On the other hand, if coupling and bypass capacitors are to serve their purpose, their values have to be extremely large. Such capacitors are not only very expensive, but also are inconveniently large in size. For example, to bypass a 100 Ω emitter resistor at a frequency of 10 Hz, we need a capacitor of about 1000 μF. The lower the frequency, the worse the problem becomes. To avoid this problem, direct coupling is used. Figure 9.8 shows a two-stage direct-coupled amplifier, using transistors. Note that no coupling and bypass capacitors are used. Therefore, both dc as well as ac are coupled to the next stage. The dc voltage at the collector of the first stage reaches the base of the second stage. This should be taken into account while designing the biasing circuit of the second stage (see Example 9.2).

The direct coupling scheme has a serious drawback. The transistor parameters like V_{BE} and β vary with temperature. This causes the collector current and voltage to change. Because of the direct coupling, this voltage change appears at the final output. Such an unwanted change in output voltage which has no relationship with input voltage is called *drift*. The drift in direct-coupled amplifiers is a serious problem. It can be wrongly interpreted as a genuine output produced by the input signal. There are some specially designed direct-coupled amplifier circuits in which the problem of drift is minimized to a considerable extent.

A direct-coupled amplifier using triodes and a single dc supply is shown in Fig. 9.9. This circuit is known as the Loftin-White circuit. Since no

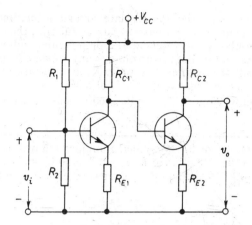

Fig. 9.8 Two-stage, direct-coupled amplifier
using transistors

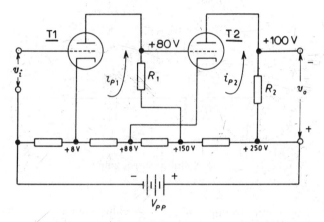

Fig. 9.9 Two-stage direct-coupled amplifier using triodes
(Loftin-White circuit)

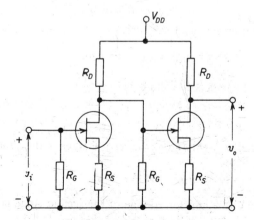

Fig. 9.10 Two-stage direct-coupled amplifier
using FETs

coupling capacitor Cc is used, the dc plate voltage of the first stage reaches the grid of the second stage. To have negative grid bias, the cathode of the second stage should be maintained at a dc potential higher than the grid potential. In the circuit, suitable dc voltages are obtained from a tapped resistor connected across the dc supply. As may be seen from the circuit, both the triodes have a negative grid bias of 8 V.

Figure 9.10 shows how FETs can be used in making a two-stage direct-coupled amplifier.

Example 9.2 A two-stage direct-coupled amplifier is shown in Fig. 9.11. The transistors used in the circuit have $V_{BE} = 0.7$ V and $\beta = 300$. If the voltage at the input is $+1.4$ V, calculate the voltage at the output terminal.

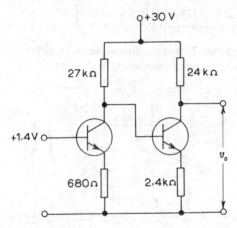

Fig. 9.11 A direct-coupled amplifier

Solution: The input voltage is $+1.4$ V. The voltage V_{BE} being 0.7 V, the voltage at the emitter terminal will be

$$V_E = 1.4 - 0.7 = 0.7 \text{ V}$$

Therefore, the emitter current (first stage)

$$I_{E1} = \frac{0.7}{680} \simeq 1 \text{ mA}$$

Since $I_{C1} \simeq I_{E1}$, the collector voltage is
$$V_{C1} = V_{CC} - I_{C1} \times 27 \times 10^3 = 30 - 1 \times 10^{-3} \times 27 \times 10^3$$
$$= 3 \text{ V}$$

For this circuit, the base voltage of the second stage is the conector voltage of the first stage. Thus,

$$V_{B2} = V_{C1} = 3 \text{ V}$$

Since, $V_{BE} = 0.7$ V, the emitter voltage of the second stage is
$$V_{B2} = 3 - 0.7 = 2.3 \text{ V}$$

The emitter current,

$$I_{E2} = \frac{2.3}{2.4 \times 10^3} \simeq 1 \text{ mA}$$

Since, $I_{C2} \simeq I_{E2}$, the voltage at the collector is

$$V_{C2} = V_{CC} - I_{C2} \times 24 \times 10^3 = 30 - 1 \times 10^{-3} \times 24 \times 10^3$$
$$= 6 \text{ V}$$

Thus, the voltage at the output terminal is

$$V_o = 6 \text{ V}$$

9.4 FREQUENCY RESPONSE CURVE OF AN RC-COUPLED AMPLIFIER

A practical amplifier circuit is meant to raise the voltage level of the input signal. This signal may be obtained from the piezoelectric crystal of a record player, the sound head of a tape recorder, the microphone in case of a PA system, or from a detector circuit of a radio or TV receiver. Such a signal is not of a single frequency. But it consists of a band of frequencies. For example, the electrical signal produced by the voice of human being or by a musical orchestra may contain frequencies as low as 30 Hz and as high as 15 kHz. Such a signal is called audio signal. If the loudspeakers are to reproduce the original sound faithfully, the amplifier used must amplify all the frequency components of the signal equally well. If it does not do so, the output of the loudspeaker will not be an exact replica of the original sound. When this happens, we say that distortion has been introduced by the amplifier.

The performance of an amplifier is judged by observing whether all frequency components of the signal are amplified equally well. This information is provided by its frequency response curve. This curve illustrates how the magnitude of the voltage gain (of amplifier) varies with the frequency of the input signal (sinusoidal). It can be plotted by measuring the voltage gain of the amplifier for different frequencies of the sinusoidal voltage fed to its input (see Fig. 9.12).

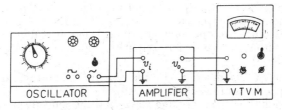

Fig. 9.12 Measurement of voltage gain for plotting
frequency-response curve

Figure 9.13 shows a frequency response curve of a typical RC-coupled amplifier. This curve is usually plotted on a semilog graph paper with frequency on logarithmic scale so as to accommodate large frequency range. Note that the gain is constant only for a limited band of frequencies. This range of frequencies is called the mid-frequency range and the gain is called mid-band gain, A_{vm}. On both sides of the mid-frequency range, the gain

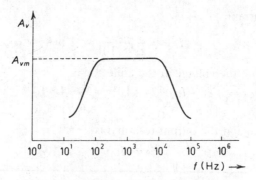

Fig. 9.13 Frequency response curve of an
RC-coupled amplifier

decreases. For very low and for very high frequencies, the gain of the amplifier reduces to almost zero.

9.4.1 Fall of Gain in Low-frequency Range

In the last section, we analysed an amplifier circuit to determine its voltage gain. This was the mid-frequency gain. In mid-frequency range, the coupling and bypass capacitors are as good as short-circuits. But, when the frequency is low, these capacitors can no longer be replaced by the short-circuit approximation. The lower the frequency, the greater is the value of reactance of these capacitors, since

$$X_C = \frac{1}{2\pi f C}$$

Let us first examine how the coupling capacitor C_C affects the voltage gain of the amplifier at low frequencies. The output section of the first stage of the two-stage RC-coupled amplifier of Fig. 9.2 is redrawn in Fig. 9.14a. The output voltage v_o of this stage is the input to the second stage. The resistors R_1 and R_2 are the biasing resistors for the second stage. From the ac point of view, this circuit is equivalent to the one drawn in Fig. 9.14b. Assume for the time being, that the capacitor C_E is replaced by a short circuit. The resistors R_1, R_2 and input impedance h_{ie} of the next stage are in parallel and are equivalent to a resistor R. This resistance forms a part of the load resistance of the previous (first) stage. It really does not matter whether the output voltage v_o is taken at the left side or at the right side of the resistor R.

The capacitor C_C is in series with the resistor R, and this series combination is in parallel with the collector resistor R_C. The whole of this impedance forms the ac load for the preceding stage. But the effective output of the stage is the ac voltage developed across the resistor R (see Fig. 9.14c). At mid-frequencies (and also at high frequencies), the reactance of the capacitor C_C is sufficiently small compared to R. We can treat it as a short-circuit so that the resistor R comes in parallel with the resistor R_C. In such a case, the voltage v_1 across resistor R_C will be the same as the voltage v_o across R. However, at low frequencies, the reactance of C_C [$= 1/(2\pi f C_C)$] becomes sufficiently large. This causes a significant voltage drop across C_C. The result is that the effective output voltage v_o decreases. The lower the frequency of this signal, higher will be the reactance of the capacitor C_C, and more will

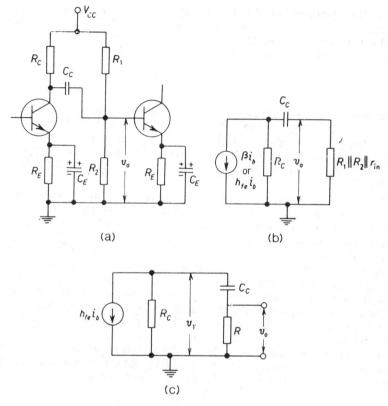

Fig. 9.14 (a) Output section of an RC-coupled amplifier
(b) Its ac equivalent; (c) The same equivalent
circuit redrawn in another way

be the reduction in output voltage v_o. At zero frequency (dc signals), the reactance of capacitor C_c is infinitely large (an open circuit). The effective output voltage v_o then reduces to zero. Thus we see that the output voltage v_o (and hence the voltage gain) decreases as the frequency of the signal decreases below the mid-frequency range.

The other component, due to which the gain decreases at low frequencies, is the bypass capacitor C_E. Figure 9.15 shows the input section of the amplifier. The capacitor C_E is connected across the emitter resistor R_E. This capacitor is meant to bypass the ac current to ground. The impedance of this capacitor is quite low (as good as a short-circuit) in the mid-frequency range as well as in high-frequency range. Therefore, at these frequencies, the emitter is effectively grounded for ac current. However, as the frequency decreases, the reactance of the capacitor C_E becomes comparable to resistance R_E. The bypassing action of the capacitor is no longer as good as at mid- and high-frequencies. The emitter is not at ground potential for ac. The emitter current i_e divides into two parts, i_1 and i_2. A part of current i_1 passes through the resistor R_E. The rest of the current i_2 ($= i_e - i_1$) passes through the capacitor C_E. Due to current i_1 in R_E, an ac voltage $i_1 \times R_E$ is developed.

When the polarity of the input signal voltage is as shown in figure, the current i_1 flows from the emitter to ground. The polarity of the voltage $i_1 R_E$ is also marked in the figure. Then, the effective input voltage to the amplifier (that is the voltage between the base and emitter of the transistor) becomes

$$v_{be} = v_s - i_1 R_E \qquad (9.6)$$

The effective input voltage is thus reduced. The output voltage v_o of the amplifier will now naturally be reduced. In other words, the gain of the amplifier ($= v_o/v_i$) reduces. This reduction in gain occurs due to the inability of the capacitor C_E to bypass ac current. The lower the frequency, the higher is the impedance of the capacitor C_E, and greater is the reduction in gain.

Note that the resistor R_E is not only a part of the input section, but also is a part of the output section. The voltage $i_1 R_E$ developed across the resistor R_E depends upon the output ac current. In this way, the effective input to the amplifier depends on the output current. The reduction in gain due to such a process is technically described as *negative current feedback effect*.

In Fig. 9.15, there is also a coupling capacitor C_C in the input section of the amplifier. Due to this capacitor, the effective input voltage is reduced at low frequencies in much the same way as the effective output voltage v_o is reduced due to the coupling capacitor in the output section. Thus, the coupling capacitor in the input side is also responsible for the decrease of gain at low frequencies.

In practical circuits, the value of the bypass capacitor C_E is very **large** ($\simeq$ 100 μF). Therefore, it is the coupling capacitor that has the **more** pronounced effect in reducing the gain at low frequencies.

9.4.2 Does Gain Fall at High Frequencies

As the frequency of the input signal increases, the gain of the amplifier reduces. Several factors are responsible for this reduction in gain. Firstly, the beta (β) of the transistor is frequency dependent. Its value decreases at high frequencies (see Fig. 9.16). Because of this, the voltage gain of the amplifier reduces as the frequency increases.

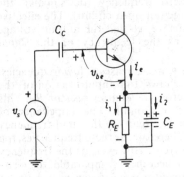

Fig. 9.15 Input section of an RC-coupled amplifier

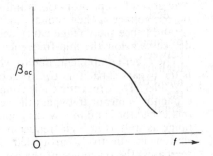

Fig. 9.16 Variation of short-circuit current gain β_{ac} with frequency

Another important factor responsible for the reduction in gain of the amplifier at high frequencies is the presence of the device. In case of a transistor, there exists some capacitance due to the formation of a depletion layer at the junctions. These inter-electrode capacitances are shown in Fig. 9.17. Note that the connection for these capacitances are shown by dotted lines. This has been done to indicate that these are not physically present in the circuit, but are inherently present with the device (whether we like it or not).

The capacitance C_{bc} between the base and collector connects the output with the input. Because of this, negative feedback takes place in the circuit and the gain decreases. This feedback effect is more, when the capacitance C_{bc} provides a better conducting path for the ac current. Such is the case at high frequencies. As the frequency increases, the reactive impedance of the capacitor becomes smaller.

The capacitance C_{be} offers a low-impedance path at high frequencies in the input side. This reduces the input impedance of the device, and the effective input signal is reduced. So, the gain falls. Similarly, the capacitance C_{ce} produces a shunting effect at high frequencies in the output side.

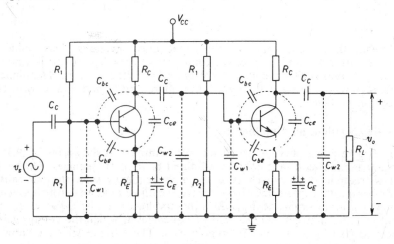

Fig. 9.17 RC-coupled amplifier. Capacitances that affect high-frequency response are shown by dotted connections

Besides the junction capacitances, there are wiring capacitances C_{w1} and C_{w2}, as shown in Fig. 9.17. The connecting wires of the circuit are separated by air which serves as a dielectric. This gives rise to some capacitance between the wires, though the capacitance value may be very small. But at high frequencies, even these small capacitances (5 to 20 pF) become important. For a multi-stage amplifier, the effect of the capacitance C_{ce}, C_{w2}, and the input capacitance C_i of the next stage can be represented by a single shunt capacitance

$$C_s = C_{ce} + C_{w2} + C_i \qquad (9.7)$$

The output section of the amplifier is shown in Fig. 9.18 from the ac point of view, for high-frequency considerations. The capacitance C_s is the equivalent shunt capacitance as given by Eq. 9.7. Note that the coupling and bypass capacitors do not appear in the figure, because they effectively represent short circuits at these frequencies.

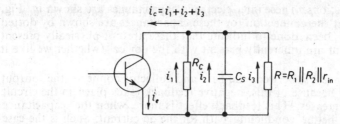

Fig. 9.18 Output section of an *RC*-coupled amplifier
at high frequencies

As can be seen from Fig. 9.18, the collector current i_c is made up of three currents i_1, i_2 and i_3. As the frequency of the input signal increases, the impedance of the shunt capacitance C_s decreases, since

$$X_{C_s} = \frac{1}{2\pi f C_s}$$

As a result, the current i_2 through this capacitance increases. This reduces both the currents i_1 and i_3, since the total current $i_c\ (=i_1+i_2+i_3)$ is almost constant. It means that the output voltage $v_o (=i_3 R)$ decreases. The higher the frequency, the lower is the impedance offered by C_s and lower will be the output voltage v_o.

9.4.3 Bandwidth of an Amplifier

Frequency response curve of an *RC*-coupled amplifier is shown in Fig. 9.19. The gain remains constant only for a limited band of frequencies. On both the low-frequency side as well as on the high-frequency side, the gain falls. Now an important question arises—where exactly should we fix the frequency limits (of input signal) within which the amplifier may be called a good amplifier? The limit is set at those frequencies at which the voltage gain reduces to 70.7 % of the maximum gain A_{vm}. These frequencies are known as the *cut-off frequencies* of the amplifier. These frequencies are marked in Fig. 9.19. The frequency f_1 is the *lower cut-off frequency* and the frequency f_2 is the *upper cut-off frequency*. The difference of the two frequencies, that is f_2-f_1, is called the *bandwidth* (BW) of the amplifier. The mid-frequency range of the amplifier is from f_1 to f_2. Usually, the lower cut-off frequency f_1 is much lower than the upper cut-off frequency f_2, so that we have

$$\text{BW} = f_2 - f_1 \simeq f_2 \qquad (9.8)$$

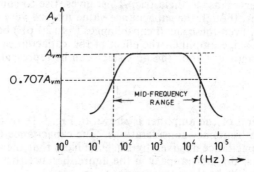

Fig. 9.19 Bandwidth of an *RC*-coupled amplifier

Why limit is set at 70.7 % of maximum gain At the cut-off frequencies the voltage gain is

$$0.707 \, A_{vm} \, [(= 1/\sqrt{2})A_{vm}]$$

where A_{vm} is the maximum gain or the mid-frequency gain of the amplifier. It means at these frequencies, the output voltage is $1/\sqrt{2}$ times the maximum voltage. Since the power is proportional to the square of the voltage, the output power at these cut-off frequencies becomes one-half of the power at mid-frequencies. On the dB scale this is equal to a reduction in power by 3 dB. For this reason, these frequencies are also called *3 dB frequencies*.

We have taken a difference of 3 dB in power to define the cut-off frequencies, because this represents an audio-power difference that can just be detected by the human ear. For the frequencies below f_1 and above f_2, the output power will reduce by more than 3 dB.

Example 9.3 An *RC*-coupled amplifier has a voltage gain of 100 in the frequency range of 400 Hz to 25 kHz. On either side of these frequencies, the gain falls so that it is reduced by 3 dB at 80 Hz and 40 kHz. Calculate gain in dB at cut-off frequencies and also construct a plot of frequency response curve.

Solution: The gain in dB is

$$A_{dB} = 20 \, \log_{10}A = 20 \, \log_{10}100 = 40 \text{ dB}$$

This is the mid-band gain. The gain at cut-off frequencies is 3 dB less than the mid-band gain, i.e.

$$(A_{dB}) \text{ (at cut-off frequencies)} = 40 - 3 = 37 \text{ dB}$$

The plot of the frequency response curve is given in Fig. 9.20.

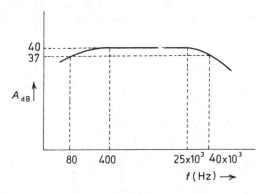

Fig. 9.20

Effect on bandwith when stages are cascaded A number of stages are cascaded to obtain higher values of voltage gain. But then the bandwidth of the amplifier does not remain the same. It decreases. The upper cut-off frequency decreases and the lower cut-off frequency increases. It happens because greater number of stages means a greater number of capacitors in the circuit. And each capacitor affects the frequency response adversely.

If n identical stages are cascaded, the overall mid-band voltage gain becomes

$$A'_v = (A_{vm})^n \qquad (9.9)$$

where A_{vm} is the mid-band voltage gain of an individual stage. If f_1 and f_2 are the lower and upper cut-off frequencies of an individual stage, the over-all cut-off frequencies are given by

$$f'_1 = \frac{1}{\sqrt{(2^{1/n}-1)}} f_1 \qquad (9.10)$$

and

$$f'_2 = \sqrt{(2^{1/n}-1)} f_2 \qquad (9.11)$$

9.5 ANALYSIS OF TWO-STAGE *RC*-COUPLED AMPLIFIER

We have complete information about an amplifier if following parameters are known :

    (i) Mid-band voltage gain A_{vm}
    (ii) Bandwidth, i.e. $f_2 - f_1$
    (iii) Input impedance Z_i
    (iv) Output impedance Z_o

The analysis of a two-stage amplifier will depend upon what active device is used. We shall analyse a two-stage amplifier circuit using transistors as well as triodes.

9.5.1 Two-Stage Transistor *RC*-Coupled Amplifier

A two-stage *RC*-coupled amplifier circuit using identical transistors is shown in Fig. 9.21. For its analysis, we first draw the circuit from the ac point of view This is shown in Fig. 9.22. Since we wish to determine the gain in

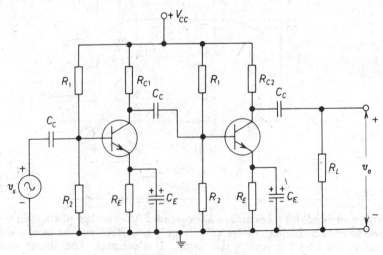

Fig. 9.21 Two-stage transistor *RC*-coupled amplifier

the mid-frequency range, all the coupling and bypass capacitors are replaced by short-circuits. The dc power supply is also replaced by a short-circuit. Next, we replace the transistors by their h-parameter approximate models. The result is the Fig. 9.23. In the approximate model of the transistor, small parameters h_{re} and h_{oe} are neglected.

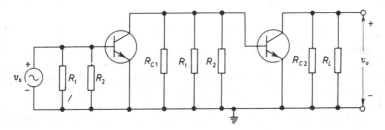

Fig. 9.22 Circuit of Fig. 9.21 from ac point of view

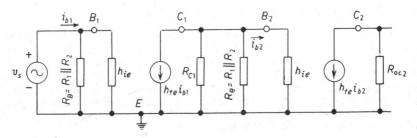

Fig. 9.23 Transistors of Fig. 9.22 replaced by their approximate
h-parameter model

The parallel combination of resistors R_1 and R_2 is replaced by a single resistor R_B, i.e.

$$R_B = R_1 \| R_2 = \frac{R_1 R_2}{R_1 + R_2} \tag{9.12}$$

For finding the overall gain of the two-stage amplifier, we should know the gains of the individual stages. Let us first find the gain A_2 of the second stage. We can use the formula derived in Unit 8 (see Eq. 8.9) and write

$$A_2 = -\frac{h_{fe} R_{ac2}}{h_{ie}} \tag{9.13}$$

where R_{ac2} is the ac load resistance for the second stage, given by

$$R_{ac2} = R_{C2} \| R_L = \frac{R_{C2} R_L}{R_{C2} + R_L} \tag{9.14}$$

For determining the gain of the first stage, let us have a closer look into what constitutes its ac load resistance, R_{ac1}. From Fig. 9.23, the resistance R_{ac1} is the parallel combination of R_{C1}, R_B and h_{ie} (the input impedance of the second stage). That is,

$$R_{ac1} = R_{C1} \| R_B \| h_{ie}$$

The voltage gain of the first stage is then given as

$$A_1 = -\frac{h_{fe}R_{ac_1}}{h_{ie}} \qquad (9.15)$$

Using Eqs. 9.13 and 9.15, the overall gain A_{vm} can be easily determined, since

$$A_{vm} = A_1 \times A_2 \qquad (9.16)$$

It should be noted that the voltage gain of the first stage A_1 is always less than the voltage gain A_2 of the second stage. This is because R_{ac_1} is very much reduced, as the input impedance h_{ie} is in parallel with R_{C_1} and R_B. This effect is called the *loading effect in multi-stage amplifiers.*

The overall input impedance is simply the input impedance of the first stage. If the biasing resistors R_1 and R_2 are large compared to h_{ie}, the overall impedance Z_i is simply h_{ie}.

The output impedance of the amplifier is R_{ac_2}.

9.5.2 Two-Stage *RC*-Coupled Amplifier Using Triodes

A two-stage vacuum triode amplifier appears in Fig. 9.24. Unlike transistors, the input impedance of a vacuum triode is very high. Also the grid-leak resistor R_G connected in the input of a stage is much greater (of the order of 1 MΩ) than the load resistance R_L (of the order of 10 kΩ) of the preceding stage. Therefore, a stage does not load its preceding stage. Each of the stages can be considered quite independently. For determining the overall gain of the amplifier, we shall consider the first stage only (shown within the dotted lines in Fig. 9.24).

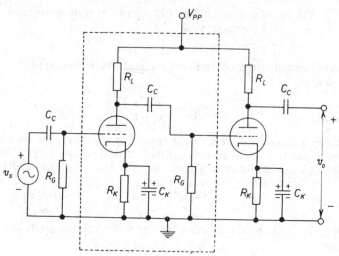

Fig. 9.24 Two-stage *RC*-coupled amplifier using vacuum triodes

In Fig. 9.25*a* the ac equivalent of the first stage of the *RC*-coupled amplifier of Fig. 9.24 is shown. The triode is replaced by its voltage-source equivalent. From the ac point of view, the dc supply V_{PP} offers a short-circuit. The parallel combination $R_K C_K$ does not appear in the equivalent

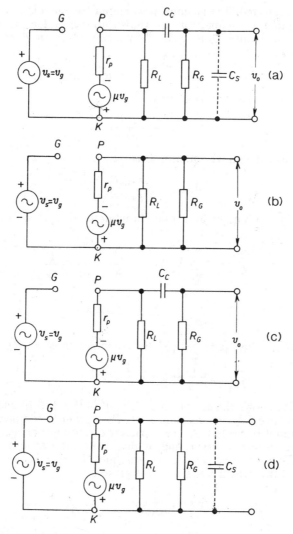

Fig. 9.25 (a) The ac equivalent of one stage of RC-coupled
triode amplifier; (b) The same at mid-frequency
(c) The same at low frequencies
(d) The same at high frequencies

circuit. The capacitor C_K is assumed to be large enough to put the cathode at ground potential, at the operating frequencies. The capacitor C_S represents the total shunt capacitance. This includes inter-electrode capacitance C_{pk}, wiring capacitance C_w and the effective input capacitance C_i of the next stage. Typical value of the shunt capacitor C_S is 200 pF.

In mid-frequency range, the frequency is high enough to make the reactance of capacitor C_C small, compared to R_G. Therefore, it may be replaced by a short-circuit. For example, if $C_C = 0.05 \, \mu F$, at a frequency of 1 kHz, the reactance $X_C = 1/2\pi f C_C = 3.18 \, k\Omega$. This is much smaller than the value of R_G (usually 1 MΩ). In this frequency range, the shunt capacitor C_S offers very high impedance in parallel with the resistors R_L and R_G.

Hence it is neglected. The ac equivalent circuit for the mid-frequency range does not contain any capacitor, as shown in Fig. 9.25b. If the parallel combination of R_L and R_G is replaced by a single resistor R, so that

$$R = R_L \parallel R_G = \frac{R_L R_G}{R_L + R_G}, \text{ or } \frac{1}{R} = \frac{1}{R_L} + \frac{1}{R_G}$$

The mid-band voltage gain is given as

$$A_{vm} = -\frac{\mu R}{r_p + R} \tag{9.17}$$

Since $\mu = r_p g_m$, the above equation becomes

$$A_{vm} = -\frac{r_p g_m R}{r_p + R}$$

We now divide both numerator and denominator by $r_p R$ so as to get

$$A_{vm} = -\frac{g_m}{\dfrac{1}{R} + \dfrac{1}{r_p}} = -\frac{g_m}{\dfrac{1}{R_L} + \dfrac{1}{R_G} + \dfrac{1}{r_p}}$$

$$= -\frac{g_m}{\dfrac{1}{R_{eq}}} = -g_m R_{eq} \tag{9.18}$$

Obviously, R_{eq} is the equivalent parallel combination of r_p, R_L and R_G, i.e.

$$\frac{1}{R_{eq}} = \frac{1}{r_p} + \frac{1}{R_L} + \frac{1}{R_G} \tag{9.19}$$

Figure 9.25c shows the ac equivalent circuit in the low-frequency range. For low frequencies, the reactance of C_C becomes comparable to R_G. Hence, it cannot be neglected. The capacitor C_S need not be considered, as its reactance becomes much higher than what it was in mid-frequency range. The frequency response of the amplifier in low-frequency depends upon C_C, and the cut-off frequency f_1 is given by

$$f_1 = \frac{1}{2\pi C_C R'} \tag{9.20}$$

where,

$$R' = (r_p \parallel R_L) + R_G \tag{9.21}$$

In the high-frequency range, the ac equivalent circuit becomes as shown in Fig. 9.25d. Now, capacitor C_C does not appear, but the capacitor C_S will have a shunting effect at the output. The fall in gain at high frequencies is due to the shunting effect of this capacitor C_S The upper cut-off frequency is given by

$$f_2 = \frac{1}{2\pi C_S R_{eq}} \tag{9.22}$$

where R_{eq} is given by Eq. 9.19.

Equations 9.17, 9.20 and 9.22 give the performance of a single-stage RC-coupled amplifier. Using these values the overall performance of the two-stage RC-coupled amplifier can be determined with the help of Eqs. 9.9, 9.10 and 9.11.

Example 9.4 A two-stage RC-coupled amplifier is shown in Fig. 9.26. Calculate (*a*) input impedance Z_i, (*b*) output impedance Z_o, and (*c*) voltage gain A_{vm}. For both the transistors, h_{fe} or $\beta = 120$ and r_{in} or $h_{ie} = 1.1$ kΩ.

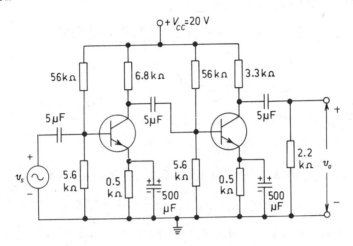

Fig. 9.26 A two-stage RC-coupled amplifier

Solution:

(*a*) *Input impedance*: From the knowledge of single-stage analysis, it is clear that for ac response, both 5.6-kΩ and 56-kΩ resistors will appear in parallel if the network is redrawn from an ac point of view. They are also in parallel with the input impedance of the first transistor, which is approximately $h_{ie} = 1.1$ kΩ, since the emitter resistor is bypassed by C_E.

$$\therefore \qquad Z_i = 5.6 \text{ k}\Omega \parallel 56 \text{ k}\Omega \parallel 1.1 \text{ k}\Omega = \textbf{0.905 k}\Omega$$

(*b*) *Output impedance*: Recall that the approximate collector-to-emitter equivalent circuit of a transistor is simply a current source $h_{fe}i_b$ (or $\beta\, i_b$). The dynamic output resistance r_o (which is $1/h_{oe}$), being very large, was neglected in the approximate analysis. Therefore, R_{eq2}, which is a parallel combination of 3.3 kΩ and 2.2 kΩ, is the effective output impedance.

$$\therefore \qquad Z_o = 3.3 \text{ k}\Omega \parallel 2.2 \text{ k}\Omega = \textbf{1.32 k}\Omega$$

(*c*) *Voltage gain*: For calculating the voltage gain, the ac equivalent circuit of the given two-stage amplifier is drawn (Fig. 9.27).

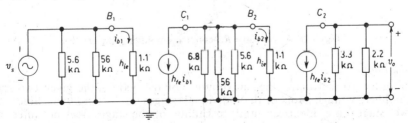

Fig. 9.27 AC equivalent circuit of a two-stage RC-coupled amplifier using approximate h-parameter model

The voltage gain of the second stage is given by

$$A_2 = \frac{-h_{fe}R_{ac2}}{h_{ie}}$$

Here $h_{fe} = 120$; $h_{ie} = 1.1$ kΩ; and

$$R_{ac} = 3.3 \text{ k}\Omega \parallel 2.2 \text{ k}\Omega = \frac{3.3 \times 10^3 \times 2.2 \times 10^3}{(3.3+2.2) \times 10^3}$$

$$= 1.32 \text{ k}\Omega$$

$$\therefore \qquad A_2 = \frac{-120 \times 1.32 \times 10^3}{1.1 \times 10^3} = -144$$

The voltage gain of the first stage is given by

$$A_1 = \frac{-h_{fe}R_{ac1}}{h_{ie}}$$

Here, $R_{ac1} = 6.8 \text{ k}\Omega \parallel 56 \text{ k}\Omega \parallel 5.6 \text{ k}\Omega \parallel 1.1 \text{ k}\Omega = 0.798 \text{ k}\Omega$

$$\therefore \qquad A_1 = \frac{-120 \times 0.798 \times 10^3}{1.1 \times 10^3} = -87.05$$

$$\therefore \qquad \text{Overall gain } A = A_1 \times A_2 = -87.05 \times (-144)$$

$$= \textbf{12 535}$$

Example 9.5 For a two-stage triode amplifier shown in Fig. 9.28, calculate the maximum voltage gain and the bandwidth. Assume the following:

$$R_L = 10 \text{ k}\Omega; \ R_G = 470 \text{ k}\Omega; \ C_S = 100 \text{ pF}; \ \mu = 25; \ r_p = 8 \text{ k}\Omega$$

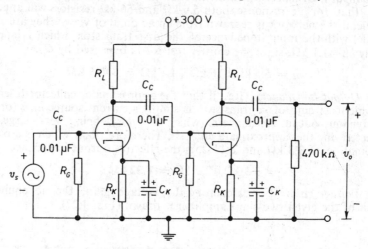

Fig. 9.28 A two-stage RC-coupled amplifier using triodes

Solution:

In order to calculate the maximum voltage gain of the given two-stage amplifier, voltage gains of the individual stages are calculated. Since the two stages are identical, and cascading of the stages does not affect the performance of the preceding stage, the voltage gain of both the stages is same. The voltage gain of a triode amplifier is given by

$$A_{vm} = g_m R_{eq} \angle 180°$$

Here, $$g_m = \frac{\mu}{r_p} = \frac{25}{8 \times 10^3} \text{ S}$$

and $$R_{eq} = r_p \parallel R_L \parallel R_G = 8 \text{ k}\Omega \parallel 10 \text{ k}\Omega \parallel 470 \text{ k}\Omega$$

$$\simeq 8 \text{ k}\Omega \parallel 10 \text{ k}\Omega = 4.44 \text{ k}\Omega$$

$\therefore$ $$A_{vm} = \frac{25 \times 4.4 \times 10^3}{8 \times 10^3} = 13.75$$

The voltage gain of the two-stages, A'_{vm}, is given by

$$A'_{vm} = (A_{vm})^2 = (13.75)^2 = \textbf{189.06}$$

The lower cut-off frequency of a triode amplifier is given by

$$f_1 = \frac{1}{2\pi C_C R'}, \text{ where } R' = (r_p \parallel R_L) + R_G$$

Here, $C_C = 0.01 \ \mu\text{F}$ and $R' = \left(\frac{10 \times 8}{18} \text{ k}\Omega + 470 \text{ k}\Omega \right) = 474.44 \text{ k}\Omega$

$\therefore$ $$f_1 = \frac{1}{2\pi \times 0.01 \times 10^{-6} \times 474.44 \times 10^3} = \textbf{33.5 Hz}$$

The lower cut-off frequency of the two stages is given by

$$f'_1 = \frac{f_1}{\sqrt{(2^{1/2} - 1)}} = \frac{33.5}{\sqrt{1.414 - 1}} = \textbf{52.05 Hz}$$

The upper cut-off frequency of a triode amplifier is given by

$$f_2 = \frac{1}{2\pi C_S R_{eq}}, \text{ where } R_{eq} = r_p \parallel R_L \parallel R_G$$

Here, $C_S = 100 \text{ pF}$ and $R_{eq} = 4.44 \text{ k}\Omega$. Therefore.

$$f_2 = \frac{1}{2\pi \times 100 \times 10^{-12} \times 4.44 \times 10^3} = \textbf{358.5 kHz}$$

The upper cut-off frequency for a two-stage amplifier is given by

$$f'_2 = \sqrt{(2^{1/2} - 1)} \times f_2 = 0.6436 \times 358.5 \text{ kHz} = 230.72 \text{ kHz}$$

$\therefore$ Bandwidth $= f'_2 - f'_1 = 230.72 \text{ kHz} - 52.05 \text{ Hz}$

$$\cong \textbf{230 kHz}$$

9.6 DISTORTION IN AMPLIFIERS

The purpose of an amplifier is to boost up the voltage or power level of a signal. During this process, the waveshape of the signal should not change. If the waveshape of the output is *not an exact replica* of the waveshape of the input, we say that *distortion* has been introduced by the amplifier. An *ideal amplifier* will amplify a signal without changing its waveshape at all. Such an amplifier *faithfully* amplifies the signal, and we say it has a good *fidelity*. Such an amplifier is called Hi-Fi (high fidelity) amplifier.

A number of factors may be responsible for causing distortion. It may be caused either due to the reactive components of the circuit, or due to imperfect (nonlinear) characteristics of the transistor or vacuum tube. There are three types of distortion. These may exist either separately or simultaneously in an amplifier.

   (i) Frequency distortion

  (ii) Phase or time-delay distortion

 (iii) Harmonic, amplitude, or nonlinear distortion

9.6.1 Frequency Distortion

In practical situations, the signal is not a simple sinusoidal voltage. It has a complex waveshape. Such a signal is equivalent to a signal obtained by adding a number of sinusoidal voltages of different frequencies. These sinusoidal voltages are called the frequency components of the signal. If all the frequency components of the signal are not amplified equally well by the amplifier, frequency distortion is said to occur. The cause for this distortion is non-constant gain for different frequencies. This occurs due to the inter-electrode capacitances of the active devices and other reactive components of the circuit. For example, an *RC*-coupled amplifier can amplify signals whose frequency lie within its bandwidth (Fig. 9.20). Let us arbitrarily state that the input signal to this amplifier contains many equal-amplitude frequency components spread over a large band (even beyond f_2). The higher frequency components will not be amplified to the same extent as the lower frequency components (see Fig. 9.29). Due to such a distortion, the speech or music (produced by the amplified electrical signal) appears to be quite different from the original one.

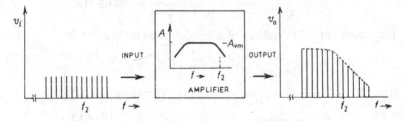

Fig. 9.29 Illustration of frequency distortion

9.6.2 Phase Distortion

Phase distortion is said to occur if the phase relationship between the various frequency components making up the signal waveform is not the same in the output as in the input. It means that the time of transmission or the delay introduced by the amplifier is different for various frequencies. The reactive components of the circuit are responsible for causing this type of distortion.

   This distortion is not important in audio amplifiers. Our ears are not capable of distinguishing the relative phases of different frequency components. But this distortion is objectionable in video amplifiers used in television.

9.6.3 Harmonic Distortion

This type of distortion is said to occur when the output contains new frequency components that are not present in the input signal. These new frequencies are the harmonics of the frequencies present in the input. For example, the input signal may consist of two frequency components. say 400 Hz and 500 Hz. If the amplifier gives rise to harmonic distortion, the output will contain

400 Hz (f_1, the fundamental)

500 Hz (f_2, the fundamental)

800 Hz (second harmonic of $f_1 = 2 \times 400$ Hz)

1000 Hz (second harmonic of $f_2 = 2 \times 500$ Hz)

1200 Hz (third harmonic of $f_1 = 3 \times 400$ Hz)

1500 Hz (third harmonic of $f_2 = 3 \times 500$ Hz)

1600 Hz (fourth harmonic of $f_1 = 4 \times 400$ Hz)

...   .

...   .

This is illustrated in Fig. 9.30.

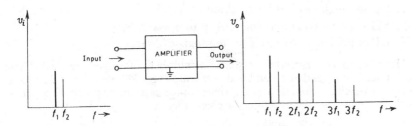

Fig. 9.30 Illustration of harmonic distortion

Harmonic distortion in an amplifier occurs because of the *nonlinearity* in the dynamic transfer characteristic curve. Hence this distortion is also called *nonlinear distortion*. In small-signal amplifiers (voltage amplifiers), the amplified signal is small. Only a small part of the transfer characteristic curve is used. Because of this, the operation takes place over an almost linear part of the curve. It is called linear because changes in the input voltage produce proportionate changes in the output current. It means that the shape of the amplified waveform is the same as the shape of the input wavetorm. Thus, in case of voltage amplifiers, where small-signals are handled, no harmonic distortion occurs.

In power amplifiers, the input signal is large. The change in the output current is no longer proportional to the changes in input voltage (*see* Fig. 9.31). If the input is a sinusoidal voltage, the output is no longer a pure sine wave. This type of distortion is also sometimes called *amplitude distortion*.

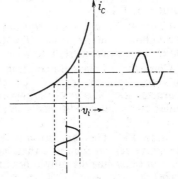

Fig. 9.31 Harmonic distortion occurs due to nonlinearity of the transfer characteristics

9.7 CLASSIFICATION OF AMPLIFIERS

An amplifier is a circuit meant to amplify a signal with a minimum of distortion, so as to make it more useful. The classification of amplifiers is somewhat involved. A complete classification must include information about the following:

   (i) Active device used.
  (ii) Frequency range of operation.
 (iii) Coupling scheme used.
  (iv) Ultimate purpose of the circuit.
   (v) Condition of dc bias and magnitude of signal.

An amplifier may use either a semiconductor device (such as BJT or FET) or a vacuum-tube device (such as triode, pentode, or beam power tube).

Based on frequency range of operation, the amplifiers may be classified as follows:

1. DC amplifiers (from zero to about 10 Hz)
2. Audio amplifiers (30 Hz to about 15 kHz)
3. Video or wide-band amplifiers (up to a few MHz)
4. RF amplifiers (a few kHz to hundreds of MHz)

Usually, in an amplifier system, a number of stages are used. These stages may be cascaded by either direct coupling, *RC* coupling, or transformer coupling. Sometimes, *LC* (inductance capacitance) coupling is also used. Accordingly, the amplifiers are classified as:

1. Direct-coupled amplifiers
2. *RC*-coupled amplifiers
3. Transformer-coupled amplifiers
4. *LC*-coupled amplifiers

Depending upon the ultimate purpose of an amplifier, it may be broadly classified as either voltage (small-signal), or power (large-signal) amplifier. Till now, we had considered the voltage amplifiers. In the next unit, we shall discuss the power amplifiers.

The amplifiers may also be classified according to where the quiescent point is fixed and how much the magnitude of the input signal is. Accordingly, four classes of operation for either transistor- or tube-amplifiers are defined as follows:

Class A: In class A operation, the transistor stays in the active region throughout the ac cycle. The Q point and the input signal are such as to make the output current flow for $360°$, as shown in Fig. 9.32a.

Class B: In class B operation, the transistor stays in the active region only for half the cycle. The Q point is fixed at the cut-off point of the characteristic. The power drawn from the dc power supply, by the circuit, under quiescent conditions is small. The output current flows only for $180°$ (see Fig. 9.32b).

Class AB: This operation is between class A and B. The transistor is in the active region for more than half the cycle, but less than the whole cycle. The output current flows for more than $180°$ but less than $360°$ (see Fig. 9.32c).

Class C: In a class C amplifier, the Q point is fixed beyond the extreme end of the characteristic. The transistor is in the active region for less than half cycle. The output current remains zero for more than half cycle, as shown in Fig. 9.32*d*. The dc current drawn from the power supply is very small.

In case of a vacuum-tube amplifier, sometimes a suffix 1 or 2 may be added to the class of operation. The suffix 1 indicates the absence of grid current, whereas suffix 2 indicates that during some part of the cycle, grid current flows. Thus, the designation class AB_1 will mean that the amplifier operates in class AB condition and that no grid current flows during any part of the cycle.

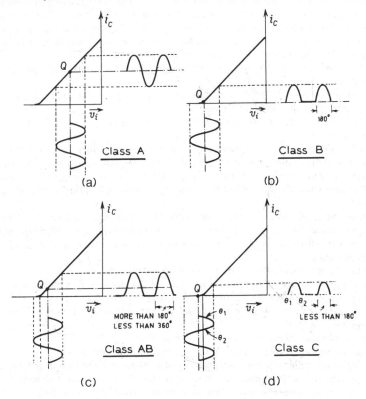

Fig. 9.32 Classification of amplifiers based on the biasing condition

In this unit, as well as in the previous unit, we have considered small-signal amplifiers under class A operation. In the units to follow, we will see different applications of other classes of operation.

REVIEW QUESTIONS

9.1 Why do you need more than one stage of amplifiers in practical circuits ?

9.2 It is known that a high value of gain can be obtained from a single stage of amplifier, if we use positive or regenerative feedback. Then, why can't we use this technique of obtaining high gain rather than using a multi-stage amplifier ?

9.3 State the reasons why we prefer expressing the gain of an amplifier on a logarithmic scale rather than on a linear scale.

9.4 Define the unit *decibel* for expressing (a) voltage; (b) current; and (c) power.

9.5 What are the various coupling schemes of two stages of amplifiers ?

9.6 (a) Draw the circuit diagram of a two-stage RC-coupled amplifier using transistors. Give the typical values of the components used; (b) Explain why RC-coupled amplifier circuits cannot be used to amplify slowly varying dc signals.

9.7 Draw the circuit diagram of a two-stage RC-coupled amplifier using vacuum triodes. Give the typical values of the components used.

9.8 State the applications where you would prefer using transformer coupling scheme.

9.9 State the advantages and disadvantages of a transformer-coupled amplifier.

9.10 What type of coupling scheme would you use for amplifying a signal obtained from a thermocouple meant to measure the temperature of a furnace ? Give reasons.

9.11 Draw the frequency response curve of a typical RC-coupled amplifier. Mark the gain-axis in dB. Why do you prefer using logarithmic scale for the frequency axis? How will you find the 3-dB frequencies from this curve ?

9.12 Why does the gain of an RC-coupled amplifier fall in (a) low-frequency range, (b) high-frequency range ?

9.13 How do you define the cut-off frequencies of an amplifier and what do you understand by the bandwidth of an amplifier ?

9.14 While defining the cut-off frequencies of an amplifier, why do we take 70.7 % of the mid-band gain ?

9.15 When more stages are cascaded to obtain high gain, does the bandwidth of the multi-stage amplifier remain the same as that of the individual stages ? If not, why ?

9.16 What do you understand by the loading effect in a multi-stage transistor amplifier? Why does such a loading effect not occur in the case of a multi-stage triode amplifier ?

9.17 Draw the ac equivalent circuit of one stage of a multi-stage RC-coupled amplifier using triodes for (a) mid-frequencies, (b) low frequencies, and (c) high frequencies. Give the expressions of lower and upper cut-off frequencies in terms of circuit components.

9.18 What do you understand by Hi-Fi amplifier system ?

9.19 What are the different types of distortions that can occur while a signal is amplified by an amplifier? Give the reasons for each type of distortion.

9.20 It is said that phase distortion does not have any importance in the case of audio amplifiers. Why ?

9.21 Harmonic distortion is also called nonlinear distortion. Why ?

9.22 State at least one typical application of each type of coupling.

9.23 "In a multi-stage amplifier, the input impedance of an amplifier stage should be very high, and output impedance must be very low". Justify this statement.

9.24 In a two-stage amplifier, each stage uses identical transistors and components; yet the voltage gain of the first stage is much less than that of the second stage. Explain why this happens. If triodes or FETs are used in this circuit, then it is not so. Why ?

OBJECTIVE-TYPE QUESTIONS

I. Four suggested alternative answers are given below each of the following questions. Tick (√) the one which you think to be correct.

1. Two identical stages of triode amplifiers are cascaded by RC-coupling. If 10 is the mid-band voltage gain of each stage, the overall gain of the cascaded amplifier will be

(a) 40 dB   (c) 20 dB

(b) 100 dB   (d) $(20 \log_{10} 20)$ dB

2. For amplifying a signal containing frequency components from 450 kHz to 460 kHz, the most appropriate amplifier is
   (a) RC-coupled amplifier using triodes
   (b) RC-coupled amplifier using FETs
   (c) direct-coupled amplifier using pentodes
   (d) transformer-coupled tuned amplifier using transistors

3. The coupling capacitor C_c in an RC-coupled triode amplifier is usually a
   (a) 5 μF; mica capacitor
   (b) 0.05 μF; paper capacitor
   (c) 0.1 μF; electrolytic capacitor
   (d) 50 μF; electrolytic capacitor

4. The main component responsible for the fall of gain of an RC-coupled amplifier in low-frequency range is
   (a) the active device itself (tube, or transistor)
   (b) stray shunt capacitance C_S
   (c) coupling capacitor C_c
   (d) the grid-leak resistor R_G

5. The overall gain of a two-stage RC-coupled amplifier is 100. A signal voltage of 10 V, 1 kHz is applied across the output terminals of this amplifier. Then, the voltage obtained across the input terminals will be
   (a) 0.1 V, 1 kHz   (c) 100 V, 1 kHz
   (b) 0 V   (d) 10 V, 1 kHz

6. Harmonic distortion of the signal is produced in an RC-coupled transistor amplifier. The probable component responsible for this distortion is
   (a) the transistor itself
   (b) the power supply V_{CC}
   (c) the coupling capacitor C_c
   (d) the biasing resistors, R_1 and R_2.

7. In Fig. O.9.1 are given the static plate characteristics of two triode tubes I and II. When used as an amplifier

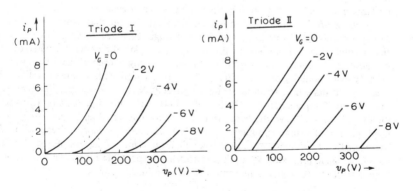

Fig. O.9.1

   (a) triode I will give more nonlinear distortion than triode II
   (b) triode II will give more nonlinear distortion than triode I
   (c) both the triodes will give same amount of nonlinear distortion
   (d) none of the triode gives any nonlinear distortion

II. **Some statements are written below. Tick those statements which you think are correct. Rewrite those statements which are false after making necessary corrections.**

1. In a multi-stage amplifier there are two or more stages.
2. The overall voltage gain of a multi-stage amplifier is obtained by adding the voltage gains of each stage when expressed as a voltage ratio.
3. When you connect an identical second-stage transistor amplifier to the first stage, the voltage gain of the first stage increases.
4. The lower cut-off frequency of a two-stage RC-coupled amplifier is higher than its value for the single-stage amplifier.
5. By cascading the second stage of an identical transistor amplifier, the upper cut-off frequency increases.
6 In a multi-stage amplifier, transformer coupling is used whenever we want to amplify very low frequency or dc signals.
7. RC-coupling is the best coupling scheme when frequency of the signal is in the range of 60 Hz to 20 kHz.
8. In a multi-stage voltage amplifier transformer coupling is usually used to amplify audio signals.
9. We always use RC coupling for amplifying a small band of rf (1400 kHz to 1410 kHz) signal.
10. While amplifying weak audio signals by a multi-stage amplifier, we should make sure that the coupling network does not disturb the biasing of the next stage.
11. From the loading point of view, the transistor is a better active device than a triode when used in a multi-stage voltage amplifier.
12. From the point of view of "loading effect" an FET is a better active device than a bipolar-junction transistor.
13. The coupling capacitors and bypass capacitors are responsible for the decrease of voltage gain at high frequencies in a multi-stage amplifier.
14. In a multi-stage amplifier, the voltage gain decreases at low frequencies mainly because of junction capacitances of the active device used.

Ans. I. 1. *a*, 2. *d*, 3. *b*, 4. *c*, 5. *b*, 6. *a*, 7. *b*.

II. 1. T. 2. The overall voltage gain of an amplifier is obtained by *multiplying* the voltage gains of each stage when expressed as voltage ratio; the overall voltage gain of an amplifier is obtained by adding the voltage gains of each stage when expressed *in terms of dB*. 3. When you connect an identical second-stage transistor amplifier to the first stage, the voltage gain of the first stage *decreases*. 4. T. 5. By cascading the second stage of an identical transistor amplifier, the upper cut-off frequency *decreases*. 6. In a multi-stage amplifier, direct coupling is used whenever we want to amplify low frequency or dc signals. 7. T. 8. In a multi-stage voltage amplifier, *RC coupling* is usually used to amplify audio signals. 9. We always use *transformer* coupling for amplifying a small band (1400 kHz to 1410 kHz) of rf signal 10. T. 11. From the "loading" point of view, the *triode* is a better active device than a *transistor*, when used in a multi-stage amplifier. 12. T. 13. The *shunt capacitance* is responsible for the decrease of voltage gain at high frequencies in a multi-stage amplifier. 14. The coupling capacitors and bypass capacitors are responsible for the decrease of voltage gain at *low* frequencies in a multi-stage amplifier.

TUTORIAL SHEET 9.1

1. A transistor multi-stage amplifier contains two stages. The voltage gain of the first stage is 50 dB and that of the second stage is 100. Calculate the overall gain of the multi-stage amplifier in dB. **[Ans. 90 dB]**

2. The overall voltage gain of a two-stage *RC*-coupled amplifier is 80 dB. If the voltage gain of the second stage is 150, calculate the voltage gain of the first stage in dB.

[**Ans. 36.47 dB**]

3. The voltage gain of a multi-stage amplifier is 65 dB. If the input voltage to the first stage is 5 mV, calculate the output voltage of the multi-stage amplifier.

[**Ans. 8.89 V**]

4. An audio signal contains 40 Hz as the lowest frequency and 10 kHz as the highest frequency. This signal is amplified by an amplifier whose maximum gain at 1 kHz is 20 dB. Draw the frequency response curve of this amplifier, indicating lower and upper cut-off frequencies.

EXPERIMENTAL EXERCISE 9.1

TITLE: Two-stage *RC*-coupled amplifier.

OBJECTIVES: To,

1. trace the given circuit and note down the values of each component;
2. measure the operating-point collector current and collector-to-emitter voltage for both the amplifier stages;
3. measure the voltage gain of the first stage with and without connecting second stage;
4. explain the loading effect of the second stage on the first stage;
5. measure the maximum signal which can be fed to the input of two-stage *RC*-coupled amplifier without causing distortion in the output;
6. measure the overall gain of the two-stage *RC*-coupled amplifier.

APPARATUS REQUIRED: Experimental board of two-stage amplifier, signal generator, CRO, ac millivoltmeter and electronic multimeter.

CIRCUIT DIAGRAM: As given in Fig. E.9.1.1 (typical values of the components are also given).

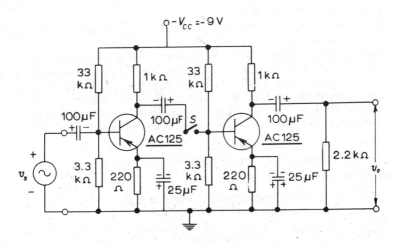

Fig. E.9.1.1

BRIEF THEORY: When the voltage gain provided by a single stage is not sufficient, we have more than one stage in the amplifier. The overall gain of the two stages is given as

$$A = A_1 \times A_2$$

where A_1 is the voltage gain of the first stage and A_2 is the voltage gain of the second stage.

In Fig. E.9.1.2, two stages are shown connected through a switch S. It is observed that the voltage gain of the first stage depends upon whether the switch S is closed or open. When the switch S is open, the voltage gain is high. If the switch S is closed the output voltage v_{o1} is very much reduced. This effect is known as *loading of first stage*. When the switch S is closed, the input impedance of the second stage comes in parallel with the load resistance of the first stage. Because of this, the effective load resistance of the first stage is reduced and hence the gain (or output voltage) also decreases.

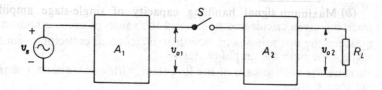

Fig. E.9.1.2

PROCEDURE:

1. Take the experimental board and trace the given circuit. Find out whether the transistor is *PNP* or *NPN*. Note down the values of all the resistors and capacitors.

2. Connect the dc voltage from the regulated power supply. Select a voltage, say 9 V. You may use 6 V or 15 V in case single-voltage IC power supply is available.

3. Note the Q point for both the transistors, i.e. find out I_C and V_{CE} for both the transistors.

4. Feed ac signal from an audio oscillator to the input of the first stage. Adjust the frequency at 1 kHz. See the output waveshape on the CRO. Go on increasing the input ac voltage till distortion starts appearing in the output waveshape. Note the value of this input signal. This is the maximum signal handling capacity. Repeat the same experiment with a single stage (open the switch S).

5. Measure the signal voltage at:
   (i) the output of the first stage
   (ii) the output of the second stage

   Calculate from these readings, the voltage gain of first stage, second stage and also overall gain of the two stages.

6. Disconnect the second stage and then measure the output voltage of the first stage. Calculate the voltage gain of the first stage under this condition and compare it with the voltage gain obtained when the second stage was connected.

OBSERVATIONS:

1. *Q point of the transistor:*

   $V_{CC} =$ _____ V

   For first stage:

   $V_{C_1} =$ _____ V; $I_{C_1} = \dfrac{V_{CC} - V_{C_1}}{R_{C_1}} =$ _____ mA

   $V_{CE_1} =$ _____ V

   For second stage:

   $V_{C_2} =$ _____ V; $I_{C_2} = \dfrac{V_{CC} - V_{C_2}}{R_{C_2}} =$ _____ mA

   $V_{CE_2} =$ _____ V

2. (a) Maximum signal handling capacity of the two-stage amplifier
      = _____ mV

   (b) Maximum signal handling capacity of single-stage amplifier =
      _____ mV.

3. *Voltage gain:*

S. no.	Input voltage	Output of first stage	Output of second stage	A_1	A_2	$A = A_1 \times A_2$
1.						
2.						

4. *Voltage gain with second stage disconnected:*

S. no.	Input voltage	Output voltage	Gain A_1
1.			
2			

RESULT:

1. Both the transistors are operating in active region, as shown by the Q-point readings.
2. The two-stage amplifier can handle a signal of _____ mV only. This is so, because the overall gain of the two stages is very high. The single-stage amplifier can handle a signal of _____ mV.
3. Loaded gain of the first stage ( ) is much less than its unloaded gain ( ).
4. The overall gain $A =$ _____.

   $= 20 \log_{10} ($ _____ $)$

   $=$ _____ dB

EXPERIMENTAL EXERCISE 9.2

TITLE: Frequency response curve of a two-stage RC-coupled amplifier.

OBJECTIVES: To,

1. identify the values of all the resistors and capacitors used in the given circuit.
2. make sure that the transistors are working in active region by measuring the Q point of the amplifier.
3. make sure that excessive signal is not fed to the input by seeing undistorted output waveshape on the CRO.
4. plot the frequency response curve of the single-stage amplifier.
5. determine the values of upper and lower cut-off frequencies of single-stage amplifier.
6. plot the frequency response curve of two-stage RC-coupled amplifier.
7. determine the values of upper and lower cut-off frequencies for a two-stage RC-coupled amplifier.
8. compare the bandwidth of a two-stage amplifier with that of single-stage amplifier.

APPARATUS REQUIRED: Experimental board, signal generator, electronic multimeter, ac millivoltmeter, CRO, and transistorized (or IC) power supply.

CIRCUIT DIAGRAM: This is same as Fig. E.9.1.1.

BRIEF THEORY: The voltage gain of an RC-coupled amplifier is maximum around 1 kHz. As frequency decreases, the gain starts falling. This decrease in gain at low frequencies is mainly because of coupling capacitors C_C. At low frequencies, coupling capacitors offer sufficiently high impedance. There occurs a voltage drop across these capacitors and hence the output voltage decreases. Also, the bypass capacitors at very low frequencies are no longer effective short-circuits. Because of this, the ac current passes through the resistor R_E. This gives rise to negative feedback, and the voltage gain reduces. The voltage gain also decreases at high frequencies because of (i) the shunt capacitances made up of junction capacitances and wiring capacitances, and (ii) the decrease in β at such frequencies.

The frequencies where voltage gain falls by 3 dB or becomes 70.7 % of the maximum value are called cut-off frequencies. These frequencies can be determined from the frequency response curve.

PROCEDURE:

1. You are already familiar with the experimental board. After making sure that transistors are biased in the active region, feed the input signal such that the output is undistorted at 1 kHz. Find the overall gain of the two-stage amplifier.
2. For plotting the frequency response curve of the first stage, disconnect the second stage. Now find the voltage gain of the amplifier at 1 kHz. Change the signal generator frequency on the lower frequency side. You will observe that the voltage gain is decreasing. Find such a frequency (f_1) where gain becomes 70.7 % of maximum gain. In a similar manner, find the signal frequency on the high frequency side (f_2) where voltage gain is reduced to 70.7 % of the maximum gain.

Calculate $f_2 - f_1$. This gives the value of bandwidth of the single-stage amplifier.

3. Now connect the second stage with the first stage. By feeding a 1 kHz signal, find the maximum gain. Now decrease the frequency such that the gain is reduced to 70.7 % of the maximum gain. This gives the value of lower cut-off frequency (f_1') of the two-stage amplifier. By changing the frequency of the signal generator on the higher frequency side, determine the value of the upper cut-off frequency (f_2'). The bandwidth of the two-stage amplifier $(f_2' - f_1')$ can now be calculated.

4. Most of the times, the signal generator does not supply constant output when the frequency is changed. It is possible that when measurements at f_1, f_2, f_1' and f_2' are made, the input would have changed. Now measure the voltage gain of a single stage as well as the two-stage amplifier near these frequencies. Take less number of readings of the voltage gain when it is constant. Make use of these readings to plot the frequency response curve (graph between voltage gain and frequency) on a semilog graph paper.

OBSERVATIONS:

1. *Q point of the transistors*:

   $V_{CC} = $ _____ V

   $V_{C_1} = $ _____ V; $I_{C_1} = $ _____ mA

   $V_{C_2} = $ _____ V; $I_{C_2} = $ _____ mA

2. (*a*) Input signal to the two-stage amplifier for which output is undistorted = _____ mV.

   (*b*) Input signal to single-stage amplifier for which output is undistorted = _____ mV.

3. *Frequency response of single stage*:

   Input signal = 5 mV

   (*a*) Voltage gain at 1 kHz = _____
   (*b*) Approximate value of f_1 = _____ Hz
   (*c*) Approximate value of f_2 = _____ kHz

4. *Frequency response of two-stage amplifier*:

   Input signal = 2 mV

   (*a*) Voltage gain at 1 kHz = _____
   (*b*) Approximate value of f_1' = _____ Hz
   (*c*) Approximate value of f_2' = _____ kHz

5. *Frequency response curve*.

S. no.	Frequency of input	Output voltage		Voltage gain	
		Single-stage	Two-stage	Single-stage	Two-stage
1.					
2.					

RESULTS:

1. The maximum voltage gain is _____ . It occurs at a frequency of _____ kHz.
2. The lower cut-off frequency (single-stage), $f_1 =$ _____
3. The upper cut-off frequency (single-stage), $f_2 =$ _____
4. The lower cut-off frequency (two-stage) $f_1' =$ _____
5. The upper cut-off frequency (two-stage) $f_2' =$ _____

It is observed that the voltage gain of the two-stage amplifier is much higher than the voltage gain of a single stage. But the bandwidth is reduced by cascading the second stage of the amplifier.

Power Amplifiers

OBJECTIVES: After completing this unit, you will be able to: ○ Explain the significance of power amplifier circuits in various electronic systems. ○ Draw single-ended and push-pull amplifier circuits. ○ Calculate the output ac power developed in a single-ended power amplifier, when supplied with various parameters. ○ Explain the working of a push-pull amplifier circuit. ○ Explain the advantages of push-pull amplifier over a single-ended power amplifier circuit. ○ Explain class-A, class-B and class-C operation; collector efficiency, collector dissipation and collector dissipation curve; harmonic distortion, with reference to power amplifiers. ○ Draw complementary symmetry push-pull amplifier circuit. ○ State the advantages and disadvantages of complementary symmetry circuits over conventional push-pull amplifier.

10.1 NEED FOR POWER AMPLIFIERS

In almost all electronic systems, the last stage has to be a power amplifier. For example, in a public-address system, it is the power amplifier that drives the loudspeakers. When a person speaks into a microphone, the sound waves are converted by it into electrical signal. This electrical signal is of very low voltage (a few mV). This signal, if fed directly, cannot drive the loudspeakers, to give sound (audio) output. The voltage level of this signal is first raised to sufficiently high values (a few V) by passing it through a multi-stage voltage amplifier. This voltage is then used to drive (or excite) the power amplifier. The power amplifier is capable of delivering power to the loudspeakers. The loudspeakers finally convert the electrical energy into sound energy. Thus, a large audience can hear the speech (or music from the orchestra, tape recorder, record player, or any other such gadget).

When we say, "a 3-W stereo tape recorder", it means the peak power fed to the loudspeakers is 3 W. The word "stereo" means "three dimensional". In this system, there are two separate channels for amplifying two signals originally obtained from the two microphones. When these signals drive two speakers placed at some distance in a room, the listener gets a three-dimensional audio effect. He feels as if a real orchestra is being played before him.

In case of broadcast transmitters, power amplifiers capable of feeding large power to the antenna are required. The power amplifier used in the last stage of a 1-kW transmitter develops a power of 1 kW. This power is fed to the antenna. The antenna then radiates the power, all around, in the form of electromagnetic waves (or radio waves). These waves are received by the receiving antenna. The signal developed in the antenna of the radio receiver is very small, and it cannot drive the loudspeaker directly. It has to be amplified first by a voltage amplifier and then by a power amplifier.

Thus we find that a power amplifier is an essential part of every electronic system.

10.2 DIFFERENCE BETWEEN VOLTAGE AMPLIFIER AND POWER AMPLIFIER

The primary function of the voltage amplifier is to raise the voltage level of the signal. It is designed to achieve the largest possible voltage gain. Only very little power can be drawn from its output. On the other hand, a power amplifier is meant to boost the power level of the input signal. This amplifier can feed a large amount of power to the load. To obtain large power at the output of the power amplifier, its input-signal voltage must be large. That is why, in an electronic system, a voltage amplifier invariably precedes the power amplifier. Also, that is why the power amplifiers are called large-signal amplifiers.

An important question arises—"Does a power amplifier actually amplify power ?" The answer is "no". In fact, no device can amplify power. The idea of amplifying power contradicts the basic principles of physics. What a power amplifier actually does is that it takes power from the dc power supply connected to the output circuit and converts it into useful ac signal power. This power is fed to the load (e.g. loudspeaker). The type of ac power developed at the output of the power amplifier is controlled by the input signal. Thus, we may say that a power amplifier is a dc-to-ac power converter, whose action is controlled by the input signal.

10.3 WHY VOLTAGE AMPLIFIER CANNOT WORK AS A POWER AMPLIFIER

The transistor used in a voltage amplifier need not have a large power-dissipation rating. It is not required to handle large power. However, the transistor must have large power dissipation rating, if it is to work in a power amplifier circuit. As a practical rule, a small-signal (voltage) transistor has a power dissipation less than 0.5 W; a power transistor more than 0.5 W.

Let us look at the simple amplifier circuit given in Fig. 10.1. In Unit 8, we had seen that this is a good voltage amplifier circuit. The question is

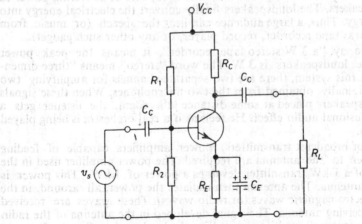

Fig. 10.1 Simple amplifier circuit

whether the circuit would work as a good power amplifier if we replace the transistor with another of higher power dissipation. We will see that this circuit is no good as a power amplifier.

In the circuit in Fig. 10.1, the whole of the voltage V_{CC} does not appear across the transistor. Under quiescent conditions, the dc collector current is $I_{CQ} \simeq I_{EQ}$. Then the dc voltage available between the collector and emitter of the transistor is

$$V_{CEQ} = V_{CC} - I_{CQ}(R_C + R_E) \qquad (10.1)$$

Thus, a portion of the supply voltage is dropped in the resistors. The dc power that goes into the transistor is

$$P_{DQ} = V_{CEQ}I_{CQ} \qquad (10.2)$$

This represents the *effective dc input power* to the amplifier. Only a portion of this power can be converted by the amplifier into the useful ac power, when the signal is applied.

Now, we can see why the circuit in Fig. 10.1 is not a good power amplifier. The total dc power drain from the supply is $V_{CC}I_{CQ}$. Out of this, only $V_{CEQ}I_{CQ}$ is the effective dc input power to the amplifier because, at best, that is the power that could be converted into useful ac power. The difference of power is

$$
\begin{aligned}
V_{CC}I_{CQ} - V_{CEQ}I_{CQ} &= I_{CQ}(V_{CC} - V_{CEQ}) \\
&= I_{CQ}I_{CQ}(R_C + R_E) \\
&= I_{CQ}^2(R_C + R_E)
\end{aligned}
$$

(we have used Eq. 10.1). This much power goes waste in unnecessarily heating the resistors.

We can attempt to reduce this wastage of power. The resistor R_E has to be there in the circuit, because it is a part of the biasing network. If R_E is absent, the stabilization of the operating point becomes poor. This may ultimately lead to the thermal runaway of the transistor. However, we can do something about the resistance R_C. This is dealt with in the next section.

10.4 HOW TO AVOID POWER LOSS IN R_C

We can avoid the dc power loss in R_C by simply short-circuiting it. But then the load resistance R_L (which is in parallel with R_C, from the ac point of view) also gets shorted. The result is disastrous. No power is transferred to the load R_L. The amplifier becomes useless.

It is to be seen whether we can replace R_C by a component whose dc resistance is zero, but ac impedance is very high. We can do this by replacing R_C by a choke (an inductor) as shown in Fig. 10.2. Two things are achieved by doing this. First, no dc voltage drop occurs across the choke (since the dc resistance is almost zero). We can afford to use lower voltage supply V_{CC} for the same amplifier. Second, the dc power loss in the choke is almost nil. Thus, this circuit is much better as compared to the one in Fig. 10.1.

Still more improvements can be made in this circuit so that it works as a better power amplifier. It can be made to convert a greater portion of the input dc power into useful ac output power. How it happens is explained in the next section.

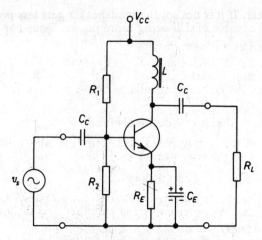

Fig. 10.2 Power amplifier having inductance-
capacitance (*LC*) coupled load

10.5 SINGLE-ENDED POWER AMPLIFIER

Figure 10.3*a* shows a typical single-ended transistor power amplifier. The term "single-ended" (denoting only one transistor) is used to distinguish this type from the push-pull amplifier (which uses two transistors; and is discussed in Sec. 10.8).

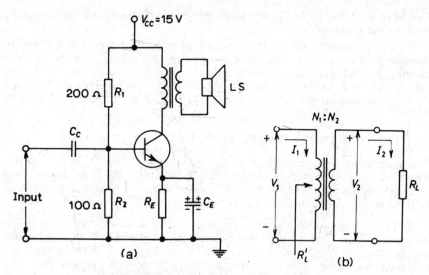

Fig. 10.3 Single-ended power amplifier

In many electronic systems, such as radio, television, tape recorder, public-address system, etc. the final output is in the form of sound. In such systems, the loudspeaker is the load for the power amplifier. The power amplifier makes the final stage and it drives the loudspeaker. We already know that maximum power will be transferred to the loudspeaker from the power amplifier, only if its output impedance is the same as the impedance of the loudspeaker. If it is not so, the loudspeaker gets less power, though the amplifier is capable of delivering more power. Since our purpose is to have maximum possible power output, we must *match* the loudspeaker impedance to the output impedance of the amplifier. This cannot be achieved in the circuit shown in Fig. 10.2 (and also Fig. 10.1). It does not have any provision for matching the load (the impedance of a loudspeaker is typically 8 Ω) with the output impedance of the transistor (of the order of a few kΩ). It is for this reason that we use a transformer in the circuit in Fig. 10.3a. This circuit has all the advantages of the inductance-capacitance coupling (Fig. 10.2). In addition, *the transformer provides impedance matching*.

10.5.1 Transformer Impedance Matching

A resistance R_L is connected across the secondary of a transformer with turns ratio N_1/N_2 (see Fig. 10.3b). The resistance seen looking into the primary of the transformer will be different than R_L. Let it be R'_L. We can find the ratio of the primary resistance to the secondary as follows:

$$\frac{R'_L}{R_L} = \frac{V_1/I_1}{V_2/I_2} = \left(\frac{V_1}{V_2}\right)\left(\frac{I_2}{I_1}\right) = \frac{N_1}{N_2}\frac{N_1}{N_2} = \left(\frac{N_1}{N_2}\right)^2$$

Therefore, the resistance looking into the primary is

$$R_L = \left(\frac{N_1}{N_2}\right)^2 R_L \qquad (10.3)$$

Thus, by using a step-down transformer of proper turns ratio, we can match a low R_L with high output impedance r_o of the transistor.

Example 10.1 We are to match a 16 Ω speaker load to an amplifier so that the effective load resistance is 10 kΩ. What should be the transformer turns ratio ?

Solution: Using Eq. 10.3, we get

$$R'_L = \left(\frac{N_1}{N_2}\right)^2 R_L$$

or

$$\left(\frac{N_1}{N_2}\right)^2 = \frac{R'_L}{R_L} = \frac{10\,000}{16} = 625$$

$\therefore$

$$\frac{N_1}{N_2} = \sqrt{625} = 25:1$$

Example 10.2 The turns ratio of a transformer is 15:1. If a load of 8Ω is connected across its secondary, what will be the effective resistance seen looking into the primary ?

Solution: Using Eq. 10.3, we get

$$R'_L = \left(\frac{N_1}{N_2}\right)^2 R_L = (15)^2 \times 8 = 1800\ \Omega = 1.8\ k\Omega$$

10.5.2 Power Considerations

In a power amplifier, the ac output power is sufficiently high, whereas the power in the base circuit is quite low. The question naturally arises where the power comes from. The only source of power in a power amplifier is the dc supply, V_{CC}. A portion of this dc input power appears as useful ac power across the load R_L. The rest of it is lost in the circuit. That is

$$\text{dc input power} = \text{ac output power} + \text{losses}$$

or
$$P_i(\text{dc}) = P_o(\text{ac}) + \text{Losses} \qquad (10.4)$$

Some of the power loss occurs in the primary of the transformer. The dc current is to pass through the primary which has got some finite resistance R_{dc}. This resistance being very small, the power loss in the primary is negligibly small. Some loss may occur in the core of the output transformer (in the form of hysteresis loss and eddy-current loss). This loss is also quite small. Some dc power is dissipated in the resistance R_E, since dc current passes through it. All these losses are small compared to the loss that occurs in the transistor itself. Of course, we cannot do without the transistor. It is because of the transistor that we have ac power available at the output.

In an earlier unit we discussed the physics behind transistor action. We had seen how the charge carriers (electrons in an NPN transistor, and holes in a PNP transistor) are made to cross the reverse-biased collector junction. When they do so, they lose some energy. This loss of energy appears as heat. The transistor must dissipate this heat to the surroundings. During the ac cycle, the collector current and voltage change. But on the average, a fixed amount of heat is produced at the collector. In other words, the transistor has to dissipate an *average power* which we denote as P_D.

In Eq. 10.4, the input dc power is obtained from the battery. It is given by the product of voltage V_{CC} and the average current drawn from the battery. If the amplifier is working in class-A operation, the average collector current will be the same as the quiescent collector current I_{CQ}. Therefore, the dc input power is

$$P_i = V_{CC} I_{CQ} , \qquad (10.5)$$

For the transformer-coupled amplifier, the only power lost is P_D which is dissipated by the transistor (other losses are negligible). We can now write Eq. (10.4) as

$$P_D = P_i - P_o$$

or
$$P_D = V_{CC} I_{CQ} - P_o \qquad (10.6)$$

This equation is simple, but it is very significant in the operation of a power amplifier. It shows that the power P_D dissipated by the transistor is the difference between the average dc input power P_i from the battery (which is constant for a fixed battery and operating point) and the output power P_o drawn by the load. In case the output power is zero, the transistor is required to dissipate or throw the maximum power, i.e. $P_D = V_{CC} \times I_{CQ}$. This is the "worst case" for the transistor. If the load does draw some power, the transistor is required to dissipate that much less power. In other words, the transistor has to work hardest (dissipate the maximum power) when the load is disconnected, or the input signal reduces to zero. The transistor dissipates least power when the load is drawing the maximum power from the circuit. Thus, the maximum power dissipation in the transistor occurs when the ac output power is zero, in which case

$$P_{D(max)} = V_{CC}I_{CQ} \qquad (10.7)$$

Obviously, this sets the safest power rating of the transistor. Under normal operation, with the load connected and the input signal present, the transistor has to dissipate less power. It is always preferable to ensure the presence of input signal and the output load as the amplifier is turned on.

If in an amplifier, the quiescent power dissipation P_{DQ} is 3.5 W, the transistor needs a power dissipation rating of at least 3.5 W. On the other hand, if we are given a power transistor with a dissipation rating of 3.5 W, we must ensure P_{DQ} does not exceed 3.5 W. This is an important factor that must be considered while designing a power amplifier. Figure 10.4 shows collector characteristics of a power transistor. Assume that its dissipation rating is 3.5 W, while designing the amplifier. We first plot its *collector dissipation curve*. We take some arbitrary values of V_{CE} and calculate corresponding values of I_C so that we always have $V_{CE}I_C = P_D = 3.5$ W. Some sample calculations are shown in Table 10.1. The curve obtained from these values is a hyperbola, as shown in Fig. 10.4. If this transistor is used in a power amplifier, its Q point must lie below this curve.

Table 10.1

V_{CE}	I_C	$P_D = V_{CE}I_C$
5 V	700 mA	3.5 W
10 V	350 mA	3.5 W
15 V	233 mA	3.5 W
20 V	175 mA	3.5 W
25 V	140 mA	3.5 W

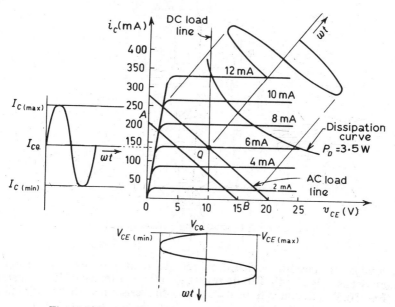

Fig. 10.4 Characteristics of a power transistor with its collector dissipation curve

A power amplifier is said to have high *efficiency*, if it can convert a greater portion of the dc input power drawn from the battery into the useful ac output power. We define output-circuit efficiency as the ratio of ac power to dc input power supplied to the collector-emitter circuit.

$$\eta = \frac{P_o \text{ (ac)}}{P_i \text{ (dc)}} = \frac{P_o}{V_{cc} I_{CQ}} \qquad (10.8)$$

Efficiency is a measure of how well an amplifier converts dc power from the battery into useful ac output power.

10.5.3 Analysis of Single-ended Power Amplifier

Power amplifiers are usually *large-signal* amplifiers. Under large-signal conditions, the transistor can no longer be considered a linear device. Its parameters such as β, r_{in}, or r_o (same as h_{fe}, h_{ie}, or $1/h_{oe}$) do not remain constant throughout its range of operation. Therefore, we cannot replace the transistor by its ac equivalent model for the analysis of this circuit, as we had done in the case of small-signal amplifiers. Here, a more appropriate method is the graphical method, though it is a little more tedious and laborious.

We shall discuss this method by taking a practical example. Figure 10.5 shows a single-ended power amplifier. The collector characteristics of the transistor are given in Fig. 10.4. The maximum dissipation rating for the transistor is specified as 3.5 W. We shall find, for this circuit, the rms values of collector current and voltage, and the ac power developed at the collector. Another quantity of interest, which is also calculated here, is the collector-circuit efficiency.

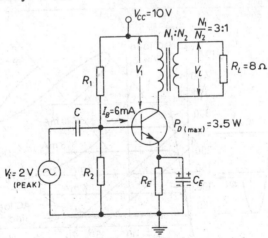

Fig. 10.5 A practical single-ended amplifier

To analyse the circuit, we first draw the dc load line on the collector characteristics (Fig. 10.4). The dc resistance of the primary of the transformer (Fig. 10.5) is assumed to be 0 Ω. Also, the resistance R_E is negligibly small. The dc load line is therefore a vertical, straight line. This is the ideal dc load line. Practical transformer windings do have small resistances. This along with the small emitter resistance R_E would provide a slight slope for the dc load line. Here, we consider the ideal case. There is no voltage drop across the dc load resistance, and the load line is drawn vertically from the voltage point $V_{CEQ} = V_{CC}$ (here, $V_{CC} = 10$ V).

The quiescent operating point of the amplifier should lie somewhere on this load line. The Q point must be located below the dissipation curve, if the transistor is not to burn out in the worst case. The biasing network R_1, R_2 and R_E fix the base current, under dc conditions. Let us say, in this case, that the dc base current is fixed at 6 mA. The Q point is then the point of intersection of the dc load line and the collector characteristic corresponding to 6-mA base current curve.

Next, we draw the ac load line. For this, we need the load resistance R_L' seen looking into the primary of the transformer. This can be calculated using Eq. 10.3. In our circuit, the loudspeaker resistance $R_L = 8$ Ω, and the transformer turns ratio is 3 : 1. Then,

$$R_L' = \left(\frac{N_1}{N_2}\right)^2 R_L = 3^2 \times 8 = 72 \ \Omega$$

We know that the slope of the ac load line is $-1/R_L'$, (negative of the reciprocal of the effective load resistance). Since the collector signal passes through the Q point, when the input ac signal goes through zero in its cycle, the ac load line must pass through the Q point. To draw such a line, we proceed as follows.

Take any voltage V_{CE}, say 15 V. Find the current I_C given by

$$I_C = \frac{V_{CE}}{R_L'}$$

Here, this current is $I_C = 15/72 = 208$ mA. Mark this current on the y axis. Join this point to the point $V_{CE} = 15$ V to get the line AB. This line will have the slope of the required ac load line. Now, we can draw the ac load line that passes through Q point and is parallel to the line AB (Fig. 10.4).

We can now see how the circuit works as a power amplifier. Suppose, at the input, we apply a sinusoidal signal of 2 V (peak). If the input impedance of the transistor is 500 Ω, the variation in the resulting base current will be $2/500 = 4$ mA (peak value). This variation in base current and the corresponding variations in collector current and collector voltage are shown in Fig. 10.4. The variation of collector voltage appears across the primary of the transformer. An ac voltage is induced in the secondary, which in turn develops ac power in R_L.

Notice that during the operation of the amplifier, the collector voltage swings much beyond the value of V_{CC}, the supply voltage. We must ensure that at no instant, the maximum voltage rating of the transistor is exceeded by the collector voltage. (For the consideration of the maximum ratings, see Sec. 7.1).

From the graph in Fig. 10.4, we can find the maximum and minimum values of the collector current and voltage, between which the signal swings. The ac power developed across the transformer primary can be calculated to be

$$P_o(\text{ac}) = V_{CE}(\text{rms}) \times I_Q(\text{rms}) = \frac{V_{CE}(\text{peak})}{\sqrt{2}} \times \frac{I_C(\text{peak})}{\sqrt{2}}$$

$$= \frac{(V_{CE(max)} - V_{CE(min)})}{2\sqrt{2}} \frac{(I_{C(max)} - I_{C(min)})}{2\sqrt{2}} \tag{10.9}$$

or $\qquad P_o = \dfrac{[V_{CE(max)} - V_{CE(min)}] \, [I_{C(max)} - I_{C(min)}]}{8}$

The same power appears across the load R_L, if the transformer is 100 % efficient. Assuming the losses in the transformer to be negligible, we may now calculate the ac power delivered to the loudspeaker R_L from Eq. 10.9. In this case,

$$V_{CE(max)} = 18.0 \text{ V}; \; V_{CE(min)} = 2.0 \text{ V}$$

$$I_{C(max)} = 245 \text{ mA}; \; I_{C(min)} = 25 \text{ mA}$$

Therefore $\qquad P_o = \dfrac{(18.0 - 2.0) \times (245 - 25) \times 10^{-3}}{8}$

$$= \frac{16.0 \times 0.220}{8}$$

$$= \textbf{0.44 W}$$

The quiescent collector current, from the graph, is $I_{CQ} = 135$ mA. The dc input power to the amplifier is given by Eq. 10.5 and in this case

$$P_i \text{ (dc)} = V_{CC} I_{CQ}$$

$$= 10 \times 0.135 = \textbf{1.35 W}$$

The output circuit efficiency can now be calculated by using Eq. 10.8.

$$\eta = \frac{P_o}{P_i(\text{dc})} = \frac{0.44}{1.35} \times 100 \text{ \%} = \textbf{32.6 \%}$$

We can use Eq. 10.6 to determine the power dissipated by the transistor.

$$P_D = P_i \text{ (dc)} - P_o = 1.35 - 0.44 = \textbf{0.91 W}$$

From the above calculations we see that, under the conditions specified, the collector-circuit efficiency of the single-ended power amplifier circuit is 32.6%. This, in fact, is low. A large part of the net input dc power is dissipated by the transistor as heat. By increasing the signal at the input we could obtain more output power. This way, the efficiency also goes up. But then the amplifier is driven into saturation (or into cut-off, or both) during a part of the ac cycle. This will cause a large distortion in the output. To keep the distortion minimum, the amplifier has to work under class-A operation. For this operation, the maximum theoretical efficiency is 50 %. In practical circuits, the efficiency is less than 35 %.

It is possible to have higher efficiency by working the amplifier under class-B (or class-AB, or even class-C) operation; and yet keep the distortion low. This is achieved by a circuit called a *push-pull amplifier*. We shall discuss this circuit in Sec. 10.8.

10.6 WHY CLASS-B AND CLASS-C OPERATION IS MORE EFFICIENT THAN CLASS A

Class-A operation provides collector (output) current during the complete cycle of the input signal. Figure 10.6a shows the output current for class-A operation. Here, the output signal does not exceed values $I_{C(\text{max})}$ or $I_{C(\text{min})}$. The collector current under no-signal conditions is I_{CQ}. Figure 10.6*b* shows class-B operation. The Q point is set at cut-off. The quiescent collector current is zero. The output current flows for only 180° of the input cycle. Since, under no-signal condition, the collector current is zero, no power is dissipated by the transistor. The transistor handles an average collector current, only when an input signal is applied.

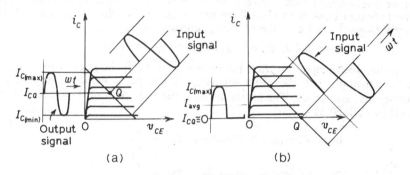

Fig. 10.6 Output of an amplifier under (*a*) class A operation
(*b*) class B operation

Average collector current I_{avg} increases for larger input signals. Thus, in class-B operation, the transistor dissipation is zero under no-signal condition; and it increases with increased input signals. This is quite contrary to class-A operation. The worst condition in class-A operation occurs when no input signal is present; the transistor then has to dissipate maximum power. Least power is dissipated by the transistor when the input signal is maximum.

Since the average current in class-B operation is less than that in class-A, the amount of power dissipated by the transistor is also less. Maximum theoretical efficiency is 78.5 % in class-B operation.

Class-AB operation is in-between class-A and class-B operation. The collector current flows for more than 180° but less than 360° of the input cycle. The operating efficiency of class-AB is between that of class-A and class-B.

In class-C operation, the collector current flows for less than 180°. The average collector current is much less, and as a result the collector losses are still less so that the efficiency is very high. Class-C operation is used with resonant or tuned circuits as for example, in radio and television transmitters where efficiency is of utmost importance. The tuned circuit helps in rejecting harmonics that are developed in the transistor due to its class-C operation.

10.7 HARMONIC DISTORTION IN POWER AMPLIFIERS

The quality of the sound given by the loudspeaker (connected in the output of the amplifier) depends very much upon the nature of the output signal. If the output is distorted, the quality of the sound will be poor. Consideration of harmonic or nonlinear distortion is therefore important in power amplifiers. Because they handle large signals, distortion is always present. If the output signal is not sinusoidal for a sinusoidal input, we say that harmonic distortion is introduced (see Sec. 9.6). As is clear from Fig. 10.6b, large amount of harmonic distortion is present in class-B operation (output waveshape is half sinusoidal). Even in class-A operation some distortion occurs; more so when the signal is large. This type of distortion is present because the transistor does not work as a linear amplifier when the signal is large. Because of this non-linearity, the signal is distorted in its waveshape. If we apply a sinusoidal signal $v_i = V \sin \omega t$ to the input of a power amplifier (whether class-A or class-B, or any other class) the waveform of the output signal can be mathematically represented as

$$i_o = I_0 + I_1 \sin \omega t + I_2 \sin 2\omega t + I_3 \sin 3\omega t + \ldots \qquad (10.10)$$

In the above equation, I_0 is the dc component, I_1 the peak value of the first harmonic (or the fundamental), I_2 the peak value of the second harmonic; and so on. The harmonic distortion for each of these components is then defined as

$$\text{Second harmonic distortion} \quad D_2 = \frac{I_2}{I_1}$$

$$\text{Third harmonic distortion} \quad D_3 = \frac{I_3}{I_1}$$

and so on $\qquad (10.11)$

When distortion occurs, the output power calculated on the basis of non-distortion is no longer correct. For example; the power calculated from Eq. 10.9 is the output power only for the non-distorted case. When distortion is present, the output power due to the fundamental component of the distorted signal is

$$P_1 = \frac{I_1^2 R_L}{2} \qquad (10.12)$$

The total power due to all the harmonic components at the output is

$$P = (I_1^2 + I_2^2 + I_3^2 + \ldots) \frac{R_L}{2}$$

$$= \left[1 + \left(\frac{I_2}{I_1} \right)^2 + \left(\frac{I_3}{I_1} \right)^2 + \ldots \right] \frac{I_1^2 R_L}{2}$$

$$= (1 + D_2^2 + D_3^2 + \ldots) P_1 \qquad (10.13)$$

We may define the total distortion or distortion factor as

$$D = \sqrt{D_2^2 + D_3^2 + \ldots} \qquad (10.14)$$

The total power can then be written from Eq. 10.13

$$P = (1 + D^2) P_1 \qquad (10.15)$$

If the total distortion is 10 %, then the total power is

$$P = [1 + (0.1)^2] P_1 = 1.01 \; P_1$$

This shows that a 10 % distortion represents a power of only 1 % of the fundamental. Thus, only a small error is made in using only the fundamental term P_1 for calculating the output power.

Example 10.3 A sinusoidal signal $v_s = 1.75 \sin 600\ t$ is fed to an amplifier. The resulting output current is of the form

$$i_o = 15 \sin 600\ t + 1.5 \sin 1200\ t + 1.2 \sin 1800\ t + 0.5 \sin 2400\ t$$

Calculate (a) second, third and fourth harmonic distortions; (b) percentage increase in power because of distortion.

Solution: The percentage harmonic distortion of each component is given as:

$$D_2 = \frac{I_2}{I_1} \times 100 = \frac{1.5}{15} \times 100 = \mathbf{10\ \%}$$

$$D_3 = \frac{I_3}{I_1} \times 100 = \frac{1.2}{15} \times 100 = \mathbf{8\ \%}$$

$$D_4 = \frac{I_4}{I_1} \times 100 = \frac{0.5}{15} \times 100 = \mathbf{3.33\ \%}$$

The distortion factor is

$$D = \sqrt{D_2^2 + D_3^2 + D_4^2} = \sqrt{(0.1)^2 + (0.08)^2 + (0.0333)^2}$$
$$= 0.1323 = \mathbf{13.23\ \%}$$

The net output power is

$$P = (1 + D^2)P_1 = [1 + (0.1323)^2]\ P_1 = 1.0175\ P_1$$

Thus the percentage increase in power is

$$= \frac{(1.0175\ P_1 - P_1)}{P_1} \times 100 = \mathbf{1.75\ \%}$$

10.8 PUSH-PULL AMPLIFIER

A push-pull amplifier circuit uses two transistors as shown in Fig. 10.7. This circuit can work in class-B, class-AB or class-A operation. Because of the special circuit connection, it provides a surprisingly low distortion, and at the same time a highly efficient operation (class-AB or class-B). Similar circuits are also made by using vacuum tubes (or FETs) as shown in Fig. 10.8.

The audio power amplifier used in transistor receivers, tape recorders, record players, PA systems, etc. make use of this circuit. These systems are usually operated by batteries (cells). Here, the efficiency of the amplifier is of prime importance; so that the battery lasts longer. However, in case of valve radio receivers which are usually operated by power mains, single-ended power amplifiers are used.

The circuit in Fig. 10.7 uses two transistors T1 and T2. The emitter terminals of the two transistors are connected together. The circuit has two transformers—one at the input and the other at the output. The input transformer has a centre-tapped secondary winding. It provides opposite polarity signals to the two transistor inputs. The primary of the output transformer is also centre-tapped. The collector terminals of the two transistors are connected to the supply V_{CC} through the primary of this transformer.

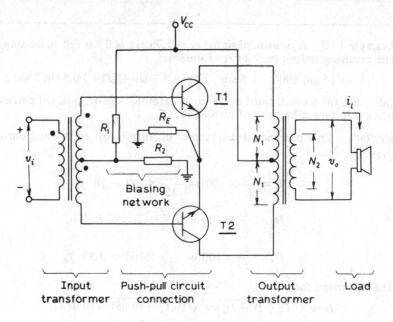

Fig. 10.7 Push-pull amplifier circuit using transistors

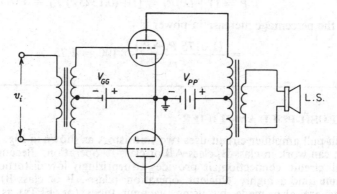

Fig. 10.8 A vacuum-triode amplifier using push-pull connection

The load resistance (usually a loudspeaker) is connected across the secondary of the output transformer. The turns ratio $2N_1 : N_2$ of the transformer is chosen so that the load R_L is matched with the output impedance of the transistor. Under matched conditions, maximum power is delivered to the load by the amplifier. Note that the resistors R_1, R_2 and R_E form the biasing network.

Let us see how opposite polarity signals appear to the two transistor inputs when we apply a signal v_i to the amplifier input (see Fig. 10.9). Assume that when we apply a sinusoidal signal v_i at the input, induced voltage of 20 V (peak) is developed across the secondary winding (i.e. across the terminals AB). Point C is the centre-tap of the secondary. This point is at ground (0 V potential) as regards ac. The voltages across each half of the

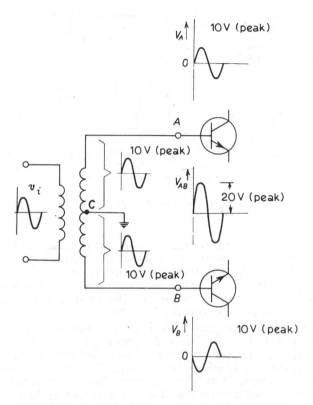

Fig. 10.9 Opposite-polarity inputs to the two transistors
obtained by using a centre-tapped transformer

secondary are 10 V (peak). The two voltages add up to a total of 20 V
across the whole winding. Let us consider the instant at which the voltage
at point A is 20 V with respect to point B, i.e. when $V_{AB} = 20$ V. At this
instant, the voltage at point A with respect to point C is 10 V, i.e. $V_{AC} =
V_A = 10$ V. At the same time, the voltage at point C with respect to point
B is also 10 V, i.e. $V_{CB} = 10$ V. That is $V_{BC} = -V_{CB} = -10$ V. This is
the voltage at point B with respect to the ground, and this voltage V_B is
appearing at the input of the transistor T2. Thus, the signals appearing at
the base of the two transistors are of opposite polarity. It can also be said
that, they are in opposite phase (phase difference of π radians). The resulting base currents of the two transistors can then be written as

$$i_{b1} = I_b \sin \omega t \qquad (10.16)$$
$$i_{b2} = I_b \sin (\omega t + \pi) \qquad (10.17)$$

We shall now see what happens at the output of the amplifier. We first
consider class-A operation (although push-pull connection can be more
efficiently used in class-B or class-AB operation). As shown in Fig. 10.10,
the quiescent collector currents (I_{CQ}) of the two transistors flow in opposite
directions through the two halves of the primary winding. These currents
produce opposite flux through the magnetic core of the transformer. If the

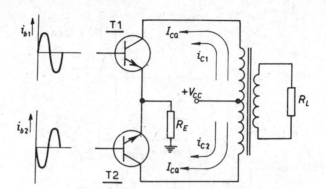

Fig. 10.10 Details of push-pull operation at the output

two transistors are perfectly matched, the net flux in the core is zero (matched or balanced-case means; the two transistors are exactly identical and the number of turns on the two sides of the centre tap are exactly the same). This is a great advantage. Now, the transformer need not handle a large flux due to the dc (quiescent) currents. We can use small-sized core, as transformers are supposed to operate near zero flux.

When an ac signal is applied to the input, opposite-phased, varying base-currents flow in the two transistors. As a result, the ac collector currents in the two transistors are also in opposite phase. The total current i_{c1} in transistor T1 and the total current i_{c2} in transistor T2 varies as shown in .Fig. 10.11a and b, respectively. These currents flow in opposite directions in two halves of the primary winding. The flux produced by these currents will also be in opposite directions. The net flux in the core will be the same as that produced by the difference of the currents i_{c1} and i_{c2}. To find the difference $i_{c1} - i_{c2}$, we first find the negative of i_{c2}. This is done in Fig. 10.11c. We can now add the currents of Fig. 10.11a and c to get the difference, since

$$i_{C1} - i_{C2} = i_{C1} + (-i_{C2})$$

The difference of the two collector currents is obtained in Fig. 10.11d. Note that during this process the quiescent currents (I_{CQ}) of the two transistors get cancelled, but the ac currents get added up. The overall operation results in a net ac current flow through the primary of the transformer. This results in a varying flux in the core. An ac voltage is induced in the secondary, and the ac power is delivered to the load resistor R_L.

From Fig. 10.11a and b, it may be seen that during the first half-cycle, the current i_{c1} increases, but at the same time the current i_{c2} decreases. In other words, when one transistor is being driven into more conduction, the other is driven into less conduction. The reverse happens in the next half-cycle. This amounts to saying that when the current in one transistor is "pushed up", the current in the other transistor is "pulled down". Hence, the name push-pull amplifier.

10.8.1 Distortion in Push-pull Amplifier

A power amplifier handles large signals. Harmonic distortion of the signal occurs due to the nonlinear characteristics of the transistors. In a push-pull

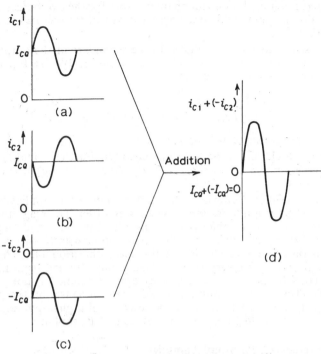

Fig. 10.11 (a) Collector current i_{c_1} in transistor T1

(b) Collector current i_{c_2} in transistor T2;

(c) The negative of i_{c_2}; (d) The difference $i_{c_1} - i_{c_2}$

amplifier, this distortion is much reduced. To prove this, we proceed as follows:

When we apply a sinusoidal signal of frequency ω at the input of the amplifier, we get opposite-phased base currents given by Eqs. 10.16 and 10.17, repeated here for convenience

$$I_{b_1} = I_b \sin \omega t \qquad\qquad (10.16)$$

$$I_{b_2} = I_b \sin (\omega t + \pi) \qquad\qquad (10.17)$$

We have seen in Sec. 10.7 that due to nonlinearity of the transistor, even if the input is sinusoidal, the output current contains harmonic terms. The output can then be represented by an equation showing the presence of harmonic terms. This is done in Eq. 10.10. Now for the push-pull amplifier, the input to transistor T1 is given by Eq. 10.16. We can represent the collector current for this transistor as

$$i_{C1} = I_0 + I_1 \sin \omega t + I_2 \sin 2\omega t + I_3 \sin 3\omega t + \ldots \qquad (10.18)$$

For transistor T2, the input is given by Eq. 10.17. Its output can be written using Eq. 10.18 and replacing ωt by $\omega t + \pi$, as

$$i_{C2} = I_0 + I_1 \sin (\omega t + \pi) + I_2 \sin 2(\omega t + \pi) + I_3 \sin 3(\omega t + \pi) + \ldots$$

$$= I_0 + I_1 \sin (\omega t + \pi) + I_2 \sin (2\omega t + 2\pi) + I_3 \sin (3\omega t + 3\pi) + \ldots$$

or $\quad i_{C2} = I_0 - I_1 \sin \omega t + I_2 \sin 2\omega t - I_3 \sin 3\omega t + \ldots \qquad (10.19)$

Here, we have assumed identical characteristics for the two transistors. Such an assumption is quite valid, since matched pair of transistors are available in the market.

We have seen that voltage induced in the secondary of the output transformer is proportional to the difference $i_{C1} - i_{C2}$. We use Eq. 10.18 and 10.19 and write the output voltage

$$v_o = k \, (i_{C1} - i_{C2})$$
$$= 2 \, k(I_1 \sin \omega t + I_3 \sin 3 \omega t + I_5 \sin 5\omega t + \ldots) \qquad (10.20)$$

where k is some constant of proportionality.

The power developed in the load R_L depends upon the voltage v_o. *Note that no even harmonic terms appear in Eq. 10.20.* The output is free from even harmonics.

In a power amplifier, all the harmonic terms produced do not have the same magnitude. Their magnitudes are in decreasing order. The magnitude of the second harmonic is more than that of the third harmonic; and the magnitude of the third harmonic is more than that of fourth harmonic, and so on. In fact, it is the second harmonic which is most objectionable. The rest of the harmonic terms are very small in magnitude. We find from Eq. 10.20, that the second harmonic (and also other even harmonics) is not present in the output. They (even harmonics) get cancelled because of the push-pull connection. The net distortion in the output of a push-pull amplifier is much less than it would have been in a single-ended amplifier.

10.8.2 Advantages of Push-pull Amplifier

A push-pull amplifier possesses many advantages over a single-ended power amplifier. For this reason, to get a given output power, we prefer using two transistors in push-pull connection rather than using a single larger power transistor in a single-ended circuit. The advantages of the push-pull circuit connection are summarized below:

(i) The dc components of the collector currents (I_{CQ}) oppose each other in the transformer. This results in zero dc flux in the core. The magnetic saturation of the core by dc current does not occur. We can use smaller size transformers. The cost thus becomes low. There is another advantage due to the core being biased near zero flux. The magnetization curve is linear near zero flux value. As a result the nonlinear distortion due to the nonlinear magnetization is avoided.

(ii) The output has much less distortion due to the cancellation of all the even harmonic components.

(iii) The dc supply V_{CC} (or V_{PP} in the case of vacuum-tube circuits) is usually obtained by rectification of the power mains. The unidirectional output of the rectifier is further smoothened by a filter circuit to give a better dc. Even after filtering, some ac ripples are left. This ripple voltage apears at the output of the amplifier as humming noise. However, in a push-pull circuit connection, these ripples get cancelled. The ripple currents flow in the opposite direction in the two halves of the primary winding (of output transformer). Thus, the ripple voltage in the voltage supply does not affect the output. We can use a poorer (less costly) smoothing filter in the voltage supply V_{CC}, without any adverse effect.

(iv) The net current flowing through the emitter resistor R_E is the sum of the two collector currents i_{C1} and i_{C2}. It may be seen from Eqs. 10.18

and 10.19, that this sum does not contain the fundamental component. The lowest frequency is 2ω. There is no need of bypassing R_E by capacitor C_E for the fundamental frequency components. And to bypass the second harmonic term, we can use a smaller capacitor. The cost is thus reduced.

(v) We can use class-AB or class-B operation to get high efficiency without producing much distortion. However, if a single-ended amplifier is used in class-AB or class-B, a large distortion results.

10.8.3 Phase Inverter Circuit

The push-pull amplifier circuit of Fig. 10.7 or 10.8 uses two transformers—one at the input and another at the output. In practical electronic circuits it is desirable to avoid the use of transformers (also of inductors) due to the following reasons:

(i) The electronic system becomes bulky with the use of transformers.

(ii) A fairly good transformer is quite expensive as compared to other components like resistors, capacitors, diodes and transistors.

(iii) The frequency response of the amplifier becomes poor because of the stray interwinding capacitances.

(iv) Unless careful shielding is used, the transformer's magnetic field may cause distortion of the signal.

(v) In integrated-circuit technology, it is very difficult to make an inductor or a transformer.

The purpose of the input transformer is to provide signals to the two stages which are equal in magnitude but opposite in phase. The same result may be achieved by using a circuit called *phase inverter*, *paraphase*, or *phase-splitter circuit*. Such a circuit, using a transistor, is shown in Fig. 10.12. In this circuit, resistors R_1 and R_2 form the voltage divider which forward-biases the emitter-base junction of the transistor. The collector resistor R_C and the emitter resistor R_E are equal in value, as are the coupling capacitors C_2 and C_3.

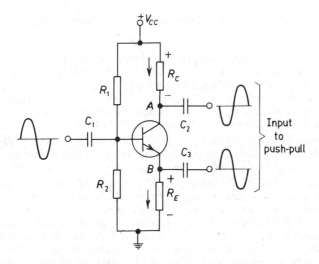

Fig. 10.12 Phase inverter circuit

Approximately the same current flows through R_C and R_E in the directions indicated by the arrows. Hence, as a result of the voltage drops, point A becomes as much negative as point B becomes positive. The circuit provides opposite-phase signals to drive the push-pull amplifier stage. The use of input transformer is thus eliminated. The use of output transformer can also be eliminated by a circuit called *complementary-symmetry* circuit. This is explained in the next section.

10.8.4 Complementary-Symmetry Push-pull Circuit

It is possible to eliminate both the input and output transformers in an ordinary push-pull amplifier circuit. One such circuit is shown in Fig. 10.13. This uses a pair of transistors having complementary symmetry; that is, one transistor is *PNP* and the other is *NPN*. Since, complementary vacuum triodes do not exist, there can be no vacuum-tube counterpart of this circuit.

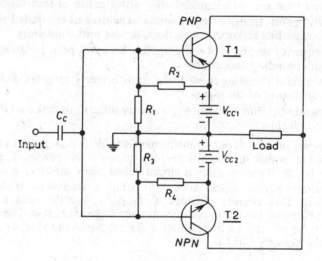

Fig. 10.13 Complementary-symmetry push-pull circuit

Note that the complementary symmetry circuit of Fig. 10.13 requires two power supplies (V_{CC_1} and V_{CC_2}), since each transistor must be biased suitably. Resistors R_1 and R_2 act as a voltage divider across V_{CC_1} to forward-bias the emitter-base junction of transistor T1 (which is *PNP* type). Similarly, the resistors R_3 and R_4 act as a voltage divider across V_{CC_2} to forward-bias the emitter-base junction of transistor T2 (which is *NPN* type).

The transistors are operated in class-B. That is, the bias is adjusted such that the operating point corresponds to the cut-off points. Hence, with no signal input, both transistors are cut-off and no collector current flows.

The signal applied at the input goes to the base of both the transistors. Since the transistors are of opposite type, they conduct in opposite half-cycles of the input. For example, during the positive half-cycle of the input signal, the *PNP* transistor T1 is reverse-biased and does not conduct. The *NPN* transistor T2, on the other hand, is forward-biased and conducts. This results in a half-cycle of output voltage across the load resistor. The other half-cycle of output across the load is provided by the conduction of transistor T1 (the transistor T2 remains cut-off) during the negative half-cycle of the input. Since the collector current from each transistor flows through the load during alternate half-cycles of the input signal, no centre-tapped output transformer is required.

The two transistors—though of opposite type—must be matched. If there is an imbalance in the characteristics of the two transistors, even harmonics will no longer be cancelled. This would result in considerable distortion. Increasing availability of complementary transistors is making the use of class-B transformer-coupled stages obsolete. All modern power-amplifier circuits are transformerless and use complementary transistors.

REVIEW QUESTIONS

10.1 Explain why a power amplifier is also known as a large-signal amplifier.
10.2 Explain the difference between a voltage amplifier and a power amplifier
10.3 Explain the following terms in connection with power amplifiers:
   (a) collector circuit efficiency;
   (b) collector dissipation rating;
   (c) class-A, class-B and class-C operation;
   (d) harmonic distortion.
10.4 Explain why a power amplifier is always preceded by a voltage amplifier.
10.5 Draw the circuit diagram of a single-ended power amplifier circuit. Explain the function of each component used in this circuit.
10.6 Explain how ac power is developed across a loudspeaker in a single-ended power amplifier.
10.7 Explain why a step-down transformer is used in the output of a power amplifier circuit.
10.8 Single-ended power amplifier is not much used in practical circuits. Instead, a push-pull amplifier circuit is used. Why ?
10.9 Although a class-B single-ended power amplifier has high efficiency, yet it is never used in practical circuits. Explain why ?
10.10 Draw the circuit diagram of a push-pull amplifier circuit. Explain:
   (a) how proper biasing is achieved in this circuit, and;
   (b) how ac power, free from even harmonics, is developed across the load.
10.11 State the advantages achieved by using a push-pull amplifier circuit, instead of connecting the two transistors (or tubes) in parallel.
10.12 State the disadvantages of using transformers in a push-pull amplifier. Explain some technique that eliminates the use of the input transformer.
10.13 Draw the circuit diagram of a phase inverter. Explain its utility in power amplifier circuits.
10.14 Explain why the complementary-symmetry power amplifier has become more popular in modern circuits. Draw a practical circuit of a complementary-symmetry push-pull amplifier circuit.

OBJECTIVE-TYPE QUESTIONS

I. Here are some incomplete statements. Four alternatives are given below each. Tick the alternative which completes the statement correctly.

1. The output power of a power amplifier is several times its input power. It is possible because
   (a) the power amplifier introduces negative resistance
   (b) the power amplifier converts a part of the input dc power into ac output
   (c) positive feedback exists in the circuit
   (d) step-up transformer is used in the circuit

2. The main function of the transformer used in the output of a power amplifier is
   (a) to step-up the voltage
   (b) to increase the voltage gain
   (c) to match the load impedance with dynamic output resistance of the transistor
   (d) to safeguard the transistor against overheating

3. Heat sinks are used in power amplifier circuits
   (a) to increase the output power
   (b) to reduce the heat losses in the transistor
   (c) to increase the voltage gain of the power amplifier
   (d) to increase the collector dissipation rating of the transistor

II. Tick the correct statements given below:
   1. Push-pull amplifier uses two transistors.
   2. Push-pull amplifier reduces odd harmonics in the output.
   3. Power amplifiers are also known as small-signal amplifiers.
   4. In class-C amplifiers, the bias voltage is beyond the cut-off voltage.
   5. The collector current in quiescent condition is zero when the amplifier is working in class-A operation.

III. Match each of the member of column A with appropriate member of column B.

Column A	Column B
1. Maximum efficiency of a class-B amplifier is	(a) 50 %
	(b) 78.5 %
2. Maximum efficiency of a class-A amplifier is	(c) 100 %
	(d) 25 %
3. Maximum efficiency of a class-C amplifier is	(e) 80 %

Ans. I. 1. (b), 2. (c), 3. (d); II. 1., 4.; III. 1. (b), 2. (a), 3. (c)

TUTORIAL SHEET 10.1

1. For the circuit shown in Fig. T.10.1.1, calculate the power across the load resistor R_L, if the load resistor R_L assumes the values: (a) 20 Ω, (b) 5 kΩ, (c) 8 kΩ, (d) 20 kΩ. For which value of this resistor is the power consumed by the load maximum ?

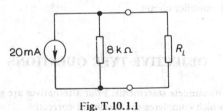

Fig. T.10.1.1

[**Ans.** (*a*) 7.96 mW, (*b*) 757.396 mW, (*c*) 800 mW, (*a*) 653.06 mW.
For maximum power, R_L should be equal to internal
resistance, i.e., 8 kΩ.]

2. In the basic power amplifier circuit given in Fig. T.10.1.2, calculate the turns ratio
of the transformer for obtaining maximum output power. Assume the loudspeaker
resistance equal to 4 Ω and the dynamic output resistance ($1/h_{oe}$) of the transistor
to be 14.4 kΩ. [**Ans.** 60 : 1]

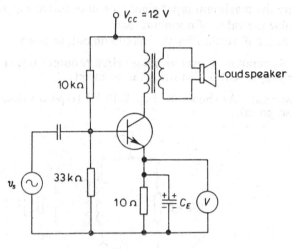

Fig. T.10.1.2

3. In the above question calculate the values of collector current (I_C) and collector-to-
emitter voltage (V_{CE}), if the voltmeter connected across the emitter resistor indicates 1 V. Assume the dc resistance of the primary winding to be 10 Ω.

[**Ans.** 100 mA, 10 V]

4. Signal power is to be delivered to a loudspeaker having a resistance of 4 Ω. The
output transformer used in the power amplifier for this purpose has a turns ratio
of 20 : 1. The primary winding of the transformer gets ac signal from a transistor
which can be represented by a current source of 5 mA and shunt resistance of 8 kΩ.
Calculate the power delivered to a loudspeaker when it is connected to the
secondary of the transformer. [**Ans.** 27.778 mW]

TUTORIAL SHEET 10.2

1. The collector circuit efficiency of a power amplifier is 20 %. This circuit is used in a
radio-receiver which is rated for an output of 500 mW. Calculate the collector
dissipating power. [**Ans.** 2 W]

2. In a power amplifier, the collector-circuit efficiency is 15 %. Calculate the ac output
power for this circuit, if $V_{CC} = 10$ V, and $I_C = 20$ mA. [**Ans.** 30 mW]

3. If power is to be developed in a 5 Ω loudspeaker coil through a transformer in an
audio-power amplifier circuit, calculate the turns-ratio of the transformer. Assume
the dynamic output resistance of the transistor to be 25 kΩ. [**Ans.** 70.71 : 1]

EXPERIMENTAL EXERCISE 10.1

TITLE: Single-ended power amplifier.

OBJECTIVES: To,
1. trace the circuit of single-ended power amplifier;
2. measure the quiescent operating point;
3. measure the maximum input signal for undistorted output;
4. determine the value of optimum load;
5. measure the distortion for the maximum output power.

APPARATUS REQUIRED: Power amplifier circuit, output power meter, transistor power supply, CRO, distortion factor meter.

CIRCUIT DIAGRAM: As shown in Fig. E.10.1.1 (typical values of components are also given).

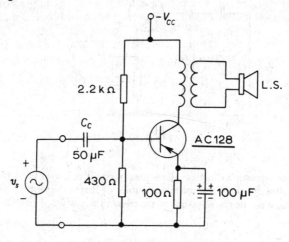

Fig. E.10.1.1

BRIEF THEORY: In almost all electronic systems, the power amplifier is the final stage. It converts dc power obtained from the battery into useful ac power. In audio systems the loudspeaker is the load for the power amplifier. For transferring maximum power from the transistor to the load, low impedance (3 Ω to 32 Ω) of the loudspeaker should be matched with the high output-impedance (12 kΩ to 20 kΩ) of the transistor. This is achieved by suitably selecting the turns ratio of the output transformer so that

$$r_o \text{ (or } 1/h_{oe}) = \left(\frac{N_1}{N_2}\right)^2 R_L$$

where R_L is the loudspeaker impedance, N_1/N_2 is primary to secondary turns ratio of the transformer and r_o is the dynamic output resistance.

PROCEDURE:
1. Find the type number of the transistor. From the data book note down whether it is *PNP* or *NPN*, and also find important ratings like maximum collector dissipation rating, maximum dc collector current, and maximum collector voltage.

2. Look into the circuit and trace it. Write down the values of resistors and capacitors used in the circuit.
3. Supply the dc voltage V_{CC} (say 9 V) to the circuit. Note the values of V_{CE} and V_E. Calculate the quiescent point current

$$I_C \simeq I_E = \frac{V_E}{R_E}$$

4. Feed the signal at the input from a low-frequency signal generator. Adjust its frequency at 1 kHz. Connect the loudspeaker at the output. If 1 kHz note is heard, it means that the circuit is working properly. Now display the output on the CRO. Increase the input voltage, till output waveshape starts showing distortion. Measure the value of the input signal. This gives the maximum signal handling capacity of the circuit.
5. In place of the loudspeaker, connect the audio power meter. By varying the impedance of the power meter note the changing output power. Plot a graph between output power and impedance. You will find that, for a particular value of load impedance, the output power is maximum. This is the value of optimum load for the circuit.
6. Connect the distortion-factor meter (DFM) to the output. Measure the distortion with the DFM.

OBSERVATIONS:

1. *Specifications from data book*:

   (*a*) Transistor used: _____
   (*b*) Maximum collector current rating: _____mA
   (*c*) Maximum collector dissipation rating: _____ W
   (*d*) Maximum collector voltage rating: _____ V

2. *Quiescent operating point*:

$$V_{CC} = \underline{\hspace{2cm}} V$$
$$V_{CE} = \underline{\hspace{2cm}} V$$
$$V_E = \underline{\hspace{2cm}} V$$

$$I_E \simeq I_C = \frac{V_E}{R_E} = \underline{\hspace{2cm}} mA$$

3. Maximum input signal for nondistorted output = _____ mV
4. Optimum load of the circuit:
   (*a*) Input signal = _____ mV
   (*b*) Frequency of the input signal = 1 kHz.

S. no.	Load impedance	Output power
1.		
2.		
3.		

5. Distortion in the output for optimum load resistor, and for given input signal (say, 100 mV) = _____ %

RESULTS:

1. Transistor is biased in the active region and it can handle a signal of _____ mV without producing distortion in the output.
2. The optimum load for the circuit is _____ Ω.
3. The percentage of the distortion in the output is _____ %.

EXPERIMENTAL EXERCISE 10.2

TITLE: Push-pull amplifier using transistors.

OBJECTIVES: To,

1. trace the circuit of the given amplifier;
2. read the values of each component used in the circuit;
3. measure the operating point (I_C and V_{CE}) for both the transistors;
4. measure the maximum signal handling capacity of the amplifier;
5. determine the value of the optimum load;
6. measure the value of distortion in the output of the amplifier.

APPARATUS REQUIRED: Experimental board, transistorized (or IC) power supply, audio signal generator, output power meter, distortion-factor meter and CRO.

CIRCUIT DIAGRAM: The circuit of a push-pull amplifier is as shown in Fig. E.10.2.1.

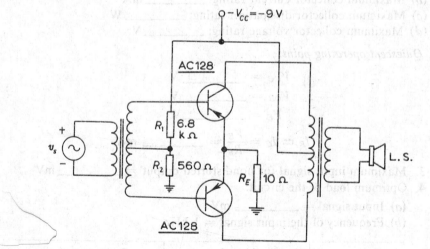

Fig. E.10.2.1

BRIEF THEORY: Push-pull connection is the most widely used in power amplifiers. It has much less distortion and high efficiency. The resistors R_1, R_2 and R_E fix up the operating point, as in any other transistor amplifier. The output transformer has a centre-tap primary, and it serves to match the load with the output impedance of the transistors. The input transformer has a centre-tap secondary. It provides input signals to the two transistors.

These signals are equal in magnitude but opposite in phase. On applying the input signal, the collector current of one transistor is 'pushed up' and that of the other transistor is 'pulled down'. Hence the name 'push-pull amplifier'. The output voltage is proportional to net current $(i_{C_1} - i_{C_2})$ flowing through the primary of the transformer.

We may connect a power meter at the output, instead of a loudspeaker. It serves the function of the load resistor as well as it measures the output power.

PROCEDURE:

1. Note down the transistor type numbers from the experimental board. Consult the data book, and find whether the transistors are *PNP* or *NPN*. Also find their maximum collector-power dissipation rating.

2. Trace the circuit of the amplifier and write down the values of the components used.

3. Connect the dc supply from the regulated power supply. Measure voltage V_{CE} for both the transistors. For a matched pair of transistors, the voltage V_{CE} for both should be the same. Measure voltage drop across R_E. This voltage drop divided by the value of the resistor R_E gives the sum of the two emitter-currents. Half of this current will be approximately the collector current of each transistor.

4. Feed the ac signal at the input. Keep the frequency at 1 kHz (mid-band frequency). Connect a loudspeaker at the output. Do you hear a sound note? If yes, the circuit is working.

5. Replace the loudspeaker with a power meter. Select a typical value of the load impedance (say, 8 Ω) in the power meter. Also connect the output to a CRO. See the waveshape of the output voltage on the CRO. Increase the input signal voltage till the output waveshape starts getting distorted. Note this input-signal voltage. Now reduce the input signal to a value slightly below (about 80 %) this voltage.

6. Change the impedance (in the power meter) in steps. For each value of impedance, note down the output power. Plot a graph between output power and load impedance. From this graph, find the impedance for which the output power is maximum. This is the value of the optimum load.

7. Select a load impedance (in the power meter) which is equal to or near about the optimum load. See the waveshape of the output on the CRO. Increase the input signal till the waveshape just shows distortion. Note the value of this input voltage. This gives the maximum signal handling capacity of the amplifier.

8. Slightly increase the input voltage (say, by 10 %). Connect a distortion-factor meter at the output and measure the distortion.

OBSERVATIONS:

1. *Operating point*:

$V_{CC} =$ _____ V; $V_{CE_1} =$ _____ V; $V_{CE_2} =$ _____ V

$V_E =$ _____ mV; $I_{C_1} = I_{C_2} = \dfrac{1}{2}\left(\dfrac{V_E}{R_E}\right) =$ _____ mA

2. Typical load connected = _____ Ω
   Input signal frequency = 1 kHz
   Maximum input signal giving undistorted output = _____ mV

3. *Optimum load*:

Input signal = _____ mV

Signal frequency = 1 kHz

S. no.	Load impedance (in Ω)	Output power (in mW)
1.		
2.		
3.		

4. *Maximum signal handling capacity*:

Selected value of optimum load = _____ Ω

Frequency of the input = 1 kHz

Maximum input signal giving undistorted output = _____ mV

5. *Distortion*:

Selected value of optimum load = _____ Ω

Frequency of the input signal = 1 kHz

% distortion in output = _____ mV

RESULTS:

1. Optimum load impedance = _____ Ω
2. Maximum signal handling capacity = _____ mV
3. Distortion = _____ %

Tuned Voltage Amplifiers

OBJECTIVES : After completing this unit, you will be able to : ○ State the applications of tuned voltage amplifiers in communication systems. ○ Explain the characteristics of series resonant circuit and parallel resonant circuit. ○ Calculate for the given tuned circuit (series or parallel) the resonant frequency, the quality factor Q, impedance at resonant frequency and bandwidth. ○ Explain the working of single-tuned voltage amplifier. ○ Explain the limitation of single-tuned voltage amplifier in terms of its frequency response characteristics. ○ Explain the working of double-tuned amplifiers.

11.1 NEED FOR TUNED VOLTAGE AMPLIFIERS

Till now we have discussed audio amplifiers. Such amplifiers are used in various audio systems, for example, record players, tape recorders, public address systems, etc. In radio broadcasting, the audio signal (voice or music) is "raised" to some high-frequency level. This high frequency is in the radio-frequency (rf) range and it serves as the carrier of the audio signal. The carrier frequencies (and corresponding wavelengths) of some of the broadcast stations are given below:

Station	Frequency	Wavelength
Delhi 'B'	1017 kHz	294.9 m
Chandigarh	1431 kHz	209.6 m
Lucknow 'A'	747 kHz	401.6 m
Bombay 'C'	1188 kHz	252.5 m

The process of raising the audio signal to rf frequencies is called *modulation*. It is the modulated wave that is transmitted by the broadcasting station. This modulated wave has a relatively narrow band of frequencies centred around the carrier frequency, as shown in Fig. 11.1. The bandwidth of the signal $(f_2 - f_1)$ is very small compared to the carrier frequency f_c.

When the rf signal (modulated wave) reaches the receiving antenna, a very weak voltage is induced in it. This voltage is of the order of a few μV. It is not possible to extract the original audio signal from this weak voltage*. It is necessary, first, to amplify the rf signal to a suitable level. This is achieved in a radio receiver with the help of a *tuned voltage amplifier*.

---

*The process of extracting the original signal from the rf signal is called demodulation. These terms (modulation and demodulation) are presented here merely by way of introduction. They will be discussed in detail when we deal with communication engineering.

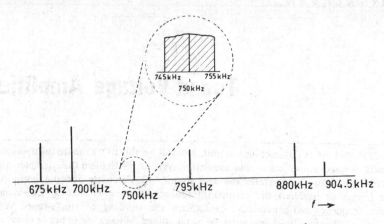

Fig. 11.1 Frequency spectrum of modulated waves that are transmitted by different broadcasting stations

The tuned voltage amplifier not only amplifies the rf signal, but it also performs another important function. The carrier frequencies assigned to different broadcasting stations may be operating simultaneously. A provision should be there in the receiver to *select* only the desired station. This function of selecting the desired rf signal, and rejecting the rest, is also achieved by the tuned voltage amplifier. Thus we see that the tuned-voltage amplifier serves two purposes:

(*i*) Selection of the desired rf signal (broadcasting station).

(*ii*) Amplification of the selected rf signal to a suitable value.

A tuned-voltage amplifier uses a tuned circuit. Such a tuned circuit is shown in Fig. 11.2. Because of the well known phenomenon called *resonance*, the tuned circuit is capable of selecting a particular frequency.

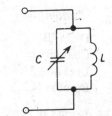

Fig. 11.2 A tuned circuit

11.2 RESONANCE

For resonance to occur, a circuit must contain inductance (*L*) and capacitance (*C*). It may also (and generally does) have some resistance. This resistance may be the effective resistance of the coil itself (all practical coils do have some resistance). Sometimes, a resistor is deliberately introduced to create some desired effect.

An inductor and a capacitor can be connected to a source in two different ways as shown in Fig. 11.3. The circuit of Fig. 11.3*a* is known as *series resonant circuit* and that of Fig. 11.3*b*, a *parallel resonant* (or *anti-resonant*, or *tank*) circuit.

Depending upon the frequency of the source voltage v_s, the circuits of Fig. 11.3 may behave either as inductive or as capacitive. *However, at a particular frequency when the inductive reactance X_L equals the capacitive reactance X_C, then the circuit behaves as a purely resistive circuit.* This phenomenon is called *resonance*; and the corresponding frequency is called *resonant frequency*.

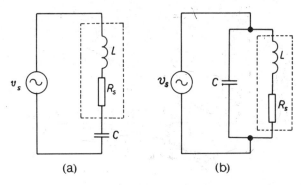

Fig. 11.3 (*a*) Series resonant circuit
(*b*) Parallel resonant circuit

11.2.1 Series Resonant Circuit

In Fig. 11.3*a*, an inductor and a capacitor are connected in series to the voltage source v_s. The frequency of this voltage source can be varied. At a given frequency f, the reactance of the inductor and the capacitor are given by

$$X_L = 2\pi fL \qquad (11.1)$$

and
$$X_C = \frac{1}{2\pi fC} \qquad (11.2)$$

The total impedance of the circuit as seen by the source is given as

$$Z_s = R + j(X_L - X_C) \qquad (11.3)$$

To understand the phenomenon of resonance, let us first see how X_L and X_C vary with frequency. From Eq. 11.1, it is clear that X_L increases linearly with frequency. But in the case of the capacitor, the reactance X_C varies inversely with frequency as can be seen from Eq. 11.2. The variation of these reactances* with frequency is shown in Fig. 11.4. At a particular frequency f_r, we find that $X_L = X_C$. This frequency is known as *resonant frequency*. Below this frequency, $X_C > X_L$. Therefore, the circuit looks *capacitive* to the source. Above resonant frequency, $X_L > X_C$, and the circuit looks *inductive* to the source. At resonance, the source sees only the resistance R, because X_L and X_C cancel.

The resonant frequency (f_r) can be found by equating the two reactance values, i.e.

$$X_L = X_C$$

or
$$2\pi f_r L = \frac{1}{2\pi f_r C}$$

or
$$f_r = \frac{1}{2\pi\sqrt{LC}} \qquad (11.4)$$

---

*As a convention, the reactance X_L of an inductor is treated as positive, **and the** reactance X_C of a capacitor as negative.

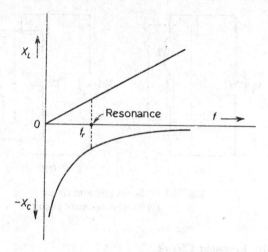

Fig. 11.4 Variation of reactances with frequency
in a series resonant circuit

At the resonant frequency, the net reactance is zero because $X_L = X_C$. The circuit impedance Z becomes minimum and is equal to the resistance R. Since the impedance is minimum, the line current will be maximum. As the impedance is purely resistive, the current is in phase with the line voltage. Summarizing these key points concerning a series resonant circuit, we have at resonance

(i) $X_L = X_C$

(ii) $f_r = \dfrac{1}{2\pi\sqrt{LC}}$

(iii) $Z_s = R$ (min)

(iv) $I = \dfrac{V_s}{R}$ (max)

Variation of Impedance with Frequency   The magnitude of the impedance of the circuit at any frequency is given as

$$Z_S = \sqrt{R^2 + (X_L - X_C)^2} \qquad (11.5)$$

At frequencies considerably below resonance, the circuit impedance is high, and is determined mainly by X_C (Fig. 11.4). As the frequency increases, the net reactance $(X_C - X_L)$ goes on decreasing. Therefore, the total impedance also decreases with frequency. At resonance, the net reactance $(X_C - X_L)$ becomes zero; the impedance Z_S is minimum and is equal to R. Above resonant frequency, the net reactance $(X_L - X_C)$ again starts increasing. Therefore, the total impedance Z_S also increases. The variation of Z_S with frequency is plotted in Fig. 11.5. Since the current in the circuit is given as $I = V_s/Z$, the curve for the current will be inverse of the impedance curve. Figure 11.5 also shows the variation of current with frequency.

From Fig. 11.5, it is clear that the current in a series resonant circuit is maximum at resonant frequency, and it decreases on either side of the resonant frequency. It is thus observed that a series resonant circuit attains the property of selecting signals of one particular frequency and rejecting

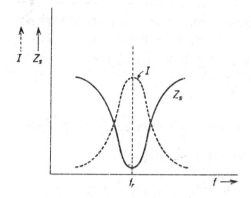

Fig. 11.5 Variation of Z_s and I with frequency,
for a series resonant circuit

those of others. The rate at which the current decreases as we move away from the resonant frequency depends upon the resistance of the circuit. If the resistance in the circuit is low, the current falls very sharply as we move away from the resonant frequency. On the other hand, high resistance makes the curve flat (see Fig. 11.6). In this way we can say that the circuit resistance R plays an important role in the selecting property (selectivity) of the resonant circuit. It is obvious from Fig. 11.6 that the current at resonance, in a series resonant circuit, is inversely proportional to the resistance in the circuit.

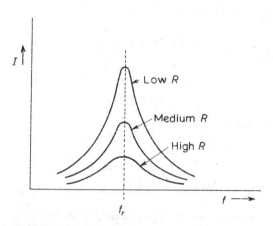

Fig. 11.6 Effect of resistance on current in a series
LC circuit

However, the resonant frequency of the circuit is not affected by the resistance of the circuit. Generally, the resistance of a resonant circuit is a part and parcel of the inductor used. The coils of same inductance L but different resistance values will have different selecting property. Thus, the resistance of a coil determines the quality of the inductor used. In technical terms, this is expressed as *quality factor Q* of the coil.

Example 11.1 A series resonant circuit has the following constants:

$$L = 220 \ \mu H, \quad C = 300 \ pF, \quad R = 20 \ \Omega$$

The supply voltage is 10 V. Calculate (a) the resonant frequency, (b) the impedance at resonance, (c) the current at resonance, and (d) the voltage across each component.

Solution: (a) The resonant frequency $(f_r) = \dfrac{1}{2\pi\sqrt{LC}}$

Here, $L = 220 \times 10^{-6}$ H; $C = 300 \times 10^{-12}$ F

Therefore, $f_r = \dfrac{1}{2\pi\sqrt{220 \times 10^{-6} \times 300 \times 10^{-12}}} =$ **620 kHz**

(b) The impedance at resonance $= R$ in the circuit. This is **20 Ω.**

(c) The current at resonance

$$I = \frac{V}{R} = \frac{10}{20} = 0.5 \text{ A}$$

(d) The voltage across the inductance

$$V_L = I \times X_L = \frac{V}{R} \times \omega_r L = V \times \frac{\omega_r L}{R}$$

$$= 10 \times \frac{2\pi \times 620 \times 10^3 \times 220 \times 10^{-6}}{20}$$

$$= 10 \times 42.8 \text{ V} = \textbf{428 V}$$

The voltage across the capacitance

$$V_C = I \times X_C = \frac{V}{R} \times \frac{1}{2\pi f_r C}$$

$$= V \times \frac{1}{20 \times 2\pi \times 620 \times 10^3 \times 300 \times 10^{-12}}$$

$$= 10 \times 42.8 = \textbf{428 V}$$

Voltage across the resistance

$$V_R = I \times R$$

$$= \frac{V}{R} \times R = V = \textbf{10 V}$$

The Quality Factor Q and Its Significance In the numerical example given above, it should be noted that the voltages across the inductor and capacitor are equal. When the line voltage is only 10 V, the voltage across the inductor and capacitor are 428 V each. This seems quite surprising but it always happens in a series resonant circuit, if the resistance is low compared to inductive and capacitive reactances (at resonant frequency). The ratio of the inductive reactance to the resistance, with reference to a coil, is a measure of the quality of the coil and is known as Q of the coil. Similarly, in a complete circuit, the ratio of the reactance to the circuit resistance is known as the Q of the circuit. If the circuit resistance comes from the coil only, then Q of the circuit and Q of the coil are the same. In calculating the circuit Q at resonance, either reactance (X_L or X_C) can be used since they are equal. In the above example, the circuit Q can be found as follows:

$$Q = \frac{X_L}{R} = \frac{\omega_r L}{R} = \frac{2\pi f_r L}{R} \qquad (11.6)$$

In the above example, it comes out to be 42.8. This can also be calculated using X_C. It again comes out to be 42.8. From this we can draw the general conclusion that in a series resonant circuit, there is a Q times rise in the voltage at resonance. This effect can also be shown mathematically as follows:

$$V_L = I X_L$$

We know that at resonance, $I = \dfrac{V}{R}$

Therefore, $V_L = \dfrac{V}{R} \times X_L = \dfrac{V}{R} \times \omega_r L = V \times \dfrac{\omega_r L}{R}$

$$= V \times Q$$

Similarly it can be shown that $V_C = V \times Q$.

Equation 11.6 indicates that the Q of the circuit varies inversely with circuit resistance. In Fig. 11.6, where the variation of the current with frequency is shown for different values of the resistance the curve labelled "low R" would represent a circuit with high Q. Circuits with medium and low Qs would be indicated by curves marked "medium R" and "high R" respectively. From Fig. 11.6 it is clear that the current falls very rapidly as we go off the resonance when Q is high. Such a circuit provides good frequency discrimination. Thus, it can be concluded that Q is a measure of the ability of a resonant circuit to select or reject a band of frequencies. The higher the Q of a series resonant circuit, the greater is its ability as a frequency selector, i.e. the narrower is the band of frequencies showing a voltage gain at resonance. This is why Q is frequently referred to as a *figure of merit* of a resonant circuit.

11.2.2 Parallel Resonant Circuit

A parallel resonant circuit is shown in Fig. 11.7. Here, an inductor and a capacitor are connected in parallel to each other, with respect to the supply source. The resistance R represents the coil resistance. Its value is usually very small, and is generally neglected compared to the other impedances. The capacitor C is assumed to be lossless. The frequency of the source V_s can be varied.

Applying Kirchhoff's current law to the junction A, we must have $I = I_L + I_C$. That is, the current I delivered by the source is divided into two branch currents I_L and I_C. The current through the inductance (neglecting the resistance R) has the value V_s/X_L and lags V_s by 90°. The current through the capacitor has the value V_s/X_C and leads V_s by 90°. Thus, as seen from the phasor diagram of Fig. 11.8, the two currents are *out of phase with each other*. Partial or complete cancellation will take place in the line current I depending on the magnitude of the branch currents.

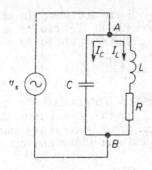

Fig. 11.7 Parallel resonant
circuit

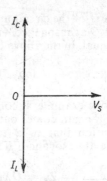

Fig. 11.8 Phasor diagram of
an ideal parallel
resonant circuit

If $X_C < X_L$, then $I_C > I_L$, and the circuit acts capacitively. On the other hand, if $X_L < X_C$; then $I_L > I_C$, and the circuit is inductive. However, when $X_L = X_C$, the inductive and capacitive currents are equal (but opposite in phase), and they cancel each other; line current in an ideal circuit is zero. Thus, an ideal parallel resonant circuit presents infinite impedance to the line and acts as an open circuit. In case of a practical circuit, the inductor coil is not free from resistance; it has some resistance, though small in value. In that case, the phasor sum of branch currents I_L and I_C is not zero but it has a minimum value. The resultant line current I is in phase with the voltage V_s, and the impedance Z_p of the circuit is maximum. Since the current and voltage are in phase, this impedance is resistive. Since $X_L = X_C$ at resonance, the resonant frequency for the parallel resonant circuit is the same as that for the series resonant circuit. The frequency at which resonance occurs in a parallel LC circuit is sometimes called the anti-resonant frequency, to distinguish it from the resonant frequency of the series LC circuit. Summarizing the key points for a parallel resonant circuit, we have at resonance

(i) $X_L = X_C$ and $I_L = I_C$

(ii) $f_r = \dfrac{1}{2\pi\sqrt{LC}}$

(iii) $I = \dfrac{V_S}{Z_p}$ (minimum and in phase with applied voltage)

(iv) Z_p is maximum and resistive

Impedance of Parallel Resonant Circuit We have seen that the impedance of a parallel resonant circuit is maximum and resistive at resonance. Furthermore, the higher the Q of the circuit (the lower the resistance R), the closer this impedance approaches infinity. The actual value of the impedance can be determined as follows. In any two-branch parallel circuit, the net impedance is given as

$$Z_p = \frac{Z_1 \times Z_2}{Z_1 + Z_2} \qquad (11.7)$$

Here, $Z_1 = 1/j\omega C = -jX_C$, $Z_2 = R + j\omega L$. If $Q = 10$ or higher, $Z_2 \simeq j\omega L = jX_L$. Also $Z_1 + Z_2$ is actually the series impedance Z_s of the three components C, L and R. Therefore,

$$Z_p = \frac{X_C \times X_L}{Z_s}. \tag{11.8}$$

But, at resonance, $X_L = X_C$, and the series impedance Z_s is equal to the resistance R of the inductive branch. Therefore, at resonance

$$Z_p = \frac{X_L \times X_C}{R} = QX_C \tag{11.9}$$

or $$Z_p = \frac{X_L^2}{R} = QX_L \tag{11.10}$$

or $$Z_p = QX_L = \frac{QX_L R}{R} = Q^2 R \tag{11.11}$$

Furthermore, Eq. 11.9 can be put in another form by replacing X_L by $2\pi f L$ and X_C by $1/(2\pi f C)$.

$$Z_p = \frac{2\pi f L}{2\pi f C \times R} = \frac{L}{CR} \tag{11.12}$$

Thus, we have a variety of equations for calculating the impedance Z_p in a parallel resonant circuit. (Actually they are merely variations of the same equation.) Sometimes one form is more convenient than another.

To see how the impedance Z_p of a parallel resonant circuit changes with frequency, let us write Eq. 11.8 as follows

$$Z_p = \frac{(2\pi f L) \times 1/(2\pi f C)}{Z_s} = \frac{(L/C)}{Z_s} \tag{11.13}$$

In Fig. 11.3, we have seen how Z_s varies with frequency. Since Z_S appears in the denominator of Eq. 11.13, the impedance curve for Z_p will be inverse of the impedance curve for Z_s. This variation is shown in Fig. 11.9. The impedance at resonance is maximum. And as we go off resonance, the impedance drops. The amount of resistance present in the circuit determines the height and steepness of the impedance curve. If the resistance is more, the height of the impedance curve reduces and the curve flattens out.

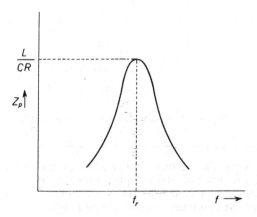

Fig. 11.9 Variation of the impedance with frequency
for a parallel resonant circuit

11.2.3 Bandwidth of Resonant Circuit

We have seen that every resonant circuit has the property of discriminating between the frequency at resonance and those not at resonance. This discriminating property of the resonant circuit is expressed in terms of its *bandwidth* (BW). The bandwidth of a circuit is the total number of hertz, above and below the resonant frequency, that gives practically the same response as at the resonant frequency itself. This band is also called the *pass-band* of the circuit. In case of a series resonant circuit, the response can be seen in terms of the line current; and for a parallel resonant circuit in terms of its impedance. Therefore, the frequency response of a resonant circuit (series or parallel) is as shown in Fig. 11.10. The effective limits of the pass-band are taken at the points on the response curve corresponding to 70.7|% of the peak value. In Fig. 11.10, the shaded area represents the band of frequencies for which the response is greater than 0.707 of the peak value. Note that one-half of the band lies above the resonant frequency (f_r to f_2) and the other half lies below the resonant frequency (f_r to f_1).

Since the Q of the circuit determines the overall steepness of the curve, the pass-band may also be found in terms of the resonant frequency and the Q of the circuit. Thus,

$$(f_2-f_1) = BW = \frac{f_r}{Q} \tag{11.14}$$

From this formula it can be seen that *the higher the Q, the smaller is the bandwidth.*

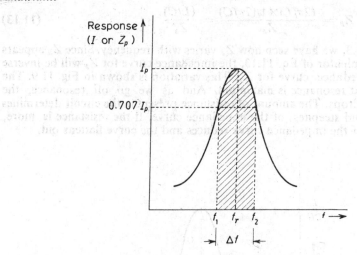

Fig. 11.10 Bandwidth of a resonant circuit

Example 11.2 A parallel resonant circuit has a capacitor of 100 pF in one branch and an inductance of 100 μH plus a resistance of 10 Ω in the parallel branch. If the supply voltage is 100 V, calculate f_r, I_L, I_C, line current and impedance of the resonant circuit at resonance.

Solution: The resonant frequency is given as

$$f_r = \frac{1}{2\pi\sqrt{LC}} = \frac{1}{2\pi \times \sqrt{100 \times 10^{-6} \times 100 \times 10^{-12}}} = 1590 \text{ kHz}$$

The inductive reactance is

$$X_L = 2\pi f_r L = 2\pi \times 1590 \times 10^3 \times 100 \times 10^{-6} = 1000 \ \Omega$$
$$R = 10 \ \Omega$$

The resistance R is very small compared to the inductive reactance X_L. Therefore, the impedance of the inductive branch is almost the same as the inductive reactance, i.e. $Z_2 \simeq X_L$. Thus, the current in the inductive branch is

$$I_L = \frac{V}{Z_2} \simeq \frac{V}{X_L} = \frac{100}{1000} = 0.1 \text{ A}$$

Now

$$X_C = \frac{1}{2\pi f_r C} = \frac{1}{2\pi \times 1590 \times 10^3 \times 100 \times 10^{-12}} = 1000 \ \Omega$$

Therefore, the current through the capacitive branch is

$$I_C = \frac{V}{X_C} = \frac{100}{1000} = 0.1 \text{ A}$$

Since

$$Z_p = \frac{L}{RC} = \frac{100 \times 10^{-6}}{100 \times 10^{-12} \times 10} = 10^5 \ \Omega$$

therefore, the line current is given as

$$I = \frac{V}{Z_p} = \frac{100}{100\ 000} = 1 \text{ mA}$$

Example 11.3 A tank circuit has a capacitor of 100 pF and an inductor of 150 μH. The series resistance is 15 Ω. Find the impedance, Q and bandwidth of the resonant circuit.

Solution:
Here, $L = 150 \times 10^{-6}$ H, $C = 100 \times 10^{-12}$ F and $R = 15 \ \Omega$. The resonant frequency f_r is given by

$$f_r = \frac{1}{2\pi\sqrt{LC}}$$

$$= \frac{1}{2\pi\sqrt{150 \times 10^{-6} \times 100 \times 10^{-12}}} = 1303 \text{ kHz}$$

The impedance of the parallel resonant circuit at resonance is

$$Z_p = \frac{L}{CR} = \frac{150 \times 10^{-6}}{100 \times 10^{-12} \times 15} = 10^5 \ \Omega$$
$$= 100 \text{ k}\Omega$$

The quality factor, $Q = \dfrac{2\pi f_r L}{R}$

$$= \frac{2\pi \times 1303 \times 10^3 \times 150 \times 10^{-6}}{15} = 81.6$$

The bandwidth,

$$\Delta f = \frac{f_r}{Q} = \frac{1303 \times 10^3}{81.6} = 15.84 \text{ kHz}$$

11.3 SINGLE-TUNED VOLTAGE AMPLIFIER

Figure 11.11 shows the circuits of a single-tuned voltage amplifier using bipolar junction transistor. In the circuit of Fig. 11.11a, the output is taken with the help of capacitive coupling, whereas in Fig. 11.11b, the output is obtained by inductive coupling. Other active devices like triode, pentode or field-effect transistors can also be used. Generally, triodes are not used at such high frequencies since the inter-electrode capacitances cause undesirable feedback at such frequencies. A single-tuned amplifier circuit using a pentode is shown in Fig. 11.12. Because of the advantages like very high input impedance and reduced inter-electrode capacitances, modern circuits make use of FETs or MOSFETs in these amplifiers. Such a circuit is shown in Fig. 11.13.

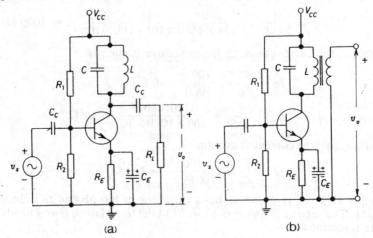

Fig. 11.11 Single-tuned amplifier circuits using bipolar junction transistor:
(a) Capacitively coupled amplifier
(b) Inductively coupled amplifier

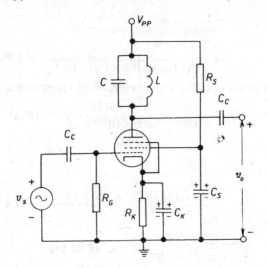

Fig. 11.12 Single-tuned amplifier using pentode

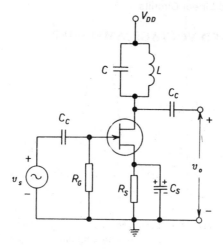

Fig. 11.13 Single-tuned amplifier using FET

In Fig. 11.11, the resistors R_1, R_2 and R_E fix up the operating point and also stabilize it. The tuned circuit consisting of inductance and capacitance acts like a load resistance of the amplifier circuit. One of the two components, that is, either inductance, or capacitance, is variable. This is for adjusting the resonant frequency of the circuit.

11.3.1 Voltage Gain and Frequency Response Curve

We have already studied that the voltage gain of an amplifier depends upon the ac load resistance. In Unit 8 the following formula for voltage gain was derived

$$A_v = \frac{\beta R_{ac}}{r_{in}} \angle 180°$$

In the tuned amplifier, R_{ac} is the impedance of the tuned circuit. This impedance is denoted by Z_p. The impedance of the tuned circuit at resonance is resistive and is equal to L/CR. Therefore, the voltage gain of the tuned amplifier at resonance is given by

$$A_v = \frac{\beta \dfrac{L}{CR}}{r_{in}} \angle 180° \tag{11.15}$$

This voltage gain is very high since the quantity L/CR is very high for a tuned circuit. The voltage gain at the frequencies away from resonance decreases, since the impedance of the tuned circuit at these frequencies also decreases. Thus, as we go away from the resonance on either side, the voltage gain of the amplifier decreases. The frequency response curve of the tuned amplifier is similar to the impedance–frequency curve for a parallel resonant circuit. This frequency response curve is plotted in Fig. 11.14. The bandwidth of the amplifier is given by

$$BW = \frac{f_r}{Q} \tag{11.16}$$

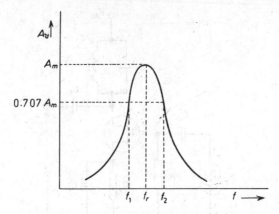

Fig. 11.14 Frequency response curve of a
single-tuned amplifier

The above expression comes from the formula for bandwidth of a parallel
resonant circuit. It was explained earlier that the sharpness of the resonant
circuit depends upon the quality factor Q of the coil used in the tuned
circuit.

11.3.2 Limitations of Single-Tuned Voltage Amplifiers

Tuned voltage amplifiers are generally used in rf stage of wireless communication systems. Here, these circuits are assigned the work of selecting the
desired carrier frequency and of amplifying the allowed pass-band around
this selected carrier frequency. The high selectivity requires a high Q-reso-
nant circuit. A high Q circuit will give high gain too, but at the same time
its bandwidth will be very much reduced. A very narrow band will result
in poor reproduction. This is the drawback with a single-tuned circuit.
However, this difficulty is overcome by using double-tuned circuits. In
double-tuned amplifiers, there are two tuned circuits coupled inductively.
A change in the coupling of the two tuned circuits results in change in the
shape of the frequency response curve. If the coupling between the two
coils of the tuned circuits is properly adjusted, the required results (high
selectivity, high gain and required bandwidth) may be obtained.

11.4 DOUBLE-TUNED VOLTAGE AMPLIFIER

In double-tuned circuits, inductive coupling is used. The primary and the
secondary coils of the transformer are shunted by capacitors, thus making
two tuned circuits. The circuit diagram of a double-tuned voltage amplifier
is shown in Fig. 11.15. In this circuit, a transistor is used as the active
device. Because of the low input impedance of a transistor, serious limita-
tions on the working of the amplifier exists. That is why, in modern solid-
state circuits, MOSFETs are generally used. Circuits with pentodes are still
in use in valve radio receivers. The frequency response curve of the double-
tuned amplifier for different coefficients of coupling is shown in Fig. 11.16.

It should be appreciated that the most suitable response curve is one
when optimum coefficient of coupling exists between the two tuned circuits.
In this condition, the circuit is highly selective and also provides sufficient
amount of gain for a particular band of frequencies.

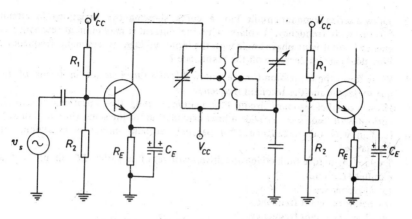

Fig. 11.15 Double-tuned transistor amplifier

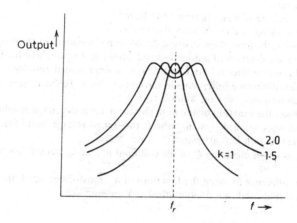

Fig. 11.16 Frequency response curve of a double-tuned
amplifier for different coefficients of coupling

REVIEW QUESTIONS

11.1 Explain the following terms in brief (within 5 to 8 lines):
   (a) Series resonance
   (b) Parallel resonance
   (c) Figure of merit of a coil
   (d) Resonant frequency

11.2 Compare the series and parallel resonant circuits from the following points of
   view:
   (a) Resonant frequency
   (b) Impedance at resonance
   (c) Voltage across each component
   (d) Current in the circuit
   (e) Bandwidth
   (f) Applications

11.3 Draw a series resonant circuit. Plot a curve showing the variations in circuit current with frequency. Explain why the current is maximum at resonant frequency ? What is its phase angle with the input voltage at resonant frequency ? Why does current decrease off the resonance ?

11.4 Write down the expression for the bandwidth of a tuned circuit in terms of its quality factor and the resonant frequency.

11.5 Draw a parallel resonant circuit. Plot a curve showing the variations of circuit current with frequency. Provide a brief explanation for the same (in 5 to 8 lines.)

11.6 Explain why the impedance of a parallel resonant circuit is maximum at resonance ?

11.7 Explain the nature (inductive/capacitive/resistive) of a series LC circuit and a parallel LC circuit.
(a) at resonance
(b) below resonant frequency
(c) above resonant frequency.

11.8 Prove that the impedance of a parallel resonant circuit at resonant frequency is approximately equal to $R_s Q^2$, where R_s is the series resistance of the coil and Q is its quality factor.

11.9 Explain the working of a single-tuned amplifier.

11.10 Explain the working of a double-tuned amplifier.

11.11 Explain in brief, the advantage in using double-tuned circuit over the single-tuned.

11.12 Draw the circuit diagram of a double-tuned amplifier. Explain how the frequency response of this amplifier is better than that of a single-tuned amplifier.

11.13 State the main difference/differences in the circuits of a tuned-voltage amplifier and an audio-frequency voltage amplifier.

11.14 Explain in brief the function of the tank circuit in a tuned-voltage amplifier.

11.15 State at least one electronic system where the tuned-voltage amplifier is used. Also state its function in that system.

11.16 Explain the effect of changing Q of the coil used in a tank circuit, on its bandwidth.

11.17 Explain the difference between the functions of a transformer used in a power amplifier and that used in a double-tuned voltage amplifier.

OBJECTIVE-TYPE QUESTIONS

I. Here are some incomplete statements. Four alternatives are given below each. Tick the alternative which completes the statement correctly.

1. A radio-frequency signal contains following three frequencies: 870 kHz, 875 kHz and 880 kHz. This signal needs to be amplified. The amplifier used should be
(a) audio-frequency amplifier
(b) wide-band amplifier
(c) tuned voltage amplifier
(d) push-pull amplifier

2. An amplifier of pass-band 450 kHz to 460 kHz will be named as
(a) wide-band amplifier
(b) audio-frequency amplifier
(c) tuned voltage amplifier
(d) video amplifier

3. A narrow-band amplifier is one that has a pass-band

    (a) limited to 2000 Hz only

    (b) limited to audio-frequency range only

    (c) approximately 10 % of its central frequency

    (d) somewhere in the region near the cut-off region of the active device used

4. Tuned voltage amplifiers are not used

    (a) in public-address systems

    (b) radio receivers

    (c) where a band of frequencies is to be selected and amplified

    (d) in television receivers

5. In a series resonant circuit, the impedance at resonance is

    (a) minimum, and is equal to the resistance of the coil

    (b) maximum and is equal to $Q \times R$ where Q is the quality factor of the coi and R is its resistance

    (c) equal to $2R$, where R is the coil resistance

    (d) equal to $R/2$, where R is the coil resistance

6. For a series or parallel LC circuit, resonance occurs when

    (a) X_L is ten times X_C (c) X_L is equal to X_C

    (b) X_L is Q times X_C (d) X_C is 10 times X_L

7. An ac circuit resonates at 1000 kHz. It has a quality factor 50. The bandwidth (BW) and half-power points shall be at

    (a) BW = 10 kHz; f_1 = 1000 kHz; f_2 = 1010 kHz

    (b) BW = 20 kHz; f_1 = 1000 kHz; f_2 = 1020 kHz

    (c) BW = 20 kHz; f_1 = 990 kHz; f_2 = 1010 kHz

    (d) BW = 10 kHz; f_1 = 995 kHz; f_2 = 1005 kHz

II. Match the items listed under Column A with those under Column B.

Column A	Column B
1. Class-A operation	(a) Current in the output flows only during a portion of the positive half of the input cycle.
2. Class-B operation	(b) Amplifies frequencies in the frequency range of 40 Hz to 15 kHz.
3. Class-C operation	(c) Current in the output side flows for the positive half of the input cycle.
4. RF amplifier	(d) Amplifies frequencies in range of 0 Hz to 5 MHz.
5. AF amplifier	(e) Amplifies frequencies in the range of 100 kHz to 5 MHz.
6. Video amplifier	(f) Current in the output side flows for the complete cycle of the input signal.

Fill your response (1. to 6.) in the space given below:

    Column A   Matches with

      1.   ————

      2.   ————

      3.   ————

      4.   ————

      5.   ————

      6.   ————

Ans. I. 1. (c), 2. (c), 3. (c), 4. (a), 5. (a), 6. (c), 7. (c).
  II. 1. (f), 2. (c), 3. (a), 4. (e), 5. (b), 6. (d).

TUTORIAL SHEET 11.1

1. Calculate the resonant frequency of a series resonant circuit in each of the following cases:
   (a) Inductance $L = 46$ mH, the series capacitor 5.5 nF
   (b) Coil of $L = 200$ μH, $R = 2$ Ω, in series with a capacitor of 200 pF.
   [**Ans.** (a) 10 kHz; (b) 0.796 MHz]

2. A series circuit of $L = 15.8$ mH, $C = 0.1$ μF and $R = 10$ Ω, is connected to a line voltage of 10 V_{rms}. Find the resonant frequency, the current at resonant frequency, and voltage across each component at resonance.
   [**Ans.** $f_r = 4$ kHz; $I_{res} = 1$ A; $V_L = V_C = 397.3$ V; $V_R = 10$ V]

3. Design a series resonant circuit to resonate at 1 MHz, using (a) 1 mH, 2 Ω coil; (b) 16 mH, 2 Ω coil, and (c) 16 mH, 5 Ω coil. Calculate the voltage available across each component in these circuits, when connected to a supply of 5 V.
   [**Ans.** (a) $C = 25.33$ pF, $V_L = V_C = 15.71$ kV, $V_R = 5$ V
   (b) $C = 1.58$ pF, $V_L = V_C = 251.3$ kV, $V_R = 5$ V
   (c) $C = 1.58$ pF, $V_L = V_C = 100.5$ kV, $V_R = 5$ V]

4. Determine the impedance at resonant frequency, in each of the circuits given in problems 1, 2 and 3 above.
   [**Ans.** 1. (a) Zero, (b) 2 Ω;
   2. 10 Ω;
   3. (a) 2 Ω, (b) 2 Ω, (c) 5 Ω]

TUTORIAL SHEET 11.2

1. Design an anti-resonant circuit using a 200 μH coil and a variable capacitor to resonate between 550 kHz and 1600 kHz. What should be the minimum and the maximum values of the capacitor to be employed for this work ?
   [**Ans.** $C_{min} = 49.5$ pF, $C_{max} = 425$ pF]

2. A parallel resonant circuit has a 100 pF capacitor in one branch and a 100 μH, 10 Ω coil in the other. Calculate f_r, Z_p, V_c, line current, and current in each branch at resonance; and the Q of the circuit. Assume line voltage to be 100 V.
   [**Ans.** $f_r = 1590$ kHz; $Z_p = 100$ kΩ; $V_c = 100$ V;
   $I_{line} = 1$ mA; $I_c = I_L = 100$ mA; $Q = 100$]

3. A tank circuit has a capacitor of 100 pF and an inductor of 150 μH. The resistance of the coil is 15 Ω. Find the impedance and Q of the circuit (a) without any shunting resistance; (b) shunted with a 2 MΩ resistor; (c) shunted with a 0.1 MΩ resistor.
   [**Ans.** (a) $Z_p = 100$ kΩ, $Q = 81.6$;
   (b) $Z_p = 95.3$ kΩ, $Q = 78.8$;
   (c) $Z_p = 50$ kΩ, $Q = 40.8$]

4. A tank circuit consists of $L = 80$ μH, $C = 400$ pF and coil resistance 15 Ω. Find the resonant frequency of the circuit, Q of the coil, bandwidth, and current circulating in the tank circuit when the circuit is put to a source of 100 V having internal resistance of 6.67 kΩ.
   [**Ans.** $f_r = 0.8897$ MHz; $Q_{coil} = 29.83$;
   BW $= 29.826$ kHz; $I_L = I_c = 149$ mA]

TUTORIAL SHEET 11.3

1. An LC tank circuit forms the load impedance of a tuned amplifier. The components of the tank circuit are $L = 15.8$ mH, $R = 10$ Ω and $C = 0.1$ µF. Calculate: (a) frequency at which the gain is maximum; (b) quality factor of the coil; (c) maximum voltage gain of the amplifier in dB; and (d) the bandwidth of the amplifier. Assume the following parameters of the transistor $\beta_{ac} = 120$ and dynamic input resistance is equal to 1.2 kΩ.   [Ans. (a) 4004 Hz; (b) 39.75; (c) 64 dB; (d) 101 Hz]

2. A variable capacitor has a range of 40 to 400 pF. The highest frequency to be tuned is 1700 kHz. Calculate the value of the inductor required. Also find the value of minimum frequency which can be tuned by this tank circuit. If a bandwidth of 10 kHz is to be obtained from the amplifier using this tank circuit, find the value of the resistance of the coil. Also calculate the maximum gain of the amplifier in dB. Assume the following parameters for the transistor $\beta_{ac} = 115$ and dynamic input resistance = 1.3 kΩ.   [Ans. $L = 219$ µH; $f_{r(min)} = 537.6$ kHz; $R_{coil} = 13.75$ Ω; gain = 91 dB]

EXPERIMENTAL EXERCISE 11.1

TITLE: Series and parallel resonant circuits.

OBJECTIVES: To,

1. determine the resonant frequency of series resonant circuit;
2. calculate the quality factor Q of the coil with the help of the formula $Q = V_L/V_s$, where V_L is the voltage across the inductor and V_s is the signal voltage;
3. plot the graph between the frequency of the source and the current in the circuit;
4. determine the resonant frequency of parallel resonant circuit;
5. plot the graph between line current and frequency of the signal;
6. calculate the quality factor Q of the coil from the graph.

APPARATUS REQUIRED: Experimental board containing coils and capacitor to make connections for series and parallel resonant circuits, electronic multimeter, signal generator.

CIRCUIT DIAGRAM: As shown in Fig. E.11.1.1.

BRIEF THEORY: At series resonant frequency, the inductive reactance is equal to the capacitive reactance, and the current is maximum. The circuit behaves as purely resistive, and the voltage and current are in phase. The series resonant frequency is given as

$$f_r = \frac{1}{2\pi\sqrt{LC}}$$

At this frequency, the voltage across the coil is Q times the supply signal V_s.

$$V_L = V_s \frac{\omega_r L}{R} = V_s Q$$

In case of the parallel resonant circuit, the line current is minimum at resonant frequency, since the impedance becomes very high. If we plot

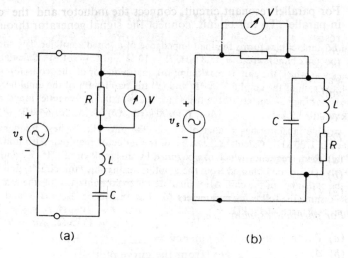

(a)   (b)

Fig. E.11.1.1

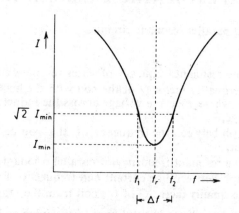

Fig. E.11.1.2

line current versus frequency, we get a curve as shown in Fig. E.11.1.2. The Q of the circuit is given by the formula

$$Q = \frac{f_r}{\Delta f}$$

PROCEDURE:

1. Connect the ac signal source to the series resonant circuit through a series resistor. Connect a voltmeter across the series resistor.

2. Vary the frequency of the ac signal and watch the voltmeter. Note the frequency at which the voltage is maximum. This is the series resonant frequency. At this frequency, measure the voltage across the capacitor and inductor. Also measure the value of the input signal. Calculate the Q of the circuit.

3. Vary the signal frequencies on both sides of resonance, and measure the voltage across the series resistor. Plot this voltage against frequency.

4. For parallel resonant circuit, connect the inductor and the capacitor in parallel. To this circuit, connect the signal generator through a line resistor as shown in the figure. Vary the frequency and watch the voltage across the line resistor. At resonant frequency, this voltage becomes minimum.

5. Vary the signal frequency on both sides of the resonance. Note the voltage across the line resistor. Plot the curve. Calculate Q.

OBSERVATIONS:

1. *For series resonant circuit*:

   (*a*) Series resonant frequency = _____ Hz.

   (*b*) V_L = _____ V; V_S = _____ V.

   (*c*) $Q = V_L/V_S$ = _____

2. *For parallel resonant circuit*:

   (*a*) Parallel resonant frequency = _____ Hz.

   (*b*) Δf = _____ Hz (from the curve plotted)

   (*c*) $Q = f_r/\Delta f$ = _____

RESULTS: As given above.

Feedback in Amplifiers

OBJECTIVES: After going through this unit, you will be able to: ○ Explain the meaning of "feedback" in amplifiers. ○ Explain different types of feedback in amplifiers. ○ Draw some typical feedback amplifier circuits. ○ Derive the mathematical formula for negative feedback amplifiers. ○ Explain the advantage of negative feedback in terms of improvement in stability of gain, reduction in distortion and noise, increase in input impedance, decrease in output impedance, increase in bandwidth. ○ Explain the effect of removing the emitter bypass capacitor in an *RC*-coupled amplifier ○ Draw the circuit diagram of the emitter follower and the cathode follower. ○ Explain the working of emitter follower. ○ State the applications of emitter-follower circuit.

12.1 CONCEPT OF FEEDBACK IN AMPLIFIERS

The important characteristics of an amplifier are its voltage gain, input impedance, output impedance, and bandwidth. These parameters are more or less constant for a given amplifier. That is, the basic amplifier has fixed values of these parameters, and the designer does not have any control over them. Quite often, the values of these parameters are required to be changed. This could be done in a number of ways. For example, voltage gain could be reduced by using a resistive network either in the input or in the output. Similarly, the input impedance of an amplifier could be increased, if desired, by simply connecting a series resistance. But these methods result in wastage of useful signals. We have with us a more powerful technique to do this. The technique is to introduce *feedback* in the amplifier circuit.

Feedback is the process of taking a part of output signal and *feeding* it *back* to the input circuit. Figure 12.1a shows a block diagram of a basic amplifier. Here, v_i is the input signal and v_o the output. If A is the voltage gain of the amplifier, the output v_o is related to the input v_i by

$$A = \frac{v_o}{v_i} \tag{12.1}$$

In this amplifier, the input does not know what is happening at the output. If due to some reason, the output changes, the net input remains unaffected. Such a system is called *open-loop* or *non-feedback* system.

In Fig. 12.1b it can be seen that the output of the amplifier is *fed back* to the input through a network. This network is called a β *network* or a *feedback network*. A fraction $\beta v_o'$ of the output voltage is going back to the input. This changes the net input voltage to the amplifier. Thus, the input is modified by the output. The input knows at every instant what the output is. Such a system is called a *closed-loop* or *feedback system*.

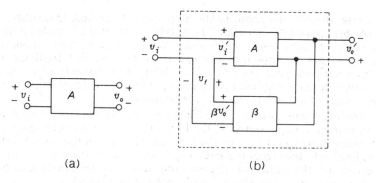

(a) (b)

Fig. 12.1 (*a*) Block diagram of a basic amplifier;
(*b*) Feedback introduced in the amplifier

Amplifiers are not the only things where the feedback is used. We use the idea of feedback in our daily life too. You may not have realised it, but even we use feedback in the process of learning. When a child is asked to write the letter A, he will probably write it as is shown in Fig. 12.2*b*. When he finds that the stroke is not going in the correct direction, the information goes to his brain through the eyes. The brain immediately orders the hand to correct the direction of the stroke. With much effort and with constant feedback, the child writes the letter 'A' as shown in Fig. 12.2*b*.

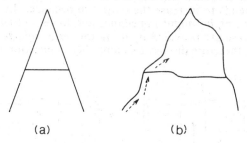

(a) (b)

Fig. 12.2 A child writes the letter "A" as in
(*b*) by seeing the letter as in (*a*),
and using feedback continuously

12.2 TYPES OF FEEDBACK

In Fig. 12.1*b*, a feedback circuit was introduced. You may have noticed that the feedback voltage v_f was in phase opposition to the input voltage v_i. In such a case, the net input voltage v_i' to the amplifier becomes the difference of the external input voltage v_i and the feedback voltage v_f. That is

$$v_i' = v_i - v_f \qquad (12.2)$$

Since the net input to the amplifier is reduced, the output of the amplifier also decreases from v_o to v_o'. In other words, the gain of this amplifier reduces because of the feedback. This type of feedback is called *negative, inverse,* or *degenerative feedback.* There exists another possibility. The feedback voltage can be in the same phase as the external input voltage. In such a case, the effective input to the amplifier is increased. Obviously, the gain of this amplifier increases. This type of feedback is called *positive, direct,* or *regenerative feedback*. You may wonder why we have emphasized negative feedback in amplifiers, whereas it is the positive feedback that provides increased gain. As you will see later, the positive feedback necessarily brings in more distortion, and poor gain-stability. In practice, almost always negative feedback is employed in amplifiers.

The feedback can also be classified as *voltage feedback* or *current feedback*. In case of voltage feedback, the signal fed back is proportional to the output voltage, whatever may be the load impedance. On the other hand, in current feedback, the signal fed back is proportional to the output current irrespective of the value of the load impedance. A combination of both the voltage and current feedback may be present in a circuit.

Both voltage and current can be fed back to the input either in series or parallel. Thus, there are four basic ways of connecting the feedback signal, namely:

  (i) Series-voltage feedback (Fig. 12.3a)
 (ii) Series-current feedback (Fig. 12.3b)
(iii) Shunt-voltage feedback (Fig. 12.3c)
 (iv) Shunt-current feedback (Fig. 12.3d)

Series-feedback connections tend to increase the input impedance of the amplifier, while shunt-feedback connections tends to decrease the input impedance. Voltage feedback tends to decrease the output impedance, while current feedback tends to increase the output impedance. In most of the cascaded amplifiers, higher input impedance and lower output impedance is desired. Both of these are provided by using the *series-voltage feedback* connection. We shall therefore discuss this amplifier connection in the sections :hat follow.

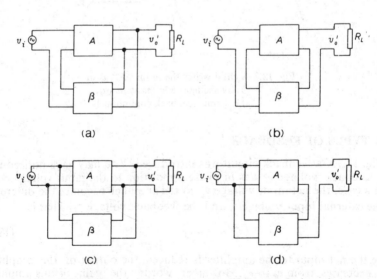

(a)   (b)

(C)   (d)

Fig. 12.3 Different connections in amplifiers: (a) Series-voltage feedback; (b) Series-current feedback; (c) Shunt-voltage feedback, and (d) Shunt-current feedback

12.3 VOLTAGE GAIN OF FEEDBACK AMPLIFIER

Figure 12.4a shows a triode amplifier using series-voltage feedback. The same circuit is drawn using a FET, in Fig. 12.4b. Here, the output of the amplifier is available across the load resistance R_L. This output voltage v_o' is fed to a β-network, consisting of resistors R_1 and R_2. This combination of resistors works as a potential divider. The values of these resistors are generally high; otherwise the effective ac load resistance in the output circuit will be very much reduced. The voltage v_f developed across resistor R_1 is fed back to the input side. This voltage v_f is in series with the input voltage v_i.

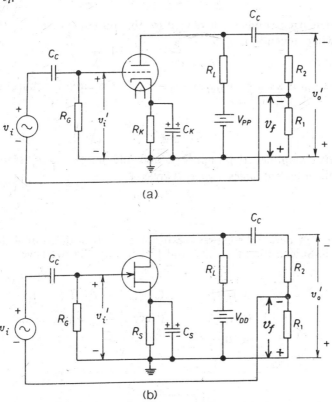

(a)

(b)

Fig. 12.4 Amplifier with series-voltage feedback using
(a) a triode, and (b) an FET

Let us now check whether the feedback is positive or negative in the amplifiers shown in Fig. 12.4. It is known to us that in a single-stage common-cathode (or common-source) amplifier, a phase reversal of the signal occurs during amplification. The polarities of the input and output voltages are shown in the figure. You may now see that the feedback voltage v_f is out of phase with the input voltage v_i. The effective input voltage v_i' is, therefore, reduced. The type of feedback is therefore negative. It is easy to see that the feedback connections in both the amplifiers in Fig. 12.4 represent a case of series-voltage feedback. We can represent such amplifiers by a block diagram, drawn in Fig. 12.5. In this block diagram, a pair of input terminals and another pair of output terminals are available outside the dotted block. The input to this feedback amplifier is v_i and the output is v_o'. The voltage gain of this feedback amplifier is then

$$A_f = \frac{v_o'}{v_i} \tag{12.3}$$

The effective input to the basic amplifier (i.e. the amplifier with no feedback connections) is v_i' and not the external input v_i. In fact, due to negative feedback, the effective input voltage gets modified and becomes

$$v_i' = v_i - v_f \tag{12.4}$$

Here, the feedback voltage v_f is related to the output voltage v_o' through the β-network. From Fig. 12.5

$$v_f = \beta v_o' \tag{12.5}$$

The constant β is known as *fedback factor*. In the circuit connections of Fig. 12.4, the value of β is

$$\beta = \frac{R_1}{R_1 + R_2} \tag{12.6}$$

For the basic amplifier, the input is v_i' and the output is v_o'. Hence, its voltage gain A (called *internal gain*) is given as

$$A = \frac{v_o'}{v_i'} \tag{12.7}$$

We shall now derive the expression of the gain A_f in terms of A (internal gain) and β. Substituting Eqs. 12.7 and 12.5 in Eq. 12.4. we have

$$v_i' = v_i - \beta v_o' \quad \text{(since } v_f = \beta v_o')$$

or $\quad (v_o'/A) = v_i - \beta v_o' \quad \text{(since } v_i' = v_o'/A)$

or $\quad v_o = A v_i - A \beta v_o'$

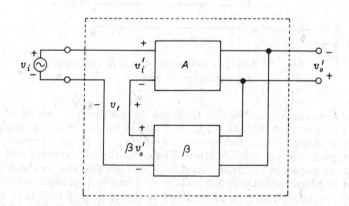

Fig. 12.5 Block diagram of a series-voltage feedback

or $\qquad (1+A\beta)\, v'_0 = Av_i$

or $\qquad \dfrac{v'_0}{v_i} = \dfrac{A}{1+A\beta}$

According to Eq. 12.3, the left hand side of the above equation is A_f.

Hence, $\qquad A_f = \dfrac{A}{1+A\beta} \qquad (12.8)$

In case of positive feedback, the feedback voltage is in the same phase as the input voltage and both are added up to make v'_i. Equation 12.4 will then become

$$v'_i = v_i + v_f$$

or $\qquad v'_i = v_i + \beta v_0$

For such a feedback amplifier, the voltage gain A_f is given by

$$A_f = \dfrac{A}{1-A\beta} \qquad (12.9)$$

As can be seen from the above equation, the gain of an amplifier increases when we apply positive feedback. However, for good performance of an amplifier, we always use negative feedback. The use of positive feedback in making oscillators will be seen in the next unit.

Example 12.1 Calculate the gain of a negative-feedback amplifier with an internal gain, $A = 100$, and feedback factor $\beta = 1/10$.

Solution: The gain of the feedback amplifier is given by

$$A_f = \dfrac{A}{1+A\beta}$$

Here, $A = 100$; $\beta = 1/10 = 0.1$. Hence

$$A_f = \dfrac{100}{1+100\times0.1} = \dfrac{100}{1+10} = \dfrac{100}{11}$$
$$= 9.09$$

Example 12.2 An amplifier with negative feedback has a voltage gain of 100. It is found that without feedback, an input signal of 50 mV is required to produce a given output; whereas with feedback, the input signal must be 0.6 V for the same output. Calculate the value of A and β.

Solution: The gain A_f of the feedback amplifier is 100. The input voltage required to produce the same output voltage as for the amplifier without feedback, is 0.6 V. Thus, the output will be

$$v'_0 = A_f V_i = 100\times0.6 = 60 \text{ V}$$

If no feedback is employed, the required input to produce 60 V output is 50 mV $= 0.05$ V. Hence, the internal gain of the amplifier is

$$A = \dfrac{V_o}{V_i} = \dfrac{60}{0.05} = 1200$$

Now using Eq. 12.8, we have

$$1+A\beta = \frac{A}{A_f}$$

or $\qquad 1+1200 \times \beta = \frac{1200}{100}$

or $\qquad \beta = \frac{12-1}{1200} = \frac{11}{1200} = \frac{11}{12} \%$

12.4 HOW NEGATIVE FEEDBACK IS ADVANTAGEOUS

We have seen that the gain of an amplifier is reduced when negative feedback is used. This appears to be a serious drawback, and one may wonder why we use negative feedback in amplifiers. The negative feedback improves the performance of the amplifier from so many other points of view. The reduction in gain due to negative feedback, can always be compensated by increasing the number of stages. The advantages of negative feedback are listed as follows (see Fig. 12.6 also):

(i) It improves the stability of amplifier gain.
(ii) It reduces the distortion and noise.
(iii) It increases the input impedance.
(iv) It decreases the output impedance.
(v) It increases the bandwidth.

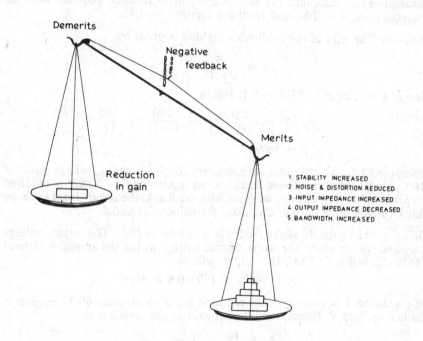

Fig. 12.6 Merits and demerits of negative feedback

12.4.1 Stabilization of Gain

The gain of an amplifier may change because of so many reasons. Change in power supply voltage, or change in the parameters of the active device (triode, FET or the transistor) may change the gain. This adversely affects the performance of the amplifier. It would have been ideal, if the gain of the amplifier was independent of these changes. Negative feedback achieves this object to a great extent.

The gain of a negative feedback amplifier is given as (see Eq. 12.8)

$$A_f = \frac{A}{1+A\beta}$$

If we make $A\beta \gg 1$, then in the denominator of the above equation we can neglect unity as compared to $A\beta$, and write

$$A_f = \frac{A}{A\beta} = \frac{1}{\beta} \tag{12.10}$$

Thus, the gain A_f of the feedback amplifier is made independent of the internal gain A. The gain A_f depends only on β, which in turn depends on passive elements such as resistors (see Fig. 12.4 where $\beta = R_1/(R_1+R_2)$).

The values of the resistors remain fairly constant, and hence the gain is stabilized. The only condition for this stabilization is that $A\beta \gg 1$. Even if this condition is not fully met, some improvement occurs in the stability of the gain.

Suppose a certain change in the internal gain of the amplifier takes place. We can find the corresponding percentage change in the overall gain of the feedback amplifier. This we do by differentiating Eq. 12.8 with respect to A.

$$\frac{dA_f}{dA} = \frac{(1+A\beta)\times 1 - A\times\beta}{(1+A\beta)^2} = \frac{1}{(1+A\beta)^2}$$

or
$$dA_f = \frac{dA}{(1+A\beta)^2}$$

Dividing the above by Eq. 12.8, we get

$$\frac{dA_f}{A_f} = \frac{dA}{(1+A\beta)^2} \times \frac{(1+A\beta)}{A}$$

or
$$\frac{dA_f}{A_f} = \frac{1}{(1+A\beta)} \frac{dA}{A} \tag{12.11}$$

Since $(1+A\beta) > 1$, the percentage change in A_f is seen to be much less than the percentage change in A.

12.4.2 Reduction in Distortion and Noise

Another desirable characteristic of negative feedback is the reduction of harmonic distortion. The reason for this may be seen in Fig. 12.7. Here, the basic amplifier is assumed to distort the sinusoidal input waveform by flattening the peaks. The feedback voltage v_f has the same waveform as the output voltage. The voltage v_f gets subtracted from the input voltage v_i to make the net input v_i' to the amplifier. Since the peaks of the voltage v_f are flattened when v_f is subtracted from v_i, the peak of resulting v_o' will become

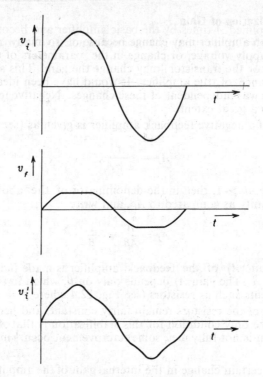

Fig. 12.7 Feedback voltage V_f predistorts the net input
V_i to partially compensate the distortion

more peaked. Thus, the net input v'_i is predistorted in such a way so as to partially compensate for the flattening caused by the amplifier. Now, when the peaked input v'_i gets amplified, the output will tend to be sinusoidal, because the amplifier tries to flatten the peaks.

We can find the amount of reduction in the distortion caused by negative feedback, with the help of Fig. 12.8. Suppose the amplifier with gain A produces a distortion D without feedback. This distortion appears at the output. After feedback is applied, the gain becomes A_f and the distortion in the output becomes D_f. Let us see how the distortion in the output changes from D to D_f. A part βD_f of the distortion D_f is fed back to the input.

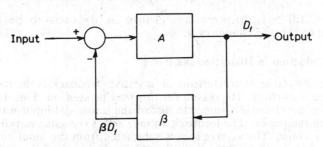

Fig. 12.8 A block diagram showing the distortion in the output
of a feedback amplifier

This gets amplified A times by the basic amplifier and becomes $A\beta D_f$. This gets added up (in reverse polarity because of negative feedback) to the original distortion D to make the net distortion D_f. Thus,

$$D_f = D - A\beta D_f$$

or
$$D_f = \frac{D}{1+A\beta} \tag{12.12}$$

Note that the distortion is reduced by the same factor as the gain.

Electrical noise may appear due to many reasons. If a noise voltage appears just at the input of the amplifier, it is amplified by the same amount as the signal voltage. However, some types of noise voltages appear inside the active device of the amplifier. This noise does not get amplified by the same amount, as the input signal. If negative feedback is employed, the net noise in the output reduces, and the performance of the amplifier is much improved.

12 4.3 Increase in Input Impedance

It is desirable to have a high input impedance for an amplifier. Then it will not load the preceding stage or the input voltage source. Such a desirable characteristic can be achieved with the help of negative series-voltage feedback. To see this, we examine the block diagram in Fig. 12.9.

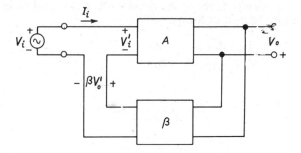

Fig. 12.9 Block diagram of a negative feedback amplifier to show increase in input impedance

In Fig. 12.9, V_I is the input voltage and I_i is the input current. Therefore, the input impedance of the feedback amplifier is

$$Z_{if} = \frac{V_i}{I_i} \tag{12.13}$$

The net input to the basic amplifier is
$$V_i' = V_i - \beta V_o' \tag{12.14}$$

The input current to the basic amplifier is also I_i'. Hence, the input impedance of the amplifier without feedback is

$$Z_i = \frac{V_i'}{I_i} \tag{12.15}$$

Since, $V_o' = AV_i'$, from Eq. 12.14, we get

$$V'_i = V_i - \beta A V'_i$$

or

$$V_i = V'_i + A\beta V'_i$$

Dividing the above equation by I_i, we get

$$\frac{V_i}{I_i} = \frac{V'_i}{I'_i}(1 + A\beta)$$

Using Eqs. 12.13 and 12.15, we get

$$Z_{if} = Z_i (1 + A\beta) \qquad (12.16)$$

Thus, we see that the input impedance is increased by a factor of $(1 + A\beta)$.

12.4.4 Decrease in Output Impedance

Just as high input impedance is advantageous to an amplifier, so is low output impedance. An amplifier with low output impedance is capable of delivering power (or voltage) to the load without much loss. Such a desirable characteristic is achieved by employing negative series-voltage feedback in the amplifier.

Figure 12.10 shows the block diagram of a feedback amplifier where the output side has been replaced by an equivalent voltage source $A\beta V_o$ in series with an impedance Z_o. The input terminals of the amplifier have been shorted. We now apply a voltage source of value V_o at the output terminals. If the input impedance of the β-network is assumed to be very high, we can write for the output loop

$$V_o + A\beta V_o = I_o Z_o$$

or

$$V_o(1 + A\beta) = I_o Z_o$$

or

$$\frac{V_o}{I_o} = \frac{Z_o}{1 + A\beta} \qquad (12.16)$$

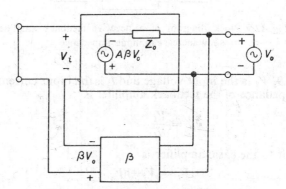

Fig. 12.10 Block diagram of a feedback amplifier to determine the output impedance

where, I_o is the current flowing into the output terminals of the amplifier from the source V_o. Thus, the output impedance with feedback is given as

$$Z_{of} = \frac{V_o}{I_o}$$

which, with the help of Eq. (12.16), becomes

$$Z_{of} = \frac{Z_o}{1+A\beta} \qquad (12.17)$$

Thus, we see that the output impedance is reduced by the factor $(1+A\beta)$.

Example 12.3   To an amplifier of 60 dB gain, a feedback (negative) of $\beta = 0.005$ is applied. What would be the change in overall gain of the feedback amplifier if the internal amplifier is subjected to a gain reduction of 12 % ?

Solution Given :

$$A = 60 \text{ dB} = 1000, \beta = 0.005$$
$$dA/A = -12 \% = -0.12$$

We know that

$$\frac{dA_f}{A_f} = \frac{1}{(1+A\beta)} \times \frac{dA}{A}$$

putting values of A, β and dA/A

$$\frac{dA_f}{A_f} = \frac{1}{(1+1000\times0.005)} \times (-0.12)$$
$$= -(0.12)/(1+5)$$
$$= -(0.12)/6 = -0.02$$

Therefore, the overall gain of the feedback amplifier will be reduced by 2 %.

Example 12.4   An amplifier with $Z_i = 1 \text{ k}\Omega$ has a voltage gain $A = 1000$. If a negative feedback of $\beta = 0.01$ is applied to it, what shall be the input impedance of the feedback amplifier ?

Solution: We know

$$Z_i' = (1+A\beta) \times Z_i$$
$$\therefore \quad Z_i' = (1+1000\times0.01)\times10^3$$
$$= (1+10)\times10^3$$
$$= (11)\times10^3 \ \Omega$$
$$= 11 \text{ k}\Omega$$

Example 12.5 We have an amplifier of 60 dB gain. It has an output impedance $Z_o = 12 \text{ k}\Omega$. It is required to modify its output impedance to 600 Ω by applying negative feedback. Calculate the value of the feedback factor. Also find the percentage change in the overall gain, for 10 % change in the gain of the internal amplifier.

Solution: We know

$$Z_u' = Z_o/(1+A\beta)$$

Given : $A = 60 \text{ dB} = 1000$, $Z_o = 12000 \ \Omega$ and $Z_o' = 600 \ \Omega^-$

$$\therefore \qquad 600 = \frac{12000}{1+1000\,\beta}$$

or
$$1+1000\beta = \frac{12000}{600}$$
$$= 20$$

or
$$1000\beta = 19$$

or
$$\beta = \frac{19}{1000} = 0.019 = 1.9\ \%$$

Again
$$\frac{dA_f}{A_f} = \frac{1}{(1+A\beta)}\,\frac{dA}{A}$$
$$= \frac{1}{1+1000\times0.019}\times(0.1)$$
$$= \frac{0.1}{1+19} = \frac{0.1}{20} = 0.005$$
$$= 0.5\ \%$$

12.4.5 Increase in Bandwidth

We know that introduction of negative feedback in amplifier reduces its gain. If A is the internal gain of the amplifier and β is the feedback factor, the gain is reduced by a factor of $(1+A\beta)$. The negative feedback also affects the cut-off frequencies. The lower cut-off frequency f_1 is lowered by a factor of $(1+A\beta)$ and the upper cut-off frequency f_2 is raised by the same factor $(1+A\beta)$. Since $f_1 \ll f_2$, the BW$(= f_2 - f_1)$ can be taken same as f_2. Thus, we may say that if negative feedback is employed in an amplifier, its BW increases by the same factor $(1+A\beta)$ by which its gain reduces. In fact, the product of gain and BW, called gain bandwidth product (GBW) remains the same. By introducing negative feedback we can trade bandwidth for gain.

12.5 AMPLIFIER CIRCUITS WITH NEGATIVE FEEDBACK

In practical amplifier circuits, it is the negative feedback that is invariably used. In Fig. 12.4, we saw an amplifier circuit in which negative series-voltage feedback was used. In this circuit, a part of the output voltage is fed back in phase opposition to the input voltage. It is also possible to have a feedback amplifier where the feedback voltage is in proportion to the output current. Such a situation arises if the emitter bypass capacitor C_E is removed from an ordinary RC-coupled CE amplifier circuit.

12.5.1 *RC*-Coupled Amplifier Without Bypass Capacitor

Figure 12.11a shows an RC-coupled amplifier circuit. Here the transistor is used in the CE mode. The effective input voltage of such an amplifier is the ac signal between the base and the emitter. In this circuit, it is the same as the voltage v_s supplied by the signal source.

Let us now consider what happens if the bypass capacitor C_E is removed. The result is the circuit it Fig. 12.11b. Now the situation becomes entirely different. The effective input voltage to the amplifier no more remains the same. At the instant when the source voltage v_s increases during its positive

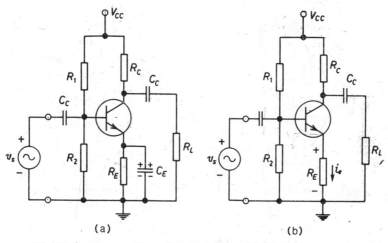

Fig. 12.11 (*a*) An ordinary *RC*-coupled amplifier; (*b*) The same circuit when emitter bypass capacitor is removed

half-cycle, the emitter-base junction becomes more forward biased. The collector current increases, and so does the ac emitter current i_e in the direction indicated in the figure. This develops an ac voltage $i_e R_E$ across the resistor R_E with a polarity as shown in the figure. Now, you may see that the effective input voltage (that is the voltage between the base and emitter) is not the same as the voltage v_s. The voltage $i_e R_E$ also appears in series with the voltage v_s, and the two may be seen in phase opposition. The effective voltage between the base and emitter is thus reduced. Hence, it is a case of negative feedback. You now see that the situation in this circuit is the same as that shown by the block diagram in Fig. 12.3*b*. The feedback voltage $i_e R_E$ is proportional to the output current ($i_e \simeq i_c$), and it appears in series with the source voltage v_s. Thus, it is a case of series-current feedback.

This type of negative feedback is very commonly used in practical amplifiers such as in public address systems, tape recorders, record players, stereo amplifiers, etc. Sometimes, only a part of the emitter resistance is bypassed, so that the gain of the amplifier is not excessively reduced, and at the same time the advantages of negative feedback are obtained.

12.5.2 Emitter Follower

An emitter follower is a very useful negative feedback circuit and is extensively used in electronic instruments. If the resistance R_C is reduced to zero, and output is taken from the emitter terminal instead of the collector terminal in the circuit in Fig. 12.11*b*, we get an emitter-follower circuit. The result is shown in Fig. 12.12*a*.

The effective input voltage in this circuit is ($v_s - v_o$). It means that the whole of the output voltage v_o is fed back to the input side. The gain of the amplifier is drastically reduced. In fact, *the voltage gain of this amplifier is less than unity.* You may wonder why to use an amplifier which has a gain less than unity. For such an amplifier the output will be less than the input.

It is found that the input impedance of this circuit is very high, and the output impedance is very low. This is where its importance lies. The circuit is used for impedance matching. The last stage of a signal generator, used in the laboratory, is an emitter follower. When you connect the output of

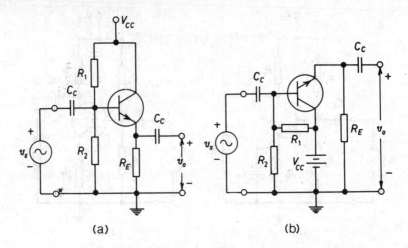

(a)   (b)

Fig. 12.12 (a) Emitter-follower circuit; (b) The same circuit drawn
in a different way to show that it is a common-collector
amplifier

the signal generator to a circuit under test, the oscillator is not loaded and
its frequency remains constant. Because of its high input impedance and
low output impedance properties, an emitter follower is capable of giving
power to a load connected to its output without requiring much power at
the input. It thus works as a *buffer amplifier*.

When the input v_s goes through its positive half-cycle, the output v_o is
also seen to go through its positive half-cycle. In other words, the output
and the input are in same phase. In magnitude, the output is almost the
same as the input (v_o is slightly less than v_s). Thus, we see that the emitter
(voltage) closely follows the input. Hence the name *emitter follower*.

The emitter-follower circuit is also called common-collector amplifier
circuit. To understand how the collector is common to the input and the
output, let us redraw the circuit by inverting the transistor as in Fig. 12.12b.
From the ac point of view, the supply V_{CC} is a short-circuit, and hence the
collector is grounded. The input is given between the base and the collector,
and the output appears between the emitter and the collector.

The vacuum-tube counterpart of the emitter follower is called *cathode
follower* (also known as *common-plate amplifier*), and is shown in
Fig. 12.13.

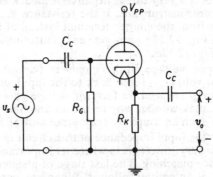

Fig. 12.13 Cathode-follower circuit

REVIEW QUESTIONS

12.1 State the merits and demerits of negative feedback in amplifiers.

12.2 Draw the block representation of the four types of negative-feedback circuits (series-voltage, series-current, shunt-voltage and shunt-current types). Which one of these types is employed to get greater input impedance and lower output impedance in an amplifier circuit.

12.3 Derive an expression to illustrate that the voltage gain in an amplifier circuit with negative feedback is somewhat stable even if the β of transistor changes due to its ageing or due to its replacement.

12.4 The voltage gain of an amplifier without feedback mainly depends on the β of the transistor used. To make the gain of the circuit less sensitive to the β of the active device, what method would you suggest ? Explain the stabilization of gain through the method suggested by you.

12.5 Explain how the negative feedback in an amplifier helps in reducing the distortion and noise.

12.6 If the bypass capacitor C_E in an RC-coupled amplifier becomes accidentally open-circuited, what happens to the gain of the amplifier ? Explain.

12.7 Draw an emitter-follower circuit. Justify that it is a common-collector amplifier circuit.

12.8 State the property of an emitter-follower circuit which justifies its wide use, even though its voltage gain is less than unity.

12.9 State and explain two specific applications of an emitter-follower circuit.

OBJECTIVE-TYPE QUESTIONS

I. Here are some incomplete statements. Four alternatives are provided below each. Tick the alternative which completes the statement correctly.

1. The negative feedback in an amplifier

    (a) reduces the voltage gain
    (b) increases the voltage gain
    (c) does not affect the voltage gain
    (d) can convert it into an oscillator if the amount of feedback is sufficient

2. Introduction of feedback in an amplifier increases the input impedance from 1 kΩ to 40 kΩ. It is due to

    (a) positive feedback
    (b) shunt-current negative feedback
    (c) series-current negative feedback
    (d) shunt-voltage negative feedback

3. One of the effects of negative feedback in amplifiers is to

    (a) increase the noise
    (b) increase the harmonic distortion
    (c) decrease the bandwidth
    (d) decrease the harmonic distortion

4. An emitter-follower circuit is widely used in electronic instruments because

    (a) its voltage gain is less than unity
    (b) its output impedance is high and input impedance is low
    (c) its output impedance is low and input impedance is high
    (d) its voltage gain is very high

5. The voltage gain of an amplifier is 100. On applying negative feedback with $\beta = 0.03$, its gain will reduce to

(a) 70   (c) 25
(b) 99.97   (d) 3

II. Write TRUE or FALSE (as the case may be) in the space provided against each of the following statements.

1. Negative feedback in an amplifier increases the stability of its voltage gain.

2. Introducing negative feedback raises the lower cut-off frequency of the amplifier.

3. Introducing negative feedback raises the upper cut-off frequency of an amplifier.

4. By introducing proper amount of negative feedback it is possible to convert an amplifier into an oscillator.

5. In a public address system, voltage amplifiers with unbypassed emitter resistors are used.

6. Emitter follower is an amplifier with positive feedback.

7. Input impedance of an emitter follower is low, and its output impedance is high.

8. Emitter follower is the same as a common-base amplifier circuit.

Ans: I. 1. *a*; 2. *c*; 3. *d*; 4. *c*; 5. *c*

II. 1. T; 2. F; 3. T; 4. F; 5. T; 6. F; 7. F; 8. F.

TUTORIAL SHEET 12.1

1. Calculate the average gain of a negative-feedback amplifier circuit having internal gain $A = 100$ and feedback factor $\beta = 0.1$. [Ans. 9.09]
2. In a certain amplifier, an output of 50 V is obtained for an input signal of 0.5 V. This circuit is now connected to a feedback (negative) network such that 20 % of the output voltage goes back to the input side. Calculate the increased value of the input if we want the same output from the circuit (i.e. 50 V). [Ans. 10.5]
3. Negative feedback reduces the gain of an amplifier from 50 to 25. Calculate the feedback factor in percentage. [Ans. 2 %]
4. The Fig. T.12.1.1 shows a vacuum-tube amplifier. The data book reveals following data for the tube used: $\mu = 20$ and $r_p = 8$ kΩ. Calculate the gain of the circuit

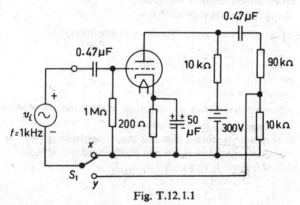

Fig. T.12.1.1

(a) when the switch S_1 is at x, (b) when the switch S_1 is at y.

[**Ans.** (a) $A = 10.64$, (b) $A_f = 5.15$]

TUTORIAL SHEET 12.2

1. A feedback amplifier is given in Fig. T.12.2.1. Calculate (a) the value of feedback factor β, (b) voltage gain A_f of the feedback amplifier, (c) voltage gain A of the amplifier without feedback, and (d) the feedback voltage (V_f).

[**Ans.** (a) $\beta = 0.05$, (b) $A_f = 19.05 = 25.6$ dB,
(c) $A = 400 = 52.04$ dB, (d) $V_f = 0.1$ V]

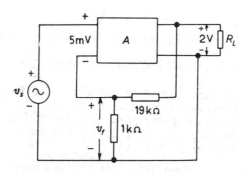

Fig. T.12.2.1

2. An amplifier with gain $A = 80$ dB is connected to a negative-feedback network. The feedback factor β is 0.02. If now, due to replacement of the active device the internal gain of the amplifier (A) is changed by 50 %, what would be the percentage change in the gain of the feedback amplifier.   [**Ans.** 0.25 %]
3. An amplifier has an internal gain A of 200. Its output impedance is 1 kΩ. Negative feedback is introduced in the circuit ($\beta = 0.02$). Calculate the output impedance of the feedback amplifier.   [**Ans.** $Z_{of} = 200 \ \Omega$]
4. If in the circuit in problem 3 above, the feedback factor is changed to 0.06, calculate the value of the new output impedance Z_{of}.   [**Ans.** $Z_{of} = 76.92 \ \Omega$]

TUTORIAL SHEET 12.3

1. A feedback amplifier has an internal gain $A = 40$ dB and feedback factor $= 0.05$. If the input impedance of this circuit is 12 kΩ, what would have been the input impedance of the amplifier if feedback was not present.   [**Ans.** 2 kΩ]
2. What should be the feedback factor of the negative feedback applied to an amplifier of internal gain $A = 180$, and $Z_i = 250 \ \Omega$, in order to increase the input impedance to 2 kΩ.   [**Ans.** 0.0389]
3. A certain amplifier has an internal gain of 80, and the harmonic distortion in the output is 12 %. To improve the performance of the amplifier from the point of view of harmonic distortion, negative feedback is introduced in the circuit. This reduces the distortion within a tolerable limit of 3%. Calculate the feedback factor in the amplifier.   [**Ans.** $\beta = 0.0375 = 3.75$ %]

4. A certain amplifier gives an output of 20 V, for an input of 200 mV. The amplifier introduces a harmonic distortion of 15 %. The tolerable limit of harmonic distortion in the output is 3 %. Calculate the feedback factor of the negative feedback to be employed for this task. Also determine the value of modified input in order to get the output unchanged, i.e. 20 V. [**Ans.** $\beta = 0.04$, v_i (modified) = 1 V]

EXPERIMENTAL EXERCISE 12.1

TITLE: Series current feedback circuit (amplifier with an unbypassed emitter-resistor).

OBJECTIVES: To,

1. trace the given circuit and to note down the values of all the components;
2. measure the quiescent operating point of the given feedback amplifier circuit;
3. measure the voltage gain of the amplifier circuit with unbypassed as well as bypassed emitter-resistor, at frequency of 1 kHz;
4. plot the frequency response curve of the feedback amplifier circuit;
5. determine the lower and upper cut-off frequencies of the amplifier circuit from the response curve.

APPARATUS REQUIRED: Amplifier circuit with an arrangement for applying feedback, signal generator, electronic multimeter, ac millivoltmeter, transistorized power supply.

CIRCUIT DIAGRAM: As given in Fig. E.12.1.1. Typical values of the components are also given.

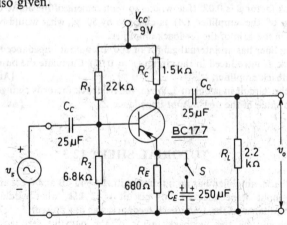

Fig. E.12.1.1

BRIEF THEORY: In an ordinary amplifier circuit, the emitter resistor is bypassed with a capacitor of large value (say, 250 μF). As long as this bypass capacitor is present, the effective input to the amplifier is the same as the signal supplied by the source. But, when the bypass capacitor is removed from the circuit (by keeping the switch S open), negative current feedback takes place in the circuit. The gain is reduced, and the bandwidth is increased.

PROCEDURE:

1. Look into the experimental board and trace the circuit. Draw the circuit diagram. Read the values of the resistors, capacitors (use colour code, if needed). Read the transistor type number. Write these in your circuit diagram.
2. Connect the dc supply voltage, of say 9 V, to the amplifier circuit. Measure V_{CC}, V_C and V_{CE}. The quiescent operating-point collector current is given by

$$I_C = \frac{V_{CC} - V_C}{R_C}$$

3. Keep the emitter resistor R_E bypassed, by keeping the switch S closed. Apply the ac signal voltage to the input of the amplifier, from the signal generator. Keep the input voltage low (say, 5 mV, otherwise distortion may result) and the frequency at 1 kHz. Measure the voltage at the output with the help of a millivoltmeter. Now open the switch S. Again measure the voltage at the output. Find the gain of the amplifier in the two cases.
4. Keep the switch S open, so that feedback is applied. Vary the frequency of the input signal from a very low value (say, 15 Hz) to very high value (say, 1 MHz), and measure the output at each frequency. Keep the input voltage constant (at, say, 5 mV) throughout. Take more readings near about the frequencies where the output varies with frequency. Draw the frequency response characteristic. A typical response characteristic is shown in Fig. E.12.1.2.

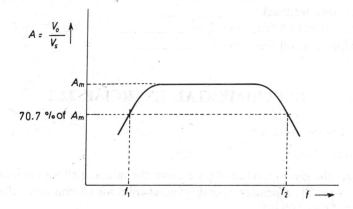

Fig. E.12.1.2

5. On the characteristic curve, note the mid-band gain. Mark a point on the gain axis of the graph corresponding to 70.7 % of the mid-band gain. Draw a horizontal line at this point. The points of intersection of this line with the response curve gives the lower and upper cut-off frequencies.

OBSERVATIONS:

1. $V_{CC} =$ _____ V; $V_C =$ _____ V; $V_{CE} =$ _____ V.

$$I_C = \frac{V_{CC} - V_C}{R_C} = \underline{\hspace{2cm}} mA$$

2. $V_s =$ _____ mV; $f = 1$ kHz

   (a) Output, when switch S is closed = _____ V

   Gain without feedback $= \dfrac{V_o}{V_s} =$ _____

   (b) Output, when switch S is open = _____ V

   Gain (with feedback) $= \dfrac{V_o}{V_s} =$ _____

3. *For frequency response curve:*

   $V_s = 5$ mV

S. No.	Frequency of input signal	Output voltage	Voltage gain
1.			
2.			
3.			

RESULTS:

1. $I_C =$ _____ mA; $V_C =$ _____ V

2. A (without feedback) = _____

   A (with feedback) = _____

3. Lower cut-off frequency $f_1 =$ _____

   Upper cut-off frequency $f_2 =$ _____

EXPERIMENTAL EXERCISE 12.2

TITLE: Emitter follower.

OBJECTIVES: To,

1. trace the given circuit and note down the values of all the components;
2. measure the quiescent operating-point collector current and collector-to-emitter voltage;
3. measure the voltage gain of the emitter-follower circuit;
4. plot the frequency response of the emitter-follower circuit;
5. determine the cut-off frequencies of emitter follower.

APPARATUS REQUIRED: Emitter-follower circuit, electronic multimeter, ac millivoltmeter, signal generator.

CIRCUIT DIAGRAM: As given in Fig. E.12.2.1. Typical values of the components are also given.

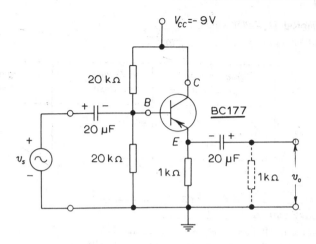

Fig. E.12.2.1

BRIEF THEORY: Emitter follower is a negative feedback amplifier circuit. The emitter resistance R_E is not bypassed by a capacitor. The collector is directly connected to the dc supply. This circuit has a voltage gain of less than unity. But its input impedance is very high and output impedance is low.

PROCEDURE:

1. Trace the given circuit of the amplifier and note down the values of all the components.
2. Connect a supply (of, say, 9 V). Note down the values of supply voltage V_{CC}, emitter voltage V_E and collector to emitter voltage V_{CE}. Calculate the emitter current by using the relation $I_E = V_E/R_E$.
3. Feed a signal of, say, 5 mV (1 kHz frequency) to the input, and measure the output voltage. Calculate the voltage gain.
4. Vary the frequency of the input signal from very low frequency to very high frequency (say from 15 Hz to 1 MHz).
5. Measure the output voltage for each frequency. Plot the frequency response curve and determine lower and upper cut-off frequencies.

OBSERVATIONS:

1. *Quiescent operating point*:

$V_{CC} =$ _____ V; $V_E =$ _____ V; $V_{CE} =$ _____ V

$I_E = \dfrac{V_E}{R_E} =$ _____ mA

2. *Maximum voltage gain* (frequency of signal = 1 kHz):

Input voltage $V_s =$ _____ mV
Output voltage $V_o =$ _____ mV

$\therefore$ Voltage gain $= \dfrac{V_o}{V_s}$

3. *Frequency response curve:*

   Input voltage = _____ mV

S. No.	Frequency of input signal	Output voltage	Voltage gain
1.			
2.			
3.			

RESULTS:

1. I_E = _____ mA; V_{CE} = _____ V
2. Voltage gain A = _____
3. Lower cut-off frequency = _____ Hz
4. Upper cut-off frequency = _____ kHz

Oscillators

OBJECTIVES: After completing this unit, you will be able to: ○ State the utility of an oscillator circuit. ○ Explain the conditions under which a feedback amplifier works as an oscillator. ○ State the classification of oscillators. ○ Explain the working of tuned-collector oscillator, tickler oscillator, Hartley oscillator, Colpitts oscillator, *RC* phase-shift oscillator, Wein bridge oscillator, crystal oscillator, astable multivibrator. ○ Derive the formula for the frequency of oscillation of Wein bridge oscillator circuits. ○ Calculate the frequency of oscillations of a tuned circuit oscillator, when supplied with the values of the circuit components. ○ Calculate the frequency of oscillations of an *RC* phase-shift oscillator, when supplied with the values of components used in the phase-shift network.

13.1 WHY DO WE NEED AN OSCILLATOR

Any circuit that generates an alternating voltage is called an *oscillator*. To generate ac voltage, the circuit is supplied energy from a dc source. The oscillators have a variety of applications. In some applications we need voltages of low frequencies; in others of very high frequencies. For example, to test the performance of a stereo amplifier, we need a signal of variable frequency in the audio range (say, from 20 Hz to 15 kHz). The commercial oscillator available in the market for this purpose is called *audio signal generator*.

Generation of high frequencies is essential in all communication systems. For example, in radio and television broadcasting, the transmitter radiates the signal using a carrier of very high frequency (say, from 550 kHz to 22 MHz in radio broadcasting, and from 47 MHz to 230 MHz in TV broadcasting). In radio and TV receivers too, there is an oscillator circuit which generates very high frequencies.

A signal generator is an important testing instrument in any electronics laboratory, whether in an educational institution or in a research and development organisation. In industry, it is frequently necessary to heat different kinds of materials. Induction and dielectric heating are very convenient ways of doing this. For such applications also, we need high frequency oscillators. These oscillators should be capable of supplying heating power.

Can an alternator (ac generating machine) serve the above purpose? Usually, an alternator is used to generate frequencies up to, say, 1000 Hz. To generate higher frequencies, many practical difficulties arise. Either the number of poles has to be made large, or the speed of rotation of the armature has to be made extremely high. Both are impracticable. This is why we have to depend upon electronic circuits to generate high frequencies.

13.2 CLASSIFICATION OF OSCILLATORS

There are mainly two types of oscillators, viz. *sinusoidal* and *nonsinusoidal*. A sinusoidal oscillator produces sine waves, whereas a nonsinusoidal oscillator can generate square waves, triangular waves, pulses, or sawtooth waves, etc. The circuits which generate square waves or pulses are usually called *multivibrators*.

Depending upon how oscillations are produced, sinusoidal oscillators are of the following types:

   (i) Tuned circuit (*LC*) oscillators,

  (ii) *RC* oscillators,

 (iii) Crystal oscillators.

13.3 HOW A TUNED CIRCUIT CAN BE MADE TO GENERATE SINE WAVES

An inductor and a capacitor connected in parallel form a tuned or tank circuit. In Fig. 13.1*a*, energy is introduced into this circuit by connecting the capacitor to a dc voltage source. The negative terminal of the battery supplies electrons to the lower plate of the capacitor. Because of the accumulation of electrons, the capacitor gets charged and there is a voltage across it. We say that energy is stored in the capacitor in the form of *electric potential energy*. When the switch *S* is thrown to position 2, current starts flowing in the circuit. The capacitor now starts discharging through the inductor. Since the inductor has the property of opposing any change in current, the current builds up slowly. Maximum current flows in the circuit when the capacitor is fully discharged. At this instant, the potential energy of the system is zero. But the *electron motion* being greatest (maximum current), the *magnetic field energy* around the coil is maximum. This condition is shown in Fig. 13.1*b*.

Once the capacitor is fully discharged, the magnetic field begins to collapse. The back emf in the inductor keeps the current flowing in the same direction. The capacitor starts charging, but with opposite polarity this time, as shown in Fig. 13.1*c*. As the charge builds up across the capacitor, the current decreases and the magnetic field decreases. In an ideal case (i.e. when there is no dissipative resistance in the circuit), the magnetic field energy drops to zero when the capacitor charges to the value it had in condition (*a*). Once again all the energy is in the form of potential energy. The capacitor now begins to discharge again. This time current flows in the opposite direction. Figure 13.1*d* shows the capacitor fully discharged, and also shows maximum current flowing in the circuit. Again, all the energy is in the magnetic field. The interchange or "oscillation" of energy between *L* and *C* is repeated again and again. This situation is similar to an oscillating pendulum, in which the energy keeps on interchanging between potential energy and kinetic energy. In a practical pendulum, because of the friction at the pivot and the air resistance, some energy is lost during each swing. The amplitude of each half-cycle goes on decreasing. Ultimately, the pendulum comes to rest, though it may take a long time. The oscillations of the pendulum are said to be *damped*.

A practical *LC* circuit deviates from the ideal one. The inductor coil will have some resistance, and the dielectric material of the capacitor will have some leakage. Because of these factors, some energy loss takes place during each cycle of the oscillation. As a result of this loss, the amplitude of

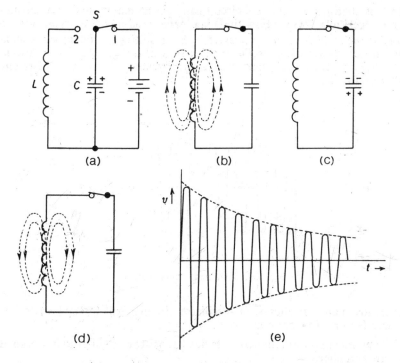

Fig. 13.1 Damped oscillation in an *LC* circuit

oscillation decreases continuously and ultimately the oscillations die down. Thus, we find that a tank circuit by itself is capable of producing oscillations, but they are damped as shown in Fig. 13.1e.

13.3.1 Frequency of Oscillations in *LC* Circuit

In case of a pendulum, the frequency of oscillation is determined by the constants of the pendulum system. The height to which the pendulum is raised has no effect on the time period of the swing. In the electrical *LC* circuit, the constants of the system are the inductance and capacitance values. The frequency of oscillation is the same as the resonant frequency of the tank circuit. It is given by

$$f_o = \frac{1}{2\pi\sqrt{LC}}$$

(13.1)*

13.3.2 Sustained Oscillations

The oscillations of a pendulum can be maintained at a constant level, if we supply additional energy to it from time to time, to overcome the effect of the losses. A light tap at the peak of its swing on either side, should be sufficient. It should be realized that the timing (or phasing) of the tap is very important. A tap downward while the pendulum is still on its upward swing can actually stop the swing.

---

*The formula for resonant frequency of a tank circuit is derived in Sec. 11.2.2.

Tne oscillations of an *LC* circuit can also be maintained at a constant level in a similar way. For this, we have to supply a spurt or pulse of energy *at the right time* in each cycle. The resulting "undamped oscillations" are called *sustained oscillations*, as shown in Fig. 13.2. Such sustained oscillations (or continuous waves) are generated by the electronic oscillator circuits.

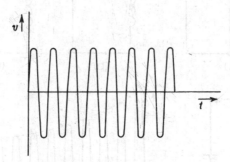

Fig. 13.2 Sustained oscillations

There are many varieties of *LC*-oscillator circuits. All of them have following three features in common:

(*i*) They must contain an active device (transistor or tube) that works as an amplifier.

(*ii*) There must be positive feedback in the amplifier.

(*iii*) The amount of feedback must be sufficient to overcome the losses.

13.4 POSITIVE FEEDBACK AMPLIFIER AS AN OSCILLATOR

The main application of positive feedback is in oscillators. An oscillator generates ac output signal without any input ac signal. A part of the output is fed back to the input; and this feedback signal is the only input to the internal amplifier.

To understand how an oscillator produces an output signal without an external input signal, let us consider Fig. 13.3*a*. The voltage source *v* drives the input terminals *YZ* of the internal amplifier (with voltage gain *A*). The amplified signal *Av*-drives the feedback network to produce feedback voltage *Aβv*. This voltage returns to the point *X*. If the phase shift due to the amplifier and feedback network is correct, the signal at point *X* will be exactly in phase with the signal driving the input terminals *YZ* of the internal amplifier.

The action of an oscillator is explained a little later. For the time being, assume that we connect points *X* and *Y*, and remove voltage source *v*. The feedback signal now drives the input terminals *YZ* of the amplifier (*see* Fig. 13.3*b*). If *Aβ* is less than unity, *Aβv* is less than *v*, and the output signal will die out as shown in Fig. 13.3*c*. This happens because enough voltage is not returned to the input of the amplifier. On the other hand, if *Aβ* is greater than unity, *Aβv* is greater than *v*, and the output voltage builds up as shown in Fig. 13.3*d*. Such oscillations are called *growing oscillations*. Finally, if *Aβ* equals unity, no change occurs in the output; we get an output whose amplitude remains constant, as shown in Fig. 13.3*e*.

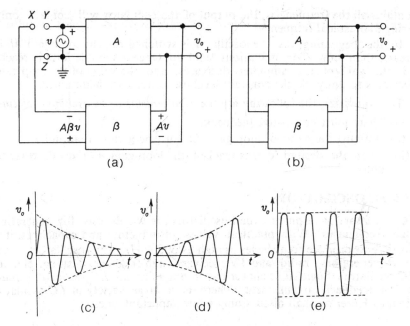

Fig. 13.3 Proper positive feedback in an amplifier makes it an oscillator

To find the necessary condition for the sustained oscillations, we can also apply the feedback theory to the circuit of Fig. 13.3*b*. In Sec. 12.3 we derived the expression for the overall voltage gain of an amplifier with negative feedback. When the *feedback is positive*, the overall gain of the amplifier can be written as

$$A_f = \frac{A}{1 - A\beta} \qquad (13.2)$$

It can now be seen that if $A\beta = 1$ $A_f = \infty$. The gain becoming infinity means that there is output without any input. In other words, the amplifier becomes an oscillator. The condition,

$$A\beta = 1 \qquad (13.3)$$

is known as *Barkhausen criterion of oscillation.*

13.4.1 The Starting Voltage

The starting voltage of an oscillator is provided by noise voltage. Such noise voltages are produced due to the random motion of electrons in resistors used in the circuit, or the active device.

This noise voltage contains almost all the sinusoidal frequencies. This voltage is very small in amplitude. It gets amplified and appears at the output terminals. The amplified noise now drives the feedback network, which is either a resonant circuit or a phase-shift network. Because of this, the feedback voltage $A\beta_v$ is maximum at a particular frequency f_o, the frequency of oscillations. Furthermore, the phase shift required for positive feedback is correct at this frequency only. Thus, although the noise voltage contains all the frequencies, the output of the oscillator will contain only a single sinusoidal frequency.

In the beginning, as the oscillator is switched on, the loop gain $A\beta$ is greater than unity. The oscillations build up. Once a suitable level is reached, the gain of the amplifier decreases, and the value of the loop gain decreases to unity. So the constant level oscillations are maintained.

To summarize, the following are the *requirements of an oscillator circuit*:

(i) There must be positive feedback.
(ii) Initially, the value of loop gain $A\beta$ must be greater than unity.
(iii) After the desired level is reached, the loop gain $A\beta$ must decrease to unity.

13.5 *LC* OSCILLATORS

LC oscillators or resonant-circuit oscillators are widely used for generating high frequencies. With practical values of inductors and capacitors, it is possible to produce frequencies as high as 500 MHz. The oscillators used in rf generators, radio and TV receivers, high-frequency heating, etc. are *LC* oscillators. Such an oscillator has an amplifier, an *LC* resonant circuit and a feedback arrangement. There is a large variety of *LC*-oscillator circuits. Here, we shall discuss only a few important ones.

13.5.1 Tuned-Collector (or Tuned-Plate) Oscillator

Figure 13.4*a* shows a basic *LC*-oscillator circuit from a purely ac point of view. It is called tuned-collector oscillator, because the tuned circuit is connected to the collector. We have used a transformer here. The primary of the transformer and the capacitor *C* form the tuned circuit (or tank circuit) which decides the frequency of the oscillation. The secondary winding is connected to the base. Since a phase difference of 180° is provided by the transistor amplifier, and an additional 180° by the transformer, the type of feedback is positive. The transistor amplifier provides sufficient gain for oscillator action to take place.

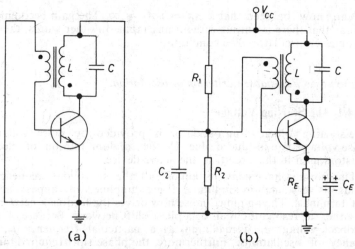

(a)　　　　　　　　(b)

Fig. 13.4 Tuned-collector oscillator: (*a*) Circuit from a purely ac point of view; (*b*) Complete circuit.

In Fig. 13.4*a*, no consideration has been shown for biasing the transistor. A practical tuned-collector oscillator with biasing is shown in Fig. 13.4*b*. Resistors R_1, R_2 and R_E provide dc bias to the transistor. The capacitors C_E and C_2 bypass resistors R_E and R_2, respectively. It is for this reason, the resistors R_E and R_2 have no effect on the ac operation of the circuit. The dc bias voltage set by the potential divider R_1 and R_2 is connected to the base through the low-resistance secondary winding of the transformer. At the same time, the secondary of the transformer provides ac feedback voltage. This voltage appears across the base-emitter junction, since the junction point of R_1 and R_2 is at ac ground (due to bypass capacitor C_2). Notice that if capacitor C_2 was not connected across resistor R_2, the feedback voltage induced in the secondary of the transformer would not be directly going to the input of the transistor; some of this voltage will drop across R_2.

A similar circuit can be made by using a vacuum tube. Figure 13.5*a* shows a tuned-plate oscillator. The parallel combination of R_G and C_G is meant to provide proper grid biasing. Figure 13.5*b* shows the same circuit using an FET. This circuit also uses $R_G C_G$ combination to provide proper gate biasing. Such a biasing arrangement when used in a triode circuit, is called *grid-leak biasing* and when used in FET circuit, is called *gate-leak biasing*.

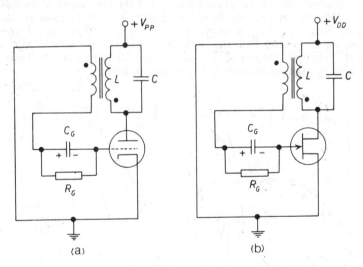

Fig. 13.5 (*a*) Tuned-plate oscillator circuit; (*b*) tuned-drain oscillator circuit

How the Oscillations Originate Let us consider the circuit in Fig. 13.5*a*. The moment we switch on the supply, the current starts building up in the plate circuit. This induces a varying voltage in the secondary of the transformer. An amplified voltage again appears in the tank circuit, which responds most to its resonant frequency. Because of the sufficient gain provided by the tube, and the proper amount of feedback in the correct phase, the oscillations grow till a certain level is reached. Thus, sustained oscillations are obtained.

Notice that in the circuit in Fig. 13.5*a*, the biasing arrangement is not the commonly used self-biasing arrangement (R_K and C_K). We use grid-leak biasing to meet specific requirements. Initially, the loop gain $A\beta$ has to be greater than unity, so that the circuit is self-starting and the oscillations grow. When a certain amplitude of oscillations is reached, the value of $A\beta$ must reduce to unity so as to maintain the oscillations at this level. These requirements are met by the use of grid-leak biasing.

Figure 13.6 shows how oscillations build up in this circuit. As the circuit is first put into operation, the grid bias is zero. The operating point is high on the transfer characteristic of the tube. At this operating point, the value of g_m is sufficiently high so as to make $A\beta$ greater than unity (the gain A directly depends upon g_m of the tube). During the first few transient cycles, as the oscillations grow, the grid potential becomes positive during the positive portion of the swing. The grid current flows and the capacitor C_G charges. This makes the grid bias negative. The operating point shifts to the left, and consequently the amplitude of oscillations increase. The amplitude of oscillations continue to increase till the operating point is taken at or beyond cut-off. At this equilibrium condition, the g_m decreases so as to make $A\beta = 1$, a condition for sustained oscillations.

It should be realized that a fixed-bias, biasing the tube beyond cut-off will not work. The plate current cannot flow and oscillations could never begin. The grid-leak biasing provides self-starting. However, the time constant $R_G C_G$ should be chosen properly. If it is too large, the bias voltage across C_G adjusts slowly to sudden changes in the amplitude of oscillations. The oscillations may then die out whenever the amplitude of oscillations decreases suddenly due to some reason. On the other hand, if the time constant $R_G C_G$ is too small, the capacitor C_G discharges almost completely during the negative portion of the swing. The operating point will not shift to the left; and the oscillations will not be stabilized.

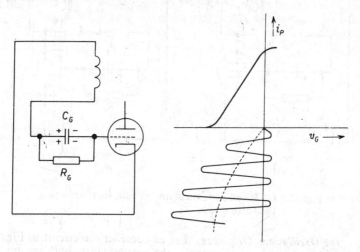

Fig. 13.6 Sketch showing building up of oscillations in an oscillator using grid-leak biasing

Where to take the Output From Simply generating sustained oscillations is not sufficient. We should be able to make use of these oscillations. These oscillations are to be coupled to the next stage in the electronic system. There are various methods of coupling the output of the oscillator to the next stage. These include capacitive, transformer and impedance-coupling networks. The choice of coupling depends on the specific application of the oscillator. Usually, the output is taken from the tank circuit using inductive coupling. The output could also be taken from any part of the circuit as the oscillations are present everywhere.

Figure 13.7 shows how the output of a tuned-collector oscillator (in Fig. 13.5) is taken by using transformer coupling. The output terminals are connected either to a power consuming device* or to the input terminals of the next stage. This has the effect of loading the circuit. To put it equivalently, the effective dissipative resistance R in the tank circuit increases. This increases the losses in the circuit. Now, to maintain the oscillations, more amount of positive feedback is required. We can increase the amount of positive feedback by simply increasing the coupling between the primary and the secondary of the transformer.

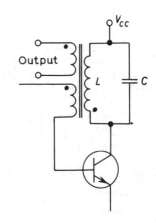

Fig. 13.7 Output of an LC oscillator taken through transformer coupling

If too high a load is connected across the output terminals of the oscillator, it may damp the oscillations, causing the oscillations to die out. It is for this reason, in a signal generator, we connect the output of the oscillator first to the input of an amplifier (such as cathode follower or emitter follower) and then take the output from the amplifier. Such an amplifier is called a *buffer amplifier*.

Example 13.1 The tuned-collector oscillator circuit used in the local oscillator of a radio receiver makes use of an LC-tuned circuit with $L = 58.6$ μH, and $C = 300$ pF. Calculate the frequency of oscillation.

Solution: The frequency of oscillation is given by

$$f_0 = \frac{1}{2\pi\sqrt{LC}}$$

Here, $L = 58.6$ μH $= 58.6 \times 10^{-6}$ H; $C = 300$ pF $= 3 \times 10^{-10}$ F

$$f_0 = \frac{1}{2\pi\sqrt{58.6 \times 10^{-6} \times 3 \times 10^{-10}}} = \textbf{1200.35 kHz}$$

13.5.2 Tuned-base (or Tuned-grid) Oscillator

If the tuned circuit is put in the base (or grid) circuit, the oscillator is called a tuned-base (or tuned-grid) circuit. This circuit is shown in Fig. 13.8. The circuit is also known by the names—*tickler oscillator*, or *Armstrong oscillator* (named after its inventor).

---

*It is very rare that the output of the oscillator is fed directly to the power consuming device.

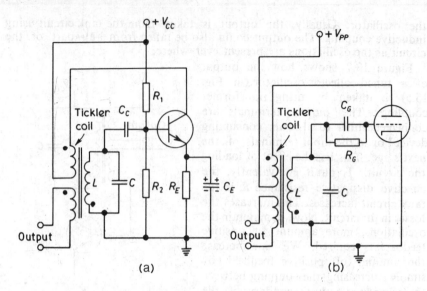

Fig. 13.8 (a) Tuned-base oscillator; (b) Tuned-grid oscillator

Let us consider the tuned-base oscillator of Fig. 13.8a. The dc bias is determined by the resistors R_1, R_2 and R_E. If resistor R_1 were omitted, then initially the transistor currents would be zero, and the circuit would not oscillate. When R_1 is present, the transistor is biased in its active region (i.e. the base-emitter junction is forward biased). When the circuit is energized, the collector current starts rising to a high value. The rising collector current, through the "tickler" coil, creates a changing magnetic field around it. This induces a voltage in the tuned circuit. Because of correct phasing (positive feedback) of the coils, and sufficient gain of the amplifier, the oscillations start growing. Dynamic self-bias is obtained from the R_2C_C combination due to the flow of base current. The voltage developed across the capacitor, due to its charging, drives the transistor towards cut-off. This reduces gain of the transistor amplifier, satisfying the condition $A\beta = 1$. As a result, sustained oscillations are obtained.

The tuned-base oscillator circuit has one drawback. Due to the low base-emitter resistance (which comes in shunt with the tuned circuit), the tank circuit gets loaded. This reduces its Q. This in turn causes drift in the oscillation frequency, and the stability becomes poorer. Due to this reason, the tuned circuit is not usually connected in the base circuit, but in the collector circuit. However, in case of an FET or a vacuum triode, the input impedance is sufficiently large. With these devices, the input does not load the tank circuit. In practice, tickler oscillator circuits are made by using either FETs or tubes.

13.5.3 Hartley Oscillator

Note that in a tuned-base oscillator (tickler oscillator) circuit, two inductor coils are used. One end of each of these coils is connected to the ground (from the ac point of view). If we make the tickler coil as an integral part of the tank circuit, the modified circuit becomes the one shown in Fig. 13.9. This is called a Hartley oscillator circuit.

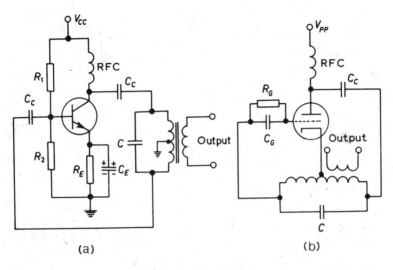

Fig. 13.9 Hartley oscillator circuit using (*a*) a transistor;
(*b*) a vacuum triode

An RFC (radio frequency choke) permits an easy flow of dc current. At the same time, it offers very high impedance to high frequency currents. In other words, an RFC ideally looks like a dc short and an ac open. The presence of the coupling capacitor C_C in the output circuit of the Hartley oscillator does not permit the dc currents to go to *the tank circuit**. The radio-frequency energy developed across the RFC is capacitively coupled to the tank circuit through the capacitor Cc.

13.5.4 Colpitts Oscillator

The Colpitts oscillator in Fig. 13.10 is a superb circuit. It is widely used in commercial signal generators above 1 MHz. This oscillator is similar to the Hartley oscillator given in Fig. 13.9. The only difference is that the Colpitts oscillator uses a split-tank capacitor instead of a split-tank inductor. The RFC has the same function as in the Hartley oscillator. The voltage developed across the capacitor C_2 provides the regenerative feedback required for the sustained oscillations. The values of L, C_1 and C_2 determine the frequency of oscillation. The frequency of oscillation is given by

$$f = \frac{1}{2\pi\sqrt{LC}} \tag{13.4}$$

where
$$C = \frac{C_1 C_2}{C_1 + C_2} \tag{13.5}$$

since C_1 and C_2 are in series.

13.6 *RC* OSCILLATORS

Till now we have discussed only those oscillators which use an *LC*-tuned circuit. These tuned circuit oscillators are good for generating high frequen-

---

*The presence of dc current in a tank circuit reduces its Q.

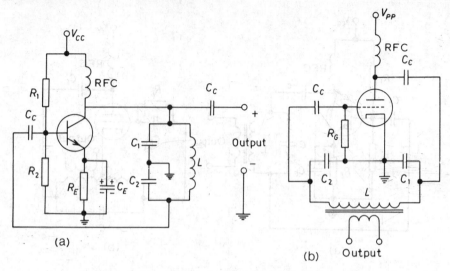

Fig. 13.10 Colpitts oscillator using (*a*) a transistor (*b*) a vacuum triode

cies. But for low frequencies (say, audio frequencies), the *LC* circuit be-comes impracticable. *RC* oscillators are more suitable. Also, with the advent of IC technology, *RC* network is the only feasible solution. It is very difficult to make an inductance, that too of high value, in an integrated circuit. Therefore *RC* oscillators are becoming increasingly popular. There are many types of *RC* oscillators, but following two are most important:

  (i) Phase-shift oscillator
  (ii) Wein bridge oscillator

13.6.1 Basic Principles of *RC* Oscillators

We know that a single stage of an amplifier not only amplifies the input signal but also shifts its phase by 180°. If we take a part of the output and feed it back to the input, a negative feedback takes place. The net output voltage then decreases. But for producing oscillations we must have positive feedback (of sufficient amount). Positive feedback occurs only when the fed back voltage is in phase with the original input signal. This condition can be achieved in two ways. We can take a part of the output of a single-stage amplifier (giving a phase shift of 180°) and then pass it through a phase-shift network giving an additional phase shift of 180°. Thus a total phase shift of 180°+180° = 360° (which is equivalent to a phase shift of 0°) occurs, as the signal passes through the amplifier and the phase-shift network. This is the principle of a phase-shift oscillator.

Another way of getting a phase shift of 360° is to use two stages of am-plifiers each giving a phase shift of 180°. A part of this output is fed back to the input through a feedback network without producing any further phase shift. This is the principle of a Wein bridge oscillator.

13.6.2 Phase-Shift Oscillator

Figure 13.11 shows a phase-shift oscillator using a vacuum triode. A bipo-lar transistor or an FET could also be used in the circuit. Here, the com-

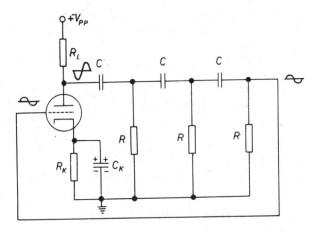

Fig. 13.11 Phase-shift oscillator

bination $R_K C_K$ provides self-bias for the amplifier. As shown in the figure, the phase of the signal at the input (for the time being, we assume the presence of the signal at the input) gets reversed when it is amplified by the amplifier. The output of the amplifier goes to a feedback network. The feedback network consists of three identical RC sections. Each RC section provides a phase shift of 60°. Thus a total of 60° × 3 = 180° phase shift is provided by the feedback network. The output of this network is now in the same phase as the originally assumed input to the amplifier, as shown in figure. If the condition $A\beta = 1$ is satisfied, oscillations will be maintained.

It may be shown by a straightforward (but a little complicated) analysis that the frequency at which the RC network provides exactly 180° phase-shift is given by

$$f_o = \frac{1}{2\pi RC\sqrt{6}} \tag{13.6}$$

This must then be the *frequency of oscillation*. Also, it can be shown that at this frequency, the feedback factor of the RC network is

$$\beta = \frac{1}{29} \tag{13.7}$$

This equation has special significance. For self starting the oscillations, we must have $A\beta > 1$. This means the gain A of the amplifier must be greater than 29; only then oscillations can start.

13.6.3 Wein Bridge Oscillator

The Wein bridge oscillator is a standard circuit for generating low frequencies in the range of 10 Hz to about 1 MHz. It is used in all commercial audio generators. Basically, this oscillator consists of two stages of RC-coupled amplifier and a feedback network. The block diagram of Fig. 1..12 explains the principle of working of this oscillator.

Here, the blocks A_1 and A_2 represent two amplifier stages. These amplifiers may use bipolar transistors, or FETs, or triodes. The output of the

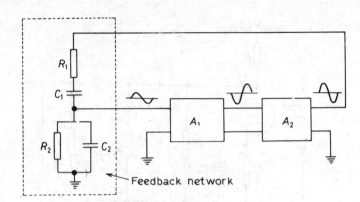

Fig. 13.12 Block diagram of a basic Wein bridge oscillator

second stage goes to the feedback network. The voltage across the parallel combination C_2R_2 is fed to the input of the first stage. The net phase shift through the two amplifiers is zero. Therefore, it is evident that for the oscillation to be maintained, the phase shift through the coupling network must be zero. It can be shown that this condition occurs at a frequency given by

$$f_o = \frac{1}{2\pi\sqrt{R_1C_1R_2C_2}} \tag{13.8}$$

Further, it can be shown that when the above condition is satisfied, we must have $\beta = 1/3$. This means that the amplifier must have a gain of at least 3. ($\because A\beta = 1$).

To have a gain of at least 3 is not difficult. On the other hand, to have a gain as low as 3 may be difficult. For this reason, we add some amount of negative feedback. As discussed earlier, the negative feedback also improves the stability of the amplifier. The addition of negative feedback modifies the circuit in Fig. 13.12 to that shown in Fig. 13.13. The same circuit is redrawn as in Fig. 13.14.

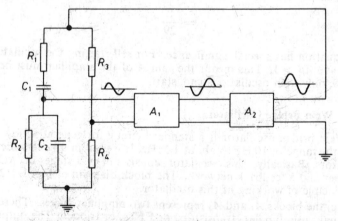

Fig. 13.13 Circuit in Fig. 13.12 modified to include **negative feedback**

Wein bridge

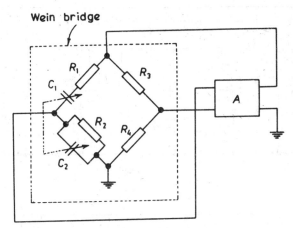

Fig. 13.14 Circuit in Fig. 13.13 is redrawn to show the presence of a "bridge" in Wein bridge oscillator

You may now see why this circuit is called a bridge oscillator. In this circuit, the resistors R_3 and R_4 provide the desired negative feedback. Note that the junction point of these resistors is connected to the lower input terminal (connecting this point to the upper terminal would mean positive feedback). The two blocks in Fig. 13.13 representing the two stages of the amplifier are replaced by a single block in Fig. 13.14.

We can have a continuous variation of frequency in this oscillator by varying the two capacitors C_1 and C_2 simultaneously. These capacitors are variable air-gang capacitors. We can change the frequency range of the oscillator by switching into the circuit different values of resistors R_1 and R_2.

Practical circuits of Wein bridge oscillators using tubes and transistors are shown in Fig. 13.15.

In Fig. 13.15*a*, the resistance R_4 is replaced by a tungsten filament lamp. This serves to stabilize the amplitude of oscillations against range-switching and against ageing of the tubes.

Example 13.2 A vacuum tube phase-shift oscillator uses three identical RC sections in the feedback network. The values of the components are $R = 100 \text{ k}\Omega$ and $C = 0.01 \ \mu\text{F}$. Calculate the frequency of oscillation.

Solution: The frequency of oscillation of a phase-shift oscillator is given as

$$f_o = \frac{1}{2\pi RC\sqrt{6}}$$

Here, $R = 100 \text{ k}\Omega = 10^5 \ \Omega$; $C = 0.01 \ \mu\text{F} = 10^{-8}$ F. Therefore,

$$f_o = \frac{1}{2 \times 3.141 \times 10^5 \times 10^{-8} \times 2.45} = \textbf{64.97 Hz}$$

Example 13.3 The RC network of a Wein bridge oscillator consists of resistors and capacitors of values $R_1 = R_2 = 220 \text{ k}\Omega$ and $C_1 = C_2 = 250$ pF. Determine the frequency of oscillations.

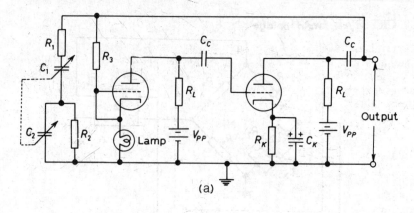

(a)

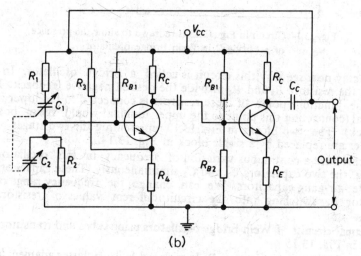

(b)

Fig. 13.15 Practical Wein Bridge circuit using (a) triodes (b) transistors

Solution: For a Wein bridge oscillator, the frequency of oscillation is given as

$$f_o = \frac{1}{2\pi\sqrt{R_1 R_2 C_1 C_2}} = \frac{1}{2\pi RC}$$

when $R_1 = R_2 = R$, and $C_1 = C_2 = C$.

Here, $R = 220\ k\Omega = 2.2\times10^5\ \Omega$; $C = 250\ pF = 2.5\times10^{-10}\ F$. Therefore,

$$f_o = \frac{1}{2\times3.141\times2.2\times10^5\times2.5\times10^{-10}}$$

$$= 2893.7\ Hz$$

$$\cong \mathbf{2.89\ kHz}$$

13.7 CRYSTAL OSCILLATORS

A crystal oscillator is basically a tuned oscillator. It uses a piezoelectric crystal as a resonant tank circuit. The crystal (usually quartz) provides a high degree of frequency stability. Therefore, the crystal oscillators are used whenever great stability is required. Examples are communication transmitters, digital clocks, etc.

13.7.1 Piezoelectric Effect

A quartz crystal exhibits a very important property known as piezoelectric effect. When an ac voltage is applied, it vibrates at the frequency of the applied voltage. Conversely, if we mechanically force it to vibrate, it generates an ac voltage. Besides quartz, the other substances that exhibit the piezoelectric effect are Rochelle salt and tourmaline.

Tourmaline is most rugged but shows the least piezoelectric activity. It is also the most expensive. It is occasionally used at very high frequencies.

Rochelle salts exhibit the greatest piezoelectric activity, but they are mechanically the weakest; they break easily. Rochelle salts are used to make microphones, headsets, and loudspeakers.

Quartz is a compromise between the piezoelectric activity of Rochelle salts and the strength of tourmaline. It is inexpensive and readily available in nature. It is mainly the quartz crystal that is used in rf oscillators.

13.7.2 Characteristics of a Crystal

For use in electronic oscillators, the crystal is suitably cut and then mounted between two metal plates, as shown in Fig. 13.16a. Let us see how the crystal behaves when an ac source is connected across it. Even when the mounted crystal in Fig. 13.16a is not vibrating, it is equivalent to a capacitance C_m because of the fact that two metal plates separated by a dielectric (viz. the crystal itself) behave like a capacitor.

However, when an ac voltage is applied to the crystal, mechanical vibrations are set up. These vibrations have a natural resonant frequency which depends upon a number of factors. Some of these factors are the dimensions of the crystal, how the surfaces are oriented with respect to its axes

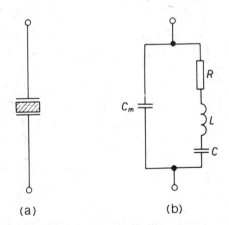

(a)   (b)

Fig. 13.16 (a) Crystal mounting
(b) Its equivalent electrical circuit

and how the crystal is mounted. Although the crystal has electromechanical resonance, we can represent its action by an equivalent electrical resonant circuit as shown in Fig. 13.16b. Typical values for a 90 kHz crystal are $L = 137$ H, $C = 0.0235$ pF, and $R = 15$ kΩ. This corresponds to a Q of 5500. The mounting capacitance C_m is much larger ($\cong 3.5$ pF) than the capacitance C.

The outstanding feature of crystals as compared with discrete LC tank circuits, is their incredibly high Q. Values of Q up to almost 10^6 can be achieved by using crystals. Whereas, a discrete LC tank circuit seldom has a Q over 100. The extremely high Q of a crystal makes the oscillation frequency very stable.

The crystal has two resonant frequencies. First, the inductance L resonates with capacitance C at a series-resonant frequency f_s. Above frequency f_s, the series branch LCR has inductive reactance. At a frequency f_p, slightly higher than f_s, the series branch has parallel resonance with capacitance C_m. This frequency f_p is called parallel resonant frequency. Above this frequency, the crystal offers capacitive reactance. The crystal as a whole behaves as an inductor only between the frequency f_s and f_p. If the crystal is used in place of an inductor, in an oscillator circuit, the frequency of oscillation must lie between f_s and f_p. The values of the components of L, C, R and C_m are such that the two frequencies f_s and f_p differ by a very small amount. This very fact gives rise to great frequency stability of a crystal oscillator. The resonant frequencies of a crystal are temperature dependent. By keeping the crystal in temperature-controlled ovens, it is possible to have frequency drifts less than 1 part in 10^{10}.

13.7.3 Crystal Oscillator Circuits

A variety of crystal-oscillator circuits is possible. A popular circuit is the Pierce crystal-controlled oscillator shown in Fig. 13.17. The circuit may use either an FET or a tube. It has no tuned circuit. The crystal is excited by a portion of the energy coming from the output. This energy is being fed back through the crystal to the input side.

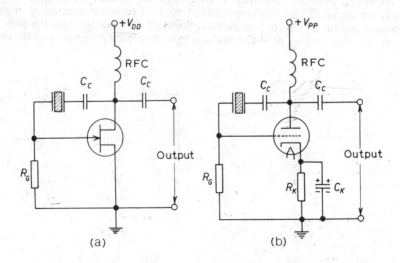

Fig. 13.17 Pierce crystal-controlled oscillator using (a) FET, (b) triode

13.8 ASTABLE MULTIVIBRATOR

All the oscillators discussed till now were sinusoidal oscillators. Sometimes we need voltages of waveshapes other than the sine wave. Among the non-sinusoidal oscillators, the square-wave generator is very important. Square waves are needed for testing video amplifiers (pulse amplifiers). Repetitive pulses find applications in radar and in triggering many digital circuits.

An astable (or free running)* multivibrator generates square waves. Figure 13.18 shows a basic collector-coupled transistor multivibrator. It is essentially a two-stage *RC*-coupled amplifier with the output of the first stage coupled to the input of the second stage; and the output of the second stage coupled to the input of the first stage. Since the phase of a signal is reversed when amplified by a single stage of the CE amplifier, it comes back to its original phase when passed through two stages. Thus, the signal fed back to the base of either transistor is in the same phase as the original signal at its input. It amounts to positive feedback. In a multivibrator (also called *relaxation oscillator*), the amount of feedback is very large—so large, that the transistors are driven between cut-off and saturation almost instantaneously. A transistor remains in either saturation or cut-off for a period determined by the time constant of the elements in the base circuit.

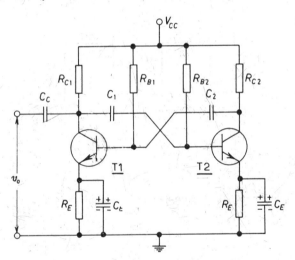

Fig. 13.18 Transistor multivibrator

13.8.1 Operation of Transistor Multivibrator Circuit

The operation of the multivibrator circuit of Fig. 13.18 is as follows: Because of circuit variations, one transistor will conduct more heavily than the other. Assume arbitrarily this to be T1. Its collector current rises rapidly. This causes its collector voltage to decrease. The resulting negative signal is fed to the base of T2 through C_1 and drives it towards cut-off. As a result, the collector voltage of T2 rises towards V_{CC}. The change in the collector voltage of T2 through C_2 (positive-going signal) is fed to the base

---

*Besides astable multivibrators there are other types such as monostable, bistable (popularly known as flip-flop) multivibrators

of transistor T1. It causes T1 to go into saturation. This happens so quickly that capacitor C_1 does not get a chance to discharge, and the decreased voltage at the collector of T1 appears across R_{B2}.

Capacitor C_1 now begins to discharge. More of the previously decreased voltage, at the collector of T1, appears across C_1 and less across R_{B2}. This decreases the reverse bias on the base of T2. Ultimately, the base-emitter junction of the transistor T2 becomes forward biased. Now T2 begins to conduct. Its collector becomes less positive. This negative-going voltage signal is fed to the base of transistor T1 through capacitor C_2. It drives the transistor T1 towards cut-off. The resulting increased voltage at the collector of T1 is coupled through C_1 and appears across R_{B2}. The collector current of T2 therefore increases. This process continues rapidly until transistor T1 is cut-off. The transistor T1 remains cut-off (and T2 in conduction) until capacitor C_2 discharges through R_{B2}, enough to decrease the reverse bias on the base of T1. The cycle then repeats itself. The output of the multivibrator can be taken from the collector of either transistor. The output is a square wave (as shown in Fig. 13.19) whose frequency is determined by the values of R_{B1}, R_{B2}, C_1 and C_2 in the circuit.

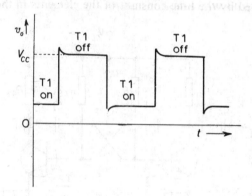

Fig. 13.19 Output of an astable multivibrator

REVIEW QUESTIONS

13.1 Explain what you understand by an electronic oscillator. State two applications of an oscillator.

13.2 State the conditions under which a feedback amplifier works as an oscillator.

13.3 Can a negative feedback amplifier work as an oscillator ? If yes, how ? If not, why ?

13.4 Explain the meaning of the terms: (a) damping oscillations, (b) growing oscillations, and (c) sustained oscillations.

13.5 Explain why an LC tank circuit, once excited, does not produce sustained oscillations.

13.6 Explain why positive feedback, and not negative feedback, is necessary to produce oscillations.

13.7 Every electronic oscillator can be considered to be an amplifier with infinite gain. Explain where you get the starting signal voltage from.

13.8 Draw the circuit diagram of the tuned-collector oscillator. Explain the following:

(a) How is the transistor biased ?

(b) How does positive feedback take place in the circuit ?

(c) How will you take the output from this circuit ?

13.9 Draw the circuit diagram of a tuned-plate oscillator using grid-leak-resistor grid-capacitor biasing. Explain how oscillations start in this circuit.

13.10 Draw the circuit diagram of a transistor Hartley oscillator. Explain the function of each component.

13.11 Draw the circuit diagram of a Colpitts oscillator using (a) a bipolar transistor, (b) a vacuum triode. Which component is responsible for positive feedback in the circuit ? Explain.

13.12 What is the basic principle of operation of an RC oscillator ?

13.13 Draw the circuit diagram of a phase-shift oscillator using a vacuum triode. Explain why we cannot have phase shift oscillator which uses only two RC networks instead of three, for obtaining a phase shift of 180°.

13.14 Explain the principle of working of the Wein bridge oscillator circuit. Draw the phase-shift network that determines the frequency of oscillations.

13.15 Explain why negative feedback in addition to the usual positive feedback is employed in the Wein bridge oscillator.

13.16 Sometimes a lamp is used as one of the resistance arms of a Wein bridge oscillator. Explain why.

13.17 Explain the properties of a quartz crystal which are responsible for its use in an oscillator.

13.18 Draw the equivalent electrical circuit of a quartz crystal. Compare this circuit with a conventional LC-tuned circuit.

13.19 Draw a crystal-controlled oscillator circuit. State some of its applications.

13.20 Draw the circuit diagram of an astable multivibrator. Justify that it is a two-stage RC-coupled amplifier using feedback. How does it generate square waves ?

OBJECTIVE-TYPE QUESTIONS

I. Here are some incomplete statements. Four alternatives are given below each. Tick the alternative which you think completes the statement most appropriately.

1. For generating a 1-kHz note, the most suitable circuit is

(a) Hartley oscillator

(b) Colpitts oscillator

(c) tuned-collector oscillator

(d) Wein bridge oscillator

2. To generate a 1-MHz signal, the most suitable circuit is

(a) Wein bridge oscillator

(b) phase-shift oscillator

(c) Colpitts oscillator

(d) none of the above

3. We use a crystal oscillator because

(a) it gives high output voltage

(b) it works at high efficiency

(c) the frequency of oscillations remains substantially constant

(d) it requires very low dc supply voltage

II. Write TRUE or FALSE (as the case may be) in the space provided against each of the following statements:

1. An audio signal generator is meant to produce sinusoidal voltages of frequencies varying from 20 Hz to 40 MHz. _____

2. The main reason why we do not use an alternator for producing sinusoidal voltages of very high frequencies is that its efficiency is very low. _____

3. For oscillations to start, the loop gain $A\beta$ of the oscillator must be greater than unity initially. _____

4. An oscillator circuit is merely a dc-to-ac converter. _____

5. An RFC (radio frequency choke) permits an easy flow of dc current, and at the same time it offers very high impedance to high-frequency currents. _____

6. A Hartley oscillator uses a split-tank capacitor, and a Colpitts oscillator uses a split-tank inductor. _____

7. A Wein bridge oscillator generates oscillations only if the gain of the two-stage amplifier is more than 3. _____

8. An RC phase-shift oscillator will not produce any oscillation until and unless the voltage gain of its internal amplifier is more than 29. _____

9. The greatest advantage of using a crystal in an oscillator is that it gives an output having a very stable frequency. _____

10. The rf signal generator used in laboratories makes use of a crystal oscillator circuit. _____

11. The output of an astable multivibrator is always a symmetrical square wave. _____

Ans: I 1. *d*, 2. *c*, 3. *c*.

II 1. F, 2. F, 3. T, 4. T, 5. T, 6. F, 7. T, 8. T, 9. T, 10. F, 11. F

TUTORIAL SHEET 13.1

1. The RC network given in Fig. T.13.1.1 is used in a low-frequency oscillator circuit. Calculate the frequency of oscillations. If the frequency of oscillation is required to be changed to 1.6 kHz, determine the new values of the capacitors.

[**Ans. 15.9 kHz, 995 pF**]

2. The tuned-collector oscillator circuit used in the local oscillator of a radio receiver makes use of an LC tank circuit with $L = 62.5\,\mu H$, and $C \cong 400$ pF. Calculate the frequency of oscillations. [**Ans. 1 MHz**]

3. The Colpitts oscillator circuit used in a TV receiver makes use of the tuned circuit given in Fig. T.13.1.2. Calculate the frequency of oscillations. [**Ans. 100 MHz**]

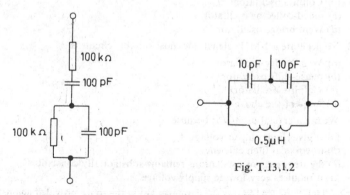

Fig. T.13.1.1

Fig. T.13.1.2

EXPERIMENTAL EXERCISE 13.1

TITLE: Wein bridge oscillator.

OBJECTIVES: To,

1. trace the given circuit and to note down the values of all the components;
2. measure the quiescent operating point of the transistors used in the oscillator circuit;
3. measure the frequency of oscillations of the Wein bridge oscillator circuit by seeing it on CRO.

APPARATUS REQUIRED: Experimental board, transistor power supply, CRO with calibrated time base or frequency counter.

CIRCUIT DIAGRAM: As given in Fig. E.13.1.1 (typical values of components are shown).

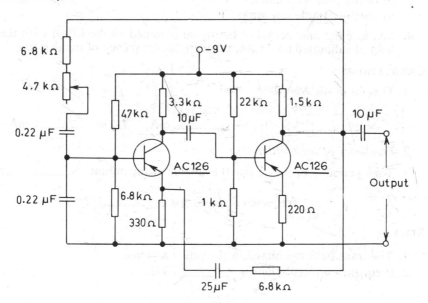

Fig. E.13.1.1

BRIEF THEORY: The method used for getting positive feedback in Wein bridge oscillator is to use two-stages of an RC-coupled amplifier. Since one stage of the RC-coupled amplifier introduces a phase shift of 180°, two stages will introduce a phase shift of 360°. At the frequency of oscillations f the positive feedback network shown in Fig. E.13.1.2, makes the input v_i and output v_o in the same phase. The frequency of oscillations is given as

$$f = \frac{1}{2\pi\sqrt{R_1 C_1 R_2 C_2}}$$

In addition to the positive feedback, the circuit is also provided with negative feedback to make the oscillations stable. This type of feedback is introduced because of unbypassed emitter resistor.

PROCEDURE:

1. Look into the given circuit. Draw it in your notebook. Note down the values of resistors and capacitors used in the circuit.

2. Connect the dc voltage, say 9 V, to the circuit. Note down the voltage available at the collector of both the transistors. Calculate the operating point collector current and collector-to-emitter voltage.

3. Feed the output of the oscillator circuit to a CRO. By making adjustments in the potentiometer connected in the positive feedback loop, try to obtain a stable sine wave.

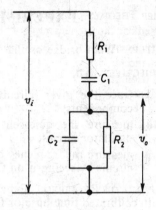

Fig. E.13.1.2

4. Measure the time period of the signal obtained on the CRO. With the help of calibrated time base, calculate the frequency of oscillations.

OBSERVATIONS:

1. *Q-point of the transistor:*

V_{CC} = _____ V; V_{C_1} = _____ V

V_{C_2} = _____ V; I_{C_1} = _____ mA; I_{C_2} = _____ mA

2. *Frequency of the oscillations:*

Time period T of the ac signal available at the output = _____ s

$\therefore$ frequency $= \dfrac{1}{T}$ Hz = _____ kHz

RESULTS:

1. The transistors are biased in the active regions.
2. Frequency of oscillations = _____ kHz.

EXPERIMENTAL EXERCISE 13.2

TITLE: Hartley oscillator circuit.

OBJECTIVES: To,

1. trace the circuit of Hartley oscillator;
2. measure the Q point of the transistor;
3. measure the frequency of oscillations by seeing the output waveshape on the CRO.

APPARATUS REQUIRED: Experimental board, transistor power supply, CRO with calibrated time-base or frequency counter.

CIRCUIT DIAGRAM: As given in Fig. E.13.2.1

BRIEF THEORY: Hartley oscillator is used for the generation of high frequencies (radio frequency). A split-inductor tank circuit used in this oscillator, provides the positive feedback. The frequency of oscillations depend upon the components of this tuned circuit.

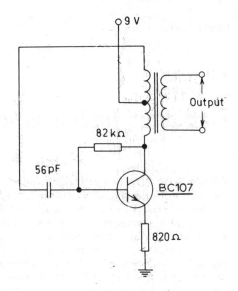

Fig. E-13-2.1

PROCEDURE:

1. Look into the circuit. Write down the circuit diagram in your note-book. Note down the values of all the resistors and capacitors used in the circuit.

2. Supply this circuit with dc voltage from the regulated power supply (say 9 V).

3. Connect the output of the oscillator circuit to the CRO. With the help of the calibrated time-base of the CRO, measure the time period of the sine wave. Calculate the frequency of oscillations.

OBSERVATIONS:

1. *Q point of the transistor:*

   $V_{CC} =$ _____V; $V_C =$ _____V

   $V_{CE} =$ _____V; $V_E =$ _____V

   $$\therefore \qquad I_E = \frac{V_E}{R_E} = \text{_____ mA}$$

2. *Frequency of oscillation:*

   Time period T of the output waveshape of the oscillator circuit

   = _____ s

   The frequency of oscillation, $f = \dfrac{1}{T} =$ _____ kHz

RESULTS:

1. From the measurements of dc voltages and current in the circuit, it is clear that the transistor is biased in _____

2. The oscillator is producing sine wave of frequency _____ kHz.

Electronic Instruments

OBJECTIVES: After completing this unit, you will be able to: ○ Name some important electronic instruments used in an electronics laboratory. ○ State the electrical quantities that can be measured by a multimeter. ○ Explain the circuits that are used to change the function and range of measurement in a multimeter. ○ Explain the loading effect of a voltmeter on the circuit under test. ○ Explain the advantages of a VTVM over an ordinary voltmeter. ○ Explain the principle of working of a VTVM. ○ Explain the principle of working of a digital multimeter (DMM). ○ Explain the working of a cathode ray tube (CRT). ○ State the function of each knob on the front panel of a cathode ray oscilloscope (CRO). ○ Explain the application of a CRO for displaying waveshapes of electrical signals, measuring voltages (ac and dc), measuring currents (ac and dc), measuring frequency of electrical signals, measuring phase difference between two sinusoidal signals. ○ Explain the working of a signal generator with the help of its block diagram. ○ Explain the principle of working of a strain gauge.

14.1 INTRODUCTION

There are many testing, measuring, and indicating instruments that have been developed to aid engineers and technicians in their work. For better understanding of electronic principles, a student is expected to do some experimental exercises. For this he goes to an electronics laboratory. There he finds a large variety of electronic instruments. Some of them are basic instruments, and some are merely improved versions of old models. Though there are many more instruments, some of the basic instruments needed in an electronics laboratory are:

  (i) Multimeter
 (ii) VTVM, or electronic multimeter (analogue or digital)
(iii) Cathode ray oscilloscope (CRO)
(iv) Signal generator
 (v) Regulated power supply

  It is not the intention of this book to give the design, or detailed description of these instruments. Only the relevant basic details are presented and explained in this chapter. This would help the student to familiarize himself with these instruments so that he can use them effectively. Since the regulated power supply has already been discussed in Unit 4 (Semiconductor Diodes), we shall not take it up here.

  A strain gauge, though not an instrument used in an electronics laboratory, is also discussed here. It illustrates the use of electronic circuitry to measure mechanical quantities.

14.2 MULTIMETER

A multimeter is used to measure many (multi) electrical quantities. It can measure voltages (ac and dc), currents (ac and dc) and resistances. A multimeter is a voltmeter, milliammeter, and ohmmeter combined together. In addition, it has various ranges of voltage, current and resistance measurements. A multimeter is a technician's constant companion. While servicing or repairing a radio receiver, or a TV set, or any other electronic system, the mechanic must have a multimeter with him. With the help of the multimeter, he can localize troubles (such as short circuits, or open circuits) in the electronic circuitry. Figure 14.1 shows the photograph of a commonly used multimeter. This instrument is available in a large number of models.

Fig. 14.1 Multimeter (Simpson-260 model)

To select the mode of measurement, a "function switch" is provided. By suitably adjusting this switch, the multimeter can be converted into a voltmeter, milliammeter, or an ohmmeter. Each position of the switch is labelled accordingly. There is another switch provided, called the "range switch". Each position of this switch is meant to fix a range of measurements. It is useful to start with the highest range and then switch down to lower ones, until the correct range is reached. This way, the instrument is saved from damage due to an excess current flowing through it.

The movement system of any multimeter is an important aspect of the instrument. Meters usually employ one of the three basic movements:

(i) The d'Arsonval movement, developed in 1881 by Jacques Arsene d'Arsonval;

(ii) iron-vane type;

(iii) electrodynamometer type.

Since most of the multimeters use the first one, we describe in brief the d'Arsonval movement in the next section.

14.2.1 The d'Arsonval Movement

Figure 14.2 illustrates the typical d'Arsonval meter movement. It consists basically of an iron core coil mounted on bearings between a permanent magnet. The coil is wound on an aluminium bobbin which is free to rotate by about 90°. An aluminium pointer is attached to the coil-and-bobbin assembly. When the coil rotates, the pointer moves on a graduated scale. There are two spiral springs attached to the coil assembly—one at the top, and the other at the bottom. These springs serve two purposes. Firstly, they provide a path for the current to reach the coil. Secondly, they keep the pointer at the low end of the scale (zero) when there is no current through the meter; and they provide a restraining or restoring torque when current flows through the coil.

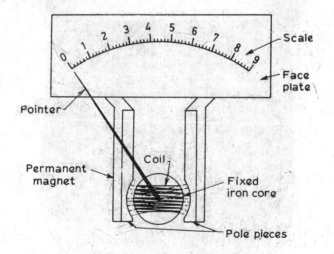

Fig. 14.2 The d'Arsonval movement

The iron core (over which the coil is mounted) helps in making the field (of the permanent magnet) radial in the air gap where the coil conductors move. This will ensure a uniform magnetic field throughout the movement of the coil. But, why should the coil move? When a current passes through the coil, a magnetic flux is produced. This flux interacts with the flux due to the permanent magnet. A torque (called deflection torque) is produced. As a result, the coil starts rotating and along with it, the pointer moves on the scale. As the coil rotates, the restoring torque provided by the spiral springs goes on increasing. The coil stops rotating at a point where the deflection torque is equal to the restoring torque. The final deflection of the pointer will depend upon the value of current passing through the coil.

An excessive current (or a current in the reverse direction) through the coil will cause the pointer to strike the end of the meter scale (or to deflect counter-clockwise); thus, the meter might be ruined. Such a situation should be avoided by taking appropriate precautions while using the instrument.

A d'Arsonval movement (or meter) is specified in terms of its "current sensitivity", CS (or its full-scale deflection current) and its "resistance". A typical movement might be having an internal resistance of 500 Ω, and full-scale deflection current of 0.1 mA. The same meter can also be rated as 50 mV (0.1 mA × 500 Ω), 500 Ω.

14.2.2 Voltage Measurement

Let us begin with a basic d'Arsonval movement having a current sensitivity of 0.1 mA and internal resistance of 500 Ω. The maximum voltage that should be applied to its terminal to have full-scale deflection is 0.1 mA × 500 Ω = 50 mV. This rating is sometimes called the *voltage sensitivity* (VS). It indicates that the maximum voltage that can be measured with this meter is only 50 mV. This is not very useful for most requirements. The voltage range can be extended by using additional circuitry. The basic construction of such a voltmeter is shown in Fig. 14.3. Here the voltage range is extended to 10 V.

The resistance R_S is connected in "series" with the movement (i.e. the main meter) to limit the current to 0.1 mA when maximum voltage is applied across the voltmeter. If the voltage across the terminals is less than the maximum, the deflection of the movement will also be less.

We can apply Kirchhoff's voltage law to determine the value of this series resistor R_S. The voltage drops, around a series circuit, must add to the total applied voltage. Therefore

$$(0.1 \times 10^{-3}) \times R_S + 500 \times 0.1 \times 10^{-3} = 10$$

$$(R_S + 500) \times 0.1 \times 10^{-3} = 10$$

or $$R_S = \frac{10}{0.1 \times 10^{-3}} - 500 = 99\,500\ \Omega = \textbf{99.5 k}\Omega$$

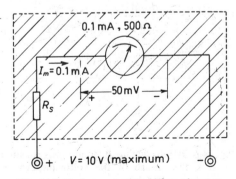

Fig. 14.3 Basic construction of a voltmeter

In general, for extending the range to voltage V,

$$R_S = \frac{V}{CS} - R_m \qquad (14.1)$$

The series resistance R_S is also called a *multiplier*, as it multiplies the voltage range. For another voltage range, we bring in another multiplier in the circuit. In a multimeter, a switch is provided to give different ranges of voltage. This switch puts in different multipliers. Such a switching device is shown in Fig. 14.4. Here, four ranges of voltage are given. For different positions of the rotary switch, net series resistances are given below:

Switch position at	Net series resistance R_S
10 V	99.5 kΩ = 99.5 kΩ
50 V	400 kΩ+99.5 kΩ = 499.5 kΩ
100 V	500 kΩ+400 kΩ+99.5 kΩ = 999.5 kΩ
500 V	4000 kΩ+500 kΩ+400 kΩ+99.5 kΩ
	= 4999.5 kΩ

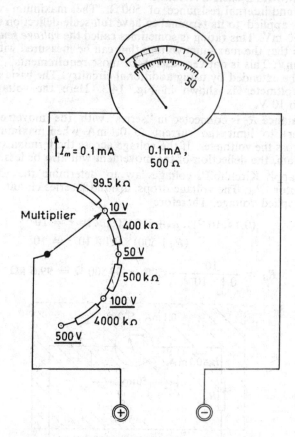

Fig. 14.4 By introducing different series resistors,
the range of the voltmeter can be
extended

Most multimeters employ the same scale for various values of maximum voltage (and/or current). For example, in Fig. 14.4 there are four voltage ranges, but only two scales. The scale $0-10$ V can be used for both the 10 V range as well as 100 V range. Similarly, the scale $0 - 50$ V is used for 50 V and 500 V ranges.

The voltmeter can be converted into an ac meter by placing a rectifier in the circuit. Usually, a full-wave bridge rectifier circuit is used It converts ac into dc which is then read on a dc meter. Different ranges of ac voltages are obtained by using a rotary switch, as in the dc voltmeter. This is illustrated in Fig. 14.5.

There are some points that must be kept in mind while using a voltmeter. Note that the voltmeter is always connected in parallel with the portion of the circuit across which the voltage is being measured. When used as dc voltmeter, you should be careful about the polarity. Usually, the positive test lead is red and the negative test lead, black.

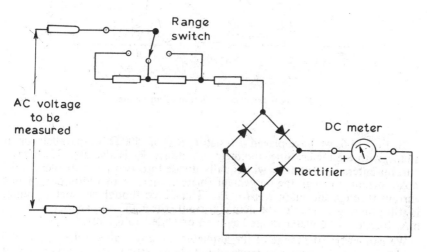

Fig. 14.5 Typical multi-range ac voltmeter

Example 14.1 You are given a d'Arsonval movement with a current sensitivity of 100 μA. Its coil resistance is 100 Ω. Determine the series resistance needed to convert it into a voltmeter with a range of 100 V.

Solution: The current sensitivity of the movement is 100 μA. The series resistance needed in the circuit to give a current of 100 μA, when 100 V is applied across its terminals; is

$$R_{total} = \frac{100}{100 \times 10^{-6}} = 10^6 \, \Omega$$

Since the coil resistance is 100 Ω, the additional series resistance required is

$$R_S = R_{total} - R_m = 10\,00\,000 - 100$$
$$= 999\,900\,\Omega = \mathbf{999.9\ k\Omega}$$

14.2.3 Current Measurement

Let us again consider the d'Arsonval movement of 0.1 mA and 500 Ω. The 0.1 mA is its current sensitivity, CS. It means that it needs 0.1 mA current to give full-scale deflection. 500 Ω is the internal resistance R_m of the movement itself. It is obvious that the maximum current this movement (meter) can measure is only 0.1 mA. For increasing its range of current measurement, we have to provide some additional circuitry. This is shown in Fig. 14.6. It illustrates the basic construction of an ammeter.

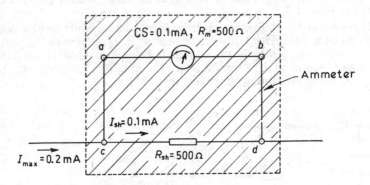

Fig. 14.6 Basic construction of an ammeter

In Fig. 14.6, we have placed a resistor R_{sh} of 500 Ω in parallel (or in shunt) with the meter. Since the meter resistance R_m is also 500 Ω, whatever current enters the ammeter will equally divide into two parts. If there is 0.1 mA current through the movement (meter), there is an additional 0.1 mA current through the shunt resistor R_{sh}. Thus, even though the meter pointer is still pointing full scale, the actual current through the ammeter is 0.2 mA. In effect, the meter range has been extended to 0.2 mA.

We can extend the range of the ammeter to any value other than 0.2 mA. If we reduce the resistance of the shunt resistor R_{sh}, a larger part of the total current is bypassed through it. The remaining part passes through the meter. We can suitably choose R_{sh}, so that this remaining part does not exceed current sensitivity, CS of the meter. Suppose we wish to extend the range to I_{max} (a current more than CS). When this current enters the ammeter, we would like only the current equal to CS to flow through the meter. The remaining current ($I_{max} - CS$) should then flow through the shunt R_{sh}. That is, the current through the shunt, $I_{sh} = I_{max} - CS$. Since the voltage across parallel elements must be the same, the potential drop across the meter (terminals a and b) must equal to that across the shunt (terminals c and d). That is,

$$CS \times R_m = R_{sh}(I_{max} - CS)$$

or
$$R_{sh} = \frac{R_m \times CS}{I_{max} - CS} \tag{14.2}$$

Figure 14.7 shows how to construct a multi-range ammeter. The rotary switch is used to select R_{sh} for the desired range. Here, three ranges are shown. For these ranges, the values of R_{sh} have been determined using Eq. 14.2.

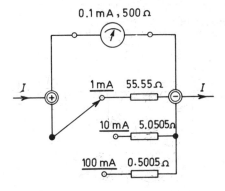

Fig. 14.7 Construction of a multi-range ammeter

Remember: *The ammeter is always placed in series with the branch in which the current is to be measured.* The conventional current should enter the positive terminal of the ammeter.

Example 14.2 A meter movement with current sensitivity 100 μA and internal resistance 100 Ω is required to measure a maximum current of 10 mA. Determine the shunt resistance required.

Solution: Here, I_{max} = 10 mA. Therefore, the current through the shunt is

$$I_{sh} = I_{max} - CS = 10\,mA - 100\,\mu A = 9.9\,mA$$

Since the voltage drop across the movement should be the same as that across the shunt, we have

$$R_{sh}I_{sh} = R_m \times CS$$

or $$R_{sh} = \frac{R_m CS}{I_{sh}} = \frac{100 \times 100 \times 10^{-6}}{9.9 \times 10^{-3}} = 1.010\,101\ \Omega$$

In the above example, a basic meter movement of 100 μA, 100 Ω was used. To extend its range to 10 mA we need a shunt resistance of 1.010 101 Ω. If we desire to extend its range to 100 mA, we would need a shunt of 0.11 001 001 Ω. (You may calculate it using Eq. 14.2.) It appears impracticable to obtain resistors of such low values, and that too so precise and accurate. But, a very novel idea is used for solving this problem This is called *universal-shunt* or *ring-shunt* method.

Universal Shunt for Extending Current Ranges The shunt resistance needed for extending the range of an ammeter is given by Eq. 14.2,

$$R_{sh} = \frac{R_m \times CS}{I_{max} - CS}$$

If I_{max} is high, the value of shunt R_{sh} required is low. If somehow we increase the value of R_m as we go for a higher range, the value of shunt resistance required need not be impractically low. This is exactly what we do in the *universal shunt*. Figure 14.8 shows a multi-range ammeter using the universal shunt. Here, the shunt resistors R_1, R_2, R_3, R_4 and R_5 form a ring with the basic meter.

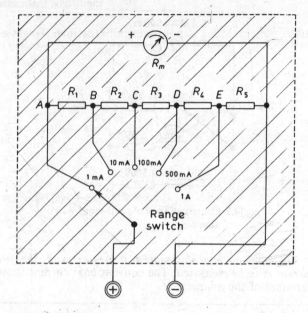

Fig. 14.8 Universal shunt for extending the current ranges

When the range switch is at position 1 (1 mA range), the current entering the meter divides into two parts at point A. The part through the meter branch is equal to its CS (if the current entering the meter is full-scale). The rest of the current bypasses through the shunt path consisting of the five resistors in series. At position 2 of the range switch, the current divides at point B. But now the meter branch consists of the meter in series with resistor R_1. The shunt branch consists of only the remaining four resistors. That is, $R_{sh} = R_2 + R_3 + R_4 + R_5$. Similarly at position 5, the current divides at point E. The meter branch is now made of the meter and the resistors R_1, R_2, R_3 and R_4. The shunt branch is made of only a single resistor. namely, R_5.

Thus we see that as we go for a higher range, the switch-cuts off a portion of the shunt-branch resistance and it gets added to the meter branch. Design of a universal shunt for a multi-range ammeter is made clear in Example 14.3.

Example 14.3 Design a universal shunt for making a multi-range milliam-meter with ranges 0–1 mA, 0–10 mA, 0–100 mA, 0–500 mA, and 0–1 A. The basic meter movement has a current sensitivity of 100 μA and nominal internal resistance of 100 Ω.

Solution: Here the current sensitivity of the basic meter is 100 μA. For 1 mA range, the total shunt resistance can be calculated using Eq. 14.2.

$$R_{sh} = R_1 + R_2 + R_3 + R_4 + R_5 = \frac{CS \times R_m}{I_{max} - CS}$$

$$= \frac{100 \times 10^{-6} \times 100}{(1 - 0.1) \times 10^{-3}} = 11.111\,111$$

To make a resistor of this value is not practically feasible. To get a suitable value for this shunt resistor, we connect a resistor R_T in series with the basic meter. Let us connect such a resistor R_T so that the total meter resistance is 900 Ω (see Fig. 14.9). Connecting resistor R_T in series with the meter serves another important purpose too. The meters that are manufactured may not all have exactly the same internal resistance. The resistance of a meter with nominal $R_m = 100$ Ω, may actually be anywhere between 90 Ω and 110 Ω. The resistor R_T provides an arrangement by which the resistance of the meter can be set to a desired value. It works as a preset adjustment. For instance, we use a 1-kΩ pot (potentiometer) for R_T. Its wiper can be set to give a total resistance of the meter at the predetermined value (900 Ω in our case). The meter resistance R_m now becomes 900 Ω.

Let us now calculate the value of the total shunt resistance needed (see Fig. 14.9).

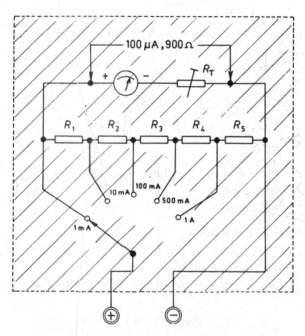

Fig. 14.9 Design of a multi-range milliammeter

(a) *Range switch at* 1 mA

Meter branch resistance, $R_{m1} = R_m = 900$ Ω
For full-scale, the shunt-path current is
$$I_{sh1} = 1 - 0.1 = 0.9 \text{ mA}$$
The shunt-path resistance is
$$R_{sh1} = R_1 + R_2 + R_3 + R_4 + R_5 = R \text{ (say)}$$
The value of R_{sh1} is calculated as
$$R_{sh1} = \frac{R_{m1} \times CS}{I_{sh1}} = \frac{900 \times 100 \times 10^{-6}}{0.9 \times 10^{-3}} = 100 \text{ Ω}$$

This value of shunt resistance is quite practicable.

(b) *Range switch at* 10 mA

$R_{m2} = R_m + R_1 = (900 + R_1)\ \Omega;\ I_{sh2} = 10 - 0.1 = 9.9$ mA

$R_{sh2} = R_2 + R_3 + R_4 + R_5 = R - R_1$

But, $R_{sh2} = \dfrac{R_{m2} \times CS}{I_{sh2}} = \dfrac{(900 + R_1) \times 100 \times 10^{-6}}{9.9 \times 10^{-3}} = \dfrac{900 + R_1}{99}$

$\therefore\ R - R_1 = \dfrac{900 + R_1}{99}$

$\therefore\quad 99 \times 100 - 99R_1 = 900 + R_1$ (since $R = 100\ \Omega$)

or $\qquad\qquad R_1 = \dfrac{9900 - 900}{100} = \mathbf{90\ \Omega}$

(c) *Range switch at* 100 mA

$R_{m3} = R_m + R_1 + R_2 = 900 + 90 + R_2 = 990 + R_2$

$I_{sh3} = 100 - 0.1 = 99.9$ mA

$R_{sh3} = R_3 + R_4 + R_5 = R - R_1 - R_2 = 100 - 90 - R_2$

$\quad = 10 - R_2$

But, $\qquad R_{sh3} = \dfrac{R_{m3} \times CS}{I_{sh3}} = \dfrac{(990 + R_2) \times 100 \times 10^{-6}}{99.9 \times 10^{-3}}$

$\therefore\qquad 10 - R_2 = \dfrac{990 + R_2}{999}$

or $\qquad\qquad 9990 - 999R_2 = 990 + R_2$

or $\qquad R_2 = \dfrac{9000}{1000} = \mathbf{9\ \Omega}$

(d) *Range switch at* 500 mA

$R_{m4} = R_m + R_1 + R_2 + R_3 = 900 + 90 + 9 + R_3 = 999 + R_3$

$I_{sh4} = 500 - 0.1 = 499.9$ mA

$R_{sh4} = R_4 + R_5 = R - R_1 - R_2 - R_3 = 100 - 90 - 9 - R_3$

$\quad = 1 - R_3$

But, $\qquad R_{sh4} = \dfrac{R_{m4} \times CS}{I_{sh4}} = \dfrac{(999 + R_3) \times 100 \times 10^{-6}}{499.9 \times 10^{-3}}$

$\therefore\qquad 1 - R_3 = \dfrac{999 + R_3}{4999}$

or $\qquad\qquad 4999 - 4999R_3 = 999 + R_3$

or $\qquad R_3 = \dfrac{4000}{5000} = \mathbf{0.8\ \Omega}$

(e) *Range switch at* 1 A

$R_{m5} = R_m + R_1 + R_2 + R_3 + R_4$

$\quad = 900 + 90 + 0.8 + R_4 = 999.8 + R_4$

$I_{sh5} = 1000 - 0.1 = 999.9$ mA

$R_{sh5} = R_5 = R - R_1 - R_2 - R_3 - R_4$

$\quad = 100 - 90 - 9 - 0.8 - R_4 = 0.2 - R_4$

But, $R_{sh5} = \dfrac{R_{m5} \times CS}{I_{sh5}} = \dfrac{(999.8 + R_4) \times 100 \times 10^{-6}}{999.9 \times 10^{-3}}$

$\therefore\ 0.2 - R_4 = \dfrac{999.8 + R_4}{9999}$

or $\qquad 1999.8 - 9999 R_4 = 999.8 + R_4$

or $\qquad R_4 = \dfrac{1000}{10\,000} = 0.1\ \Omega$

$\therefore \qquad R_5 = R - R_1 - R_2 - R_3 - R_4 = 0.2 - R_4$

$\qquad\qquad = 0.2 - 0.1 = 0.1\ \Omega$

14.2.4 Resistance Measurement

The same basic 0.1-mA meter used to measure voltage and current may also be used to measure resistance. There are three methods commonly used for the measurement of resistances. These are as follows:

(*i*) We can connect the test piece in shunt with the meter. We then determine the ability of the test piece to bypass current by this shunt path. Ohmmeters based on this principle are called *shunt-type*. Such ohmmeters are suitable for the measurement of *low*-value resistances.

(*ii*) We can connect the test piece in series with the meter. We then determine the ability of the test piece to prevent current flow in the meter path. Ohmmeters using this principle are called *series-type*. They are suitable for the measurement of *medium*-value resistances.

(*iii*) The third alternative is to apply a known voltage across the test piece and then to determine the resulting current through it. The ratio of voltage to current gives resistance. Ohmmeters using this principle are called *meggar type*. They are suitable for measurement of *high*-value resistances, such as the insulation of a cable.

In electronic circuits, we normally come across resistances of medium value. We would hardly need a resistor of less than $100\ \Omega$ or more than $10\ M\Omega$. For measurement of such medium-value resistors, we use the series-type ohmmeter.

The Series-type Ohmmeter Basically this type of ohmmeter is an ammeter. The meter movement and the resistor to be measured are kept in series. Figure 14.10 shows the basic construction of a series-type ohmmeter. It includes a battery that supplies current to energize the meter. The circuit also has a preset resistor R_T and variable resistor R_0. The resistor R_T is meant for compensating the individual differences in the meter resistances. The variable resistor R_0 works as a *zero-adjust*. It compensates for any change in battery voltage V with ageing. The series resistor R_S limits the current to full-scale deflection when the test terminals $X - X$ are shorted.

If the external terminals $X - X$ are shorted to simulate $0\ \Omega$, the current in the circuit is maximum. The series resistor R_S is suitably selected so that this current is same as the current sensitivity of the meter ($100\ \mu A$, in this case). The deflection of the meter is full-scale. (If not, it can be adjusted to full-scale by adjusting the zero-adjust.) Thus we find that the full-scale deflection corresponds to $0\ \Omega$ across the test terminals $X - X$. If the test terminals are left open, it simulates an infinite resistance. The current through the circuit reduces to zero and the pointer does not deflect at all. Thus,

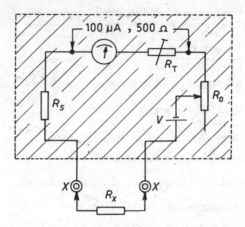

Fig. 14.10 Basic construction of a series-type
ohmmeter

zero deflection of the meter corresponds to an infinite resistance. If an unknown resistance R_X is connected across terminals $X-X$, the deflection will be less than full-scale. For this reason, the zero of the ohmmeter scale is at the right end, and the scale increases towards the left.*

Unfortunately, *the scale of the ohmmeter is not linear.* The values of resistance at the upper (or left side) end are very crowded and are hard to read. To understand this nonlinearity in scale, let us consider a specific example. In Fig. 14.10, let the battery voltage V be 1.5 V. When terminals $X-X$ are shorted, the current through the circuit should be equal to its current sensitivity (i.e. 100 μA). For this, the total resistance in the circuit should be $1.5/100 \times 10^{-6} = 15$ kΩ. If the unknown resistor R_X is 1 kΩ, the current through the meter will be $1.5/(15+1)10^3 = 93.75$ μA. Similarly, we can calculate the current corresponding to the unknown resistors of 2 kΩ, 3 kΩ, 5 kΩ, 10 kΩ, 15 kΩ, 25 kΩ, 50 kΩ, and 100 kΩ. The corresponding currents are 88.23 μA, 83.33 μA, 75 μA, 60 μA, 50 μA, 37.5 μA, 23.08 μA and 13.04 μA, respectively.

These resistance values are marked along the current scale in Fig. 14.11. You can see that the resistance scale is very much cramped in the region of higher resistances.

We avoid using the left end of the scale by providing a switching device for various ranges. To increase the resistance range, it is common practice to increase the battery voltage and the value of the series resistor R_S.

A word of caution is necessary: *An ohmmeter is never applied to an energized circuit.* If you do so, it may send excessive current through the meter; and you may damage it. Furthermore, even if excessive current does not flow through the meter, the ohmmeter scale becomes meaningless. It was calibrated only for the emf of the internal battery. Another point to be noted is that, while making resistance measurements in a circuit, ensure that there is no parallel branch across the component you are measuring. When in doubt, disconnect one terminal of the component under test.

---

*In shunt-type ohmmeters, the zero deflection corresponds to zero ohms; and full-deflection to infinite ohms. The scale is not inverted.

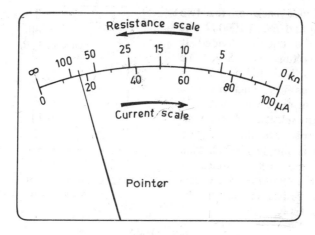

Fig. 14.11 Resistance scale of a multimeter

14.2.5 Meter Sensitivity (Ohms/Volt Rating)

An important quality of a meter is its *sensitivity*. The higher its sensitivity, the more accurate will be the measurements made with it. Meter sensitivity is measured in ohms/volt (Ω/V). If the current sensitivity of a meter is given, we can find its Ω/V rating. The meter (the basic d'Arsonval movement) discussed in this chapter has a current sensitivity CS of 100 μA. It means that it requires a current of 100 μA to give full-scale deflection. Suppose we use this meter as a voltmeter in its 1 V range. It must then have a total resistance of 10 000 Ω, so that 100 μA current flows through it when 1 V is applied across its terminals. Thus, the meter sensitivity is 10 000 Ω/V. In general,

$$\text{ohms/volt rating} = \frac{1}{\text{CS}} \qquad\qquad (14.3)$$

Thus, for a 50-μA meter,

$$\text{ohms/volt rating} = \frac{1}{50 \times 10^{-6}} = 20\ 000\ \Omega/\text{V}$$

Remember, the internal resistance of the voltmeter is not same in each of its ranges. The higher its range, the greater is its internal resistance. We can determine the resistance for each range by simply multiplying the range voltage (the maximum voltage of the scale) by the ohm/volt rating. For example, for the 100-V scale of a meter with 10 000 ohm/volt rating,

$$\text{internal resistance} = 100 \times 10\ 000 = 1\ \text{M}\Omega$$

There is another very important point. We may measure only 30 V on the 100 V scale, but the internal resistance of the above meter still remains 1 MΩ. *Whatever the deflection of the pointer, the internal resistance is determined by the full-scale voltage.*

Why is the ohms/volt rating of a meter so important? This rating gives an idea of how much the *loading effect* of the meter will be, on the circuit under test. To understand this, let us take a practical example. Suppose we wish to measure the voltage across resistor R_2 in the circuit shown in

Fig. 14.12. Since each of the two resistances, R_1 and R_2, is 1000 Ω, the battery voltage is equally divided. If the battery voltage is 1.5 V, the voltage across R_2 should be 1.5/2 = 0.75 V. Let us first measure this voltage using a meter with 1000 Ω/V rating. We can select a 1-V range, so that the internal resistance (R_i) of the voltmeter is simply 1000 Ω. This resistance of the meter comes in parallel with the resistance R_2. Therefore, the net resistance across the test terminals AB is

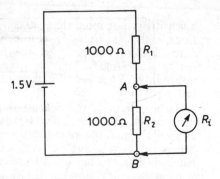

Fig. 14.12 Loading effect of a voltmeter when connected in a circuit

$$R_{net} = R_2 \| R_i = \frac{1000 \times 1000}{1000 + 1000} = 500 \ \Omega$$

Because of the modified resistance between the terminals A and B, the battery voltage does not divide equally between the two resistors any more. The voltage across A-B will now become

$$V_{AB} = \frac{1.5 \times R_{net}}{R_1 + R_{net}} = \frac{1.5 \times 500}{1000 + 500} = 0.5 \ V$$

This is the voltage that will be read by this meter. It is much below the actual value (0.75 V). This difference arises due to the *loading effect* of the voltmeter. The voltmeter loads the circuit under test.

Now, let us measure the voltage across terminals A-B (Fig. 14.12) using another meter with a higher sensitivity of 20 000 Ω/V. Again, we select the 1-V range. The internal resistance R_i of the voltmeter is now 20 000 Ω. The net resistance between the terminals A-B is now

$$R_{net} = R_2 \| R_i = \frac{1000 \times 20\ 000}{1000 + 20\ 000} = 952.38 \ \Omega$$

The resistance of 1000 Ω is modified to 952.38 Ω. In fact, it is quite a small change. The voltage across terminals A-B now becomes

$$V_{AB} = \frac{1.5 \times R_{net}}{R_1 + R_{net}} = \frac{1.5 \times 952.38}{1000 + 952.38} = 0.73 \ V$$

This is the reading given by the second meter. It is again less than the actual value (0.75 V). This too is not the most desirable solution. However, the second meter gives a reading much closer to the actual value. The loading effect due to this meter is much less. Ideally, the voltmeter should have infinite internal resistance. Only then it will give accurate readings.

Thus, we conclude that the higher the sensitivity (ohms/volt rating) of a voltmeter, the lesser will be its loading effect on the circuit under test. Generally, meters do not have sensitivities higher than 20 000 Ω/V.

14.3 ELECTRONIC MULTIMETERS

The main disadvantage of an ordinary moving-coil multimeter is its low sensitivity (i.e., ohms per volt rating). Even the best quality multimeter has a sensitivity not more than 20 000 Ω/V. For many measurement applications, this sensitivity is considered good enough. However, many times, much higher sensitivities are needed for accurate measurements. An electronic multimeter affords this quality. These meters have very high input impedance. When used for measurement, they hardly load the circuit under test.

Earlier, these instruments used vacuum tubes, and they functioned as voltmeters only. They were popularly called VTVMs (Vacuum Tube Volt-Meters). In recent years their popularity has increased because, firstly, they have become relatively inexpensive, and secondly, the function of an ohmmeter has been added. They are also referred to as "electronic volt-ohmmeters".

As the use of transistors is becoming more widespread, a transistorized version of the VTVM is also available in the market. These meters may be called "electronic solid-state volt-ohmmeters". There are two types of such instruments. One type uses analogue display of the measurement results, similar to the display of an ordinary moving-coil meter; where the pointer moves across a calibrated scale and the quantity measured is then read off the scale. The other type of solid-state volt-ohmmeter uses a digital display. It provides readings directly, in the digital form. This display is similar to the one used in digital watches.

A detailed description of all these instruments is beyond the scope of this book. Their working principles and knob functions will however be explained.

14.3.1 VTVM

A vacuum-tube voltmeter, or VTVM is a considerable improvement over the ordinary moving-coil meter. Though a VTVM uses many of the basic principles and circuitry of the moving-coil meter, it has many advantages. For example, it has a much higher input impedance (of the order of megohms) than the moving-coil meter. As a result, it does not load the circuit under test so much. A VTVM will give readings much closer to the actual value. Another advantage is the improved ac measurement. It is possible to measure low ac voltages (even of high frequency) by including electronic amplifiers in the circuitry.

A VTVM has three disadvantages, but they are not serious. Firstly, a VTVM needs a supply voltage for its operation, whereas a moving-coil meter does not. Secondly, it is relatively a costly instrument. Thirdly, in general, a VTVM cannot measure current; however, one can measure current indirectly by measuring the voltage drop across a known resistor.

Principle of Working of a VTVM Figure 14.13 shows the circuit, which is the heart of the VTVM. It uses a dual-triode tube (that is, two triodes T1 and T2 contained in one glass envelope). These triodes along with their cathode resistors, R_3 and R_4, form the lower arms of the *bridge circuit*. The anode resistors, R_1 and R_2, along with a portion of the *zero adjust* control, R_6, form the upper arms of the bridge. The moving wiper of the control R_6 is connected to the positive terminal of the dc power supply. Resistor R_5 is placed in series with the cathode resistors, and is returned to the negative terminal of the dc supply. This resistor has a comparatively higher value, so that it controls the cathode current to a large extent. The dc meter (d'Arsonval movement) is connected to the anodes of the triodes, at opposite corners of the bridge. R_S is connected in series with the meter, to limit current. The control grid of the triode T2 is grounded.

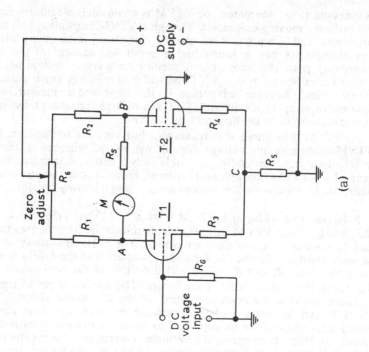

Fig. 14.13 (a) Circuit used in a VTVM; (b) Simpson VTVM

The dc voltage to be measured is applied to the control grid of the triode T1. If this voltage is positive, the plate current of T1 increases. This raises the potential of point C, the junction of R_3, R_4 and R_5. Since the grid of T2 is grounded, the effective negative grid bias of this tube is increased and its plate current decreases. The increase in the plate current of T1, causes a drop in its plate voltage. Simultaneously, the decrease in the plate current of T2 causes its plate voltage to rise. The difference in potential between the two anodes (points A and B) is registered by the meter movement. The meter deflection is linear, provided the two triodes operate in the linear region of their characteristics.

When the input dc voltage is 0 V (that is, when the control grid of T1 is grounded), the two triodes operate under identical conditions. The two anode currents should be the same; and the meter deflection should be zero. However, in practice, the two triodes may not have perfectly identical characteristics, and there may be a small difference in their plate currents. Because of this imbalance in the two plate currents, there will be a small initial deflection in the meter. The meter is initially set to zero, by the *zero adjust* resistor R_6. Adjustment of R_6 changes the relative values of the two anode resistors.

The control grid of a triode ordinarily does not draw current. It means that the circuit hardly draws any current from the source to be measured (the resistor R_G appearing across the input terminals is of very large value). The input resistance of most VTVMs is over 12 MΩ. They draw so little current from the circuit under test, that it can usually be ignored.

The range of voltages which can be applied to the control grid of T1, to maintain good linearity, is only about 0 to 3 V. The basic range can be easily expanded by using a multiplier (or attenuator) circuit, such as the one in Fig. 14.14. A number of precision resistors are used in a series circuit, with

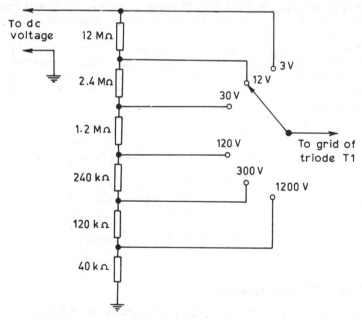

Fig. 14.14 Multiplier (or attenuator) circuit, and range switch used in a VTVM

one end grounded. The circuit works in much the same way as in an ordinary multimeter. The movable arm of the switch (rotor) can be placed at any desired resistor. This switch is called a *range switch* in most instruments.

Can We Measure ac Voltages and Resistances with VTVM A VTVM is basically a dc voltmeter. However, with additional circuitry, we can use a VTVM to measure ac voltages and resistances. When the *function switch* of the VTVM is brought in AC VOLTAGE position, a built-in rectifier is switched ahead of the basic dc instrument, as shown in Fig. 14.15. Usually, a miniature duo-diode tube (such as 6AL5) is used for the rectifier. This changes ac to dc, which is then measured on the VTVM.

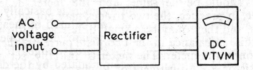

Fig. 14.15 Arrangement for measuring ac
voltages with VTVM

A VTVM measures resistances when its function switch is set to "ohm" position. In this function, the instrument includes a self-contained battery (of 1.5 V, usually) and a calibrating resistor R. As shown in Fig. 14.16, when the unknown resistor R_x is placed across the probes, a closed series circuit is formed. The battery current flows through this circuit. The voltage drop across the unknown resistor R_x is fed to the dc voltmeter. One of the scales on the instrument is calibrated directly in ohms. Note that the circuit is so arranged that the VTVM reads increasing resistances from left to right. This is opposite to the way resistance scales read on moving-coil ohmmeters.

The resistance range can be expanded by merely changing the value of the calibrating resistor R. This is easily done by taking a number of precision resistors in series, in place of a single resistor. Each resistor junction is then connected to one of the poles of the rotary switch. This switch works as *range switch*.

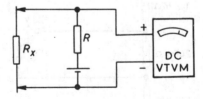

Fig. 14.16 VTVM as an ohmmeter

14.3.2 Solid-State Multimeters

These are a relatively new class of instruments which are fast becoming popular. They are transistorized · version of the VTVM. They do not contain vacuum tubes and are operated using low-voltage batteries.

The advantages of the solid-state multimeter over a regular moving-coil multimeter are: (i) much greater input impedance on ac and dc; (ii) much higher resistance measurement capability, and (iii) the ability to check transistors without the danger of damaging them (since low currents and voltages are involved). Another important advantage a solid-state multimeter has, is its overload protection.

If, by accident, input overloads are applied, the amplifier of the instrument saturates, and hence the maximum current through the meter is limited. The meter does not burn out due to overloads.

Advantages of this instrument over a VTVM are: (i) portability, as no ac voltage need be used; (ii) it can be operated using battery cells kept inside the instrument.

Similar to a VTVM, a solid-state multimeter, is basically a dc voltmeter (though, with additional circuitry, it can be used to measure ac voltages and resistances). The circuit of a solid-state dc voltmeter is shown in Fig. 14.17. The range switch selects the desired range. It attenuates the input voltage to a level that can be accommodated by the dc amplifier. The input stage of the amplifier consists of a field effect transistor (FET). The FET is a popular choice because of its high input impedance. It effectively isolates the meter from the circuit under measurement.

The two transistors, T1 and T2, form a dc amplifier. This dc amplifier drives the meter movement. Provided the amplifier operates within its linear range, the deflection of the meter movement is directly proportional to the input voltage.

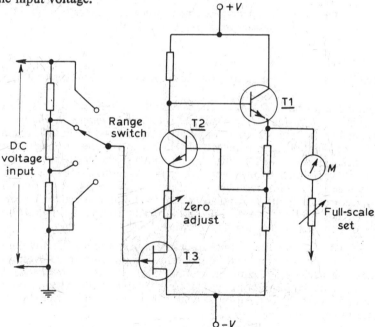

Fig. 14.17 A basic solid-state dc voltmeter circuit

Solid-state multimeters are of two types—*analogue* and *digital*. Figure 14.18 shows an electronic multimeter having an analogue display. It uses a meter movement to indicate the magnitude of the quantity under measurement.

Fig. 14.18 An electronic multimeter with analogue indication
(transistorized version of VTVM)

The reading is given on a continuous scale, hence the name *analogue*.

Digital Multimeter (*DMM*) In a digital multimeter, the result of the measurement is displayed in *discrete* intervals or *numerals*. Figure 14.19 shows such an instrument. It is popularly known as a digital voltmeter (DVM), or digital multimeter (DMM). It displays measurement of dc or ac voltages and resistances, as discrete numerals in the decimal number system. It is very convenient since we get the result of our measurement directly in the form of a number. Such instruments have many advantages. They reduce reading and interpolation errors, eliminate parallax error, and increase the reading speed. Furthermore, the output in digital form is suitable for further processing or recording.

The DMM is a versatile and accurate instrument used in many laboratories. Since the development and perfection of integrated circuit (IC) modules, the size, power requirement, and cost of the DMM has been drastically reduced. Some simple DMMs available in the market now actively compete with the conventional analogue instruments, both in portability and price.

Although there are many ways of converting the analogue reading into digital form, the most common way is to use a *ramp voltage*. The operating principle of a ramp-type DMM is simple. A ramp voltage increases linearly from zero to a predetermined level in a predetermined time interval. It simply measures the time it takes for a linear ramp voltage to change from zero volts to the level of the input dc voltage. This time interval is measured with an electronic time-interval counter. This count is displayed as a num-

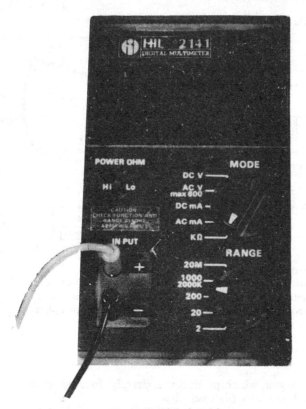

Fig. 14.19 Digital multimeter

ber of digits, using either light-emitting diodes (LEDs), or liquid-crystal display (LCD).

14.4 CATHODE-RAY OSCILLOSCOPE (CRO)

Of all the laboratory instruments available today, perhaps the most important and versatile is the cathode-ray oscilloscope (CRO). It is primarily used for the display of waveforms. It works as an "eye" for the electronics engineer. With the help of a CRO, he can "see" what is happening in each part of the electronic circuit.

A CRO is basically a very fast $X-Y$ plotter. It displays an input signal versus another signal, or versus time. The "stylus" of this "plotter" is a luminous spot which moves over the display area in response to input voltages.

The heart of the oscilloscope (or *scope* as it is sometimes called) is the cathode-ray tube (CRT). The rest of the instrument consists of circuitry necessary to operate the CRT. In the next section, we shall study the construction and working of a CRT.

14.4.1 Cathode Ray Tube (CRT)

Figure 14.20 shows the schematic diagram of a cathode-ray tube along with its control circuits. A CRT essentially consists of three basic components:

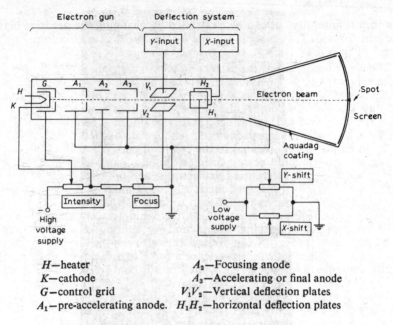

H—heater
K—cathode
G—control grid
A_1—pre-accelerating anode.

A_2—Focusing anode
A_3—Accelerating or final anode
V_1V_2—Vertical deflection plates
H_1H_2—horizontal deflection plates

Fig. 14.20 Schematic diagram of a CRT

(i) The *electron gun*, which produces a sharply focused beam of electrons, accelerated to a very high velocity.

(ii) The *deflection system*, which deflects the electrons, both in the horizontal and vertical planes electrostatically (or magnetically in TV tubes) in accordance with the waveform to be displayed.

(iii) The *fluorescent screen*, upon which the beam of electrons impinges to produce a spot of visible light.

These three essential components of a CRT are put inside a highly evacuated, funnel-shaped glass envelope. The large end of this tube is coated on the inside with a *phosphor* material. This material *fluoresces* when high-velocity electrons strike it, converting the energy of the electrons into visible light. Hence the name *fluorescent screen*. Depending upon the phosphor material used in the fluorescent screen, it is possible to have either green, orange, or white light. When the electron beam strikes the screen, besides giving out visible light, secondary emission electrons are also released. These electrons are collected by the conductive coating deposited on the inside surface of the glass bulb. The coating is usually an aqueous solution of graphite, known as *aquadag*. This is electrically connected to the final anode, as shown in Fig. 14.20.

The electron gun gets its name because it fires electrons at a very high speed, like a gun which fires high speed bullets. Electrons are emitted from the indirectly heated cathode. The control grid is a nickel cylinder surrounding the cathode. It has a small hole in the far end, opposite the cathode. The only way the emitted electrons can get past the grid is through this small hole. The action of the control grid is identical to that in a conventional vacuum triode. It controls the number of electrons passing through it. Since the brightness of the spot on the face of the screen depends upon the beam intensity, it can be controlled by changing the negative bias on the control grid. That is what is done by the *intensity control* (on the front panel of the CRO).

The electrons coming out of the control grid are accelerated by the high potential applied to the accelerating anode. These electrons, being negatively charged, have a tendency to diverge from each other. If permitted to pass as it is, the electron beam will not form a sharp image on the screen. The beam is focused to a very small dot on the screen by the focusing anode. The accelerating and focusing anodes are also cylindrical in form, with small openings located in the centre of each cylinder, along the axis of the tube. These holes permit the electrons to pass through. Note that this anode is only accelerating the electrons, and not collecting them (only a few diverging electrons are collected by the anode). The focusing anode is given a slightly lower potential than the accelerating anodes (Fig. 14.20). Because of the difference in potentials, the equipotential surfaces between the two cylinders (focusing anode and accelerating anode) form a shape like a convex lens (Fig. 14.21). When the electron beam passes through this region the electrons experience a force in a direction normal to the equipotential surfaces. As a result, the beam is converged to a sharp point on the screen. The equipotential surfaces (sometimes called *electrostatic lens*) focuses the electron beam in much the same way as a lens in an ordinary camera focuses light. By changing the potential of the focusing anode, we can change the focal length of the electrostatic lens.

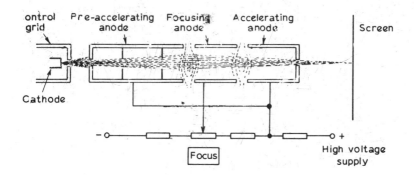

Fig 14.21 Electrostatic focusing arrangement in a CRT

The electron gun emits a very narrow (focused), highly accelerated electron beam. This beam then passes through the deflection system consisting of two pairs of parallel plates. As shown in Fig. 14.20, the Y-deflection plates (also called *vertical deflection plates*) are placed horizontally in the tube. Any voltage applied to this set of plates moves the electron beam up or down. The bright spot on the screen will move along the y-axis. The X-deflection plates (or horizontal deflection plates) are kept vertically. Any voltage applied to this set of plates moves the spot on the screen to the left or to the right (along the x-axis). If no voltage is applied externally to either set of plates, the spot should be located at the centre of the screen. The initial centering of the spot can be done by using the X-shift and Y-shift controls (see Fig. 14.20).

Before we see how a CRT is used in a CRO to display waveforms, there is one important point to be noted. In Fig. 14.20, the accelerating anodes and the deflection plates are very close to the ground potential. The cathode is given a negative potential equal to the accelerating voltage. This is done to avoid high-voltage shock hazard to the operator. The operator has to handle the vertical-input and horizontal-input terminals. He also has to handle various controls on the front panel of the CRO. The *focus control* is about 2 kV positive with respect to the cathode, but is only 100 or 150 V negative with respect to the final anode. It is, therefore, better to ground the positive side of the high-voltage supply. Imagine what would have happened if he were to touch the focus-control at high voltage (about 2 kV)!

14.4.2 How a CRO Displays Waveforms

If we apply a dc voltage to the horizontal deflection plates, the spot moves either to the left or to the right. If plate H_1 is positive, relative to plate H_2, the spot moves to the left; as the electron beam is attracted by the positive plate and repelled by the negative plate (see Fig. 14.22a). If H_2 is positive, relative to H_1, the spot moves to the right (see Fig. 14.22b). Similarly, applying a dc voltage to the vertical deflection plates moves the spot up or down. If voltages are applied to both the sets of plates together it is possible to place the spot anywhere on the screen.

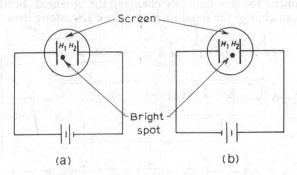

(a)   (b)

Fig. 14.22 Position of spot as dc potential is applied to
the horizontal deflection plates

Let us see how the spot moves on the screen if we apply an ac voltage (say, derived from a 50-Hz power mains) to the horizontal deflection plates. As the voltage varies sinusoidally with time, the spot also moves to the left and to the right, in the same way. It moves left and right at the rate of 50 times per second. This is too high a frequency, for the human eye to see the motion of the spot distinctly. Due to persistence of vision (and also due to the "persistence" property of the screen) we "see" a solid line on the screen (Fig. 14.23a).

Figure 14.23b shows that a solid-line trace is obtained even if the ac voltage is of the sawtooth type, instead of the sinusoidal type. But now there is one important difference. The electron spot (tracing the solid line) moves from left to right at a uniform speed, since the voltage is increasing linearly. It was not so in the case of sinusoidal voltage.

In Fig. 14.23b, when the voltage drops to the negative maximum after one cycle, the spot moves to the left almost instantaneously. In the next cycle it again starts moving to the right with uniform speed. Let us assume

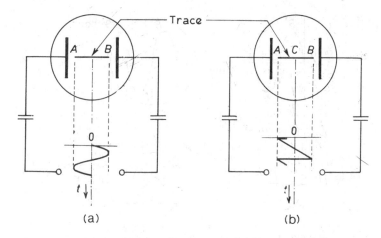

Fig. 14.23 A solid line trace is obtained when we apply to the horizontal deflection plates (*a*) a sine voltage, (*b*) a sawtooth voltage

that the time period of the sawtooth wave is 20 millisecond (corresponding to a frequency of 50 Hz), and the length of the straight line trace AB is 10 cm. At time zero (i.e. the instant when the voltage just starts rising), the spot is at point A. After 10 ms, the sawtooth voltage rises by half of its peak value; the spot must be at point C, half-way between points A and B. After 15 ms, the spot would have travelled three-fourth of the distance between points A and B. This way, the line AB is calibrated in time. The whole length AB corresponds to 20 ms; half of it to 10 ms; quarter of it to 5 ms; and so on.

Suppose we now apply a sine-wave voltage to the vertical deflection plates, and at the same time a sawtooth-wave voltage to the horizontal deflection plates. Let us further assume that the two ac voltages have the same time periods (say, 20 ms), corresponding to the frequency of 50 Hz. The trace obtained on the screen is shown in Fig. 14.24. At point A', the sine-wave voltage is zero; hence there is no deflection of the beam in the vertical deflection. The horizontal deflection plates (or X-plates) have a maximum negative voltage; hence the spot moves to the extreme left. At point B, the spot moves up since the sine wave has a positive value; at the same time it moves horizontally to the right since the sawtooth voltage is now less negative. At point C, both the voltages have zero magnitude— the spot is in the centre of the screen. We can get the figure displayed on the screen by considering the resultant displacement of the electron beam at different instants of time. Thus, points A, B, C, D and E are obtained on the screen, corresponding to the points A', B', C', D'; and E' on the vertical input wave, and the points A'', B'', C'', D'' and E'' on the horizontal input wave.

To further illustrate how a CRO displays waveforms, let us suppose that the frequency of the sawtooth wave is reduced to 25 Hz, while that of the sine wave is maintained at 50 Hz. Now, the time period of the sawtooth wave is 40 ms; the horizontal trace length* will correspond to 40 ms.

---

*The horizontal trace length can be adjusted to a suitable value by increasing or decreasing the horizontal-amplifier gain.

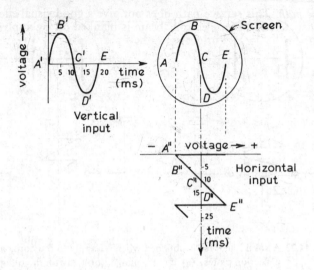

Fig. 14.24 Display of sine wave on CRO when both horizontal and vertical input waves are 50 Hz

During this time, the sine wave goes through two complete cycles. This causes two sine waveforms to be traced on the screen.

When a sawtooth wave is applied to the horizontal deflection plates, the x-axis on the screen may be taken to represent time. For this reason, the sawtooth-wave voltage is called *time-base voltage*. An ideal time-base voltage would have been the one having the waveshape shown in Fig. 14.25a. However, in practice, the time-base generator in a CRO produces the waveshape shown in Fig. 14.25b. It takes a little time for the voltage to fall from maximum positive to maximum negative. This time interval is called *flyback* or *retrace* time. Since the time-base voltage has a finite flyback period, the display of the waveform is a little distorted, as shown in Fig. 14.25c. The path traced by the spot during the flyback period is called

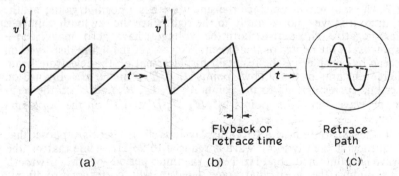

(a)   (b)   (c)

Fig. 14.25 (a) Ideal sawtooth wave; (b) Actual sawtooth wave generated by time-base generator; (c) Display of waveform gets distorted due to flyback period of time-base voltage

retrace path. This retrace path does not give a good visual effect. Therefore, in most CROs, the electron beam is blanked off by applying a negative voltage to the control grid during the flyback period.

14.4.3 Block Diagram of a CRO

A block diagram showing the various subsystems of a CRO is shown in Fig. 14.26. This diagram does not show all the possible subsystems, but the minimum stages required. These subsystems are:

  (i) the vertical deflection system;
 (ii) the horizontal-deflection system, including the time-base generator and synchronization circuitry;
(iii) the CRT; and
 (iv) the high-voltage and low-voltage power supplies.

The vertical deflection system consists of an input attenuator and a number of amplifier stages. The gain of the vertical amplifier can be controlled by the attenuator. The waveform to be displayed is fed to this Y-input. The horizontal deflection system provides the voltage for moving the beam horizontally. It includes a number of amplifier stages, the gain of which can be controlled. It has a sawtooth oscillator, or a time-base generator. Also included in this subsystem is a synchronization circuit. The purpose of this circuit is to start the horizontal sweep at a specific instant, with respect to the waveform under observation. In addition to the internal sweep, there is a provision for the external horizontal inputs (or X-inputs). One may either select the internal sweep voltage or any other voltage fed externally for deflecting the beam horizontally. Basically, the operation of the vertical section does not affect the horizontal section, and vice versa.

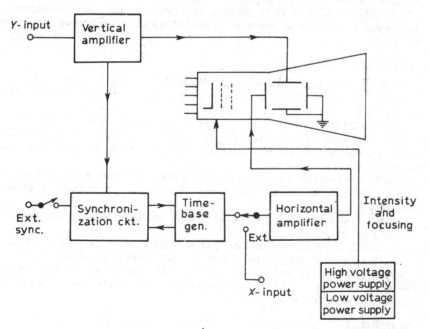

Fig. 14.26 Block diagram of CRO

When operated together, they will display the incoming signal on the screen of the CRO.

14.4.4 Front Panel Controls of a General-Purpose CRO

Figure 14.27a shows the front panel of Aplab Solid-State Oscilloscope, Type DC-15. It is a general-purpose oscilloscope manufactured by Applied Electronics Limited, Thana, India. The bandwidth of its vertical amplifier ranges from dc to 15 MHz. The time-base and trigger facilities provide a sweep speed ranging from 0.2 μs/cm to 50 ms/cm. Waveforms can be triggered from 1 Hz to 20 MHz and presented with complete stability. The scope has a provision for producing line-frequency (50 Hz) square wave of magnitude 2 V (peak-to-peak). This enables us to check the Y-calibration of the scope. The scope consumes only 55 W of power at 220 V, 50 Hz.

The front panel controls are grouped, facilitating easy use, and are clearly designated. The function of each control is explained below. The number given within brackets refers to the control number as given in Fig. 14.27a.

GENERAL:

ON POWER (1): It is a toggle switch meant for switching on power. In ON position, power is supplied to the instrument and the neon lamp (3) glows.

INTEN (2): It controls the trace intensity from zero to maximum.

FOCUS (4): It controls the sharpness of the trace. A slight readjustment of this control may be necessary after changing the intensity of the trace.

X MAG (5): It expands length of time-base from 1 to 5 times continuously, and makes maximum time-base to 40 ns/cm.

SQ. WAVE (6): This provides a square wave of 2 V (p-p) amplitude to enable one to check the Y-calibration of the scope.

SAWTOOTH WAVE (7): This provides a sawtooth-waveform output coincident to sweep-speed switch with an output of 5 V (p-p). The load resistance should not be less than 10 kΩ.

VERTICAL SECTION:

Y-POS. (10): This control enables the movement of the display along the y-axis.

Y-INPUT (13): It connects input signal to vertical amplifier through AC–DC–GND coupling switch (14).

AC–DC–GND Coupling Switch (14): It selects coupling to the vertical amplifier. In DC mode, it directly couples the signal to the input; in AC mode, it couples the signal to the input through a 0.1 μF, 400-V capacitor. In GND position, the input to the attenuator (12) is grounded, whereas Y-input is isolated.

VOLTS/cm (Attenuator) (12): It is a 10-position attenuator switch which adjusts sensitivity of vertical amplifier from 50 mV/cm to 50 V/cm in 1, 2, 5, 10 sequence. Attenuator accuracy is ±3 %.

×1-×0.1 Switch (9): When switched in ×0.1 position, it magnifies basic sensitivity to 5 mV/cm from 50 mV/cm.

CAL. Switch (8): When pressed, a dc signal of 15 mV or 150 mV is applied to a vertical amplifier depending upon the position of ×1–×0.1 switch (9) position.

DC BAL (11): It is a preset control on panel. It is adjusted for no movement of the trace when either ×1–×0.1 switch (9) is pressed, or the position of AC–DC–GND coupling switch (14) is changed.

HORIZONTAL SECTION:

X-POS (21): This control enables the movement of display along the x-axis.

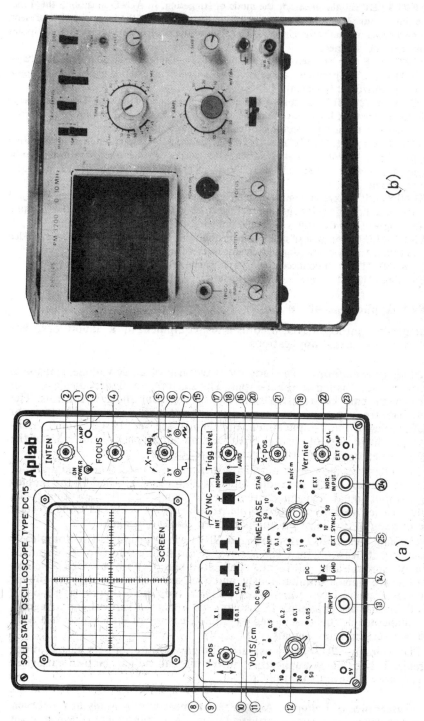

Fig. 14.27 (a) Aplab solid-state oscilloscope (Type DC-15); (b) Philips CRO

TRIGG LEVEL (18): It selects the mode of triggering. In AUTO position, the time-base line is displayed in the absence of input signal. When the input signal is present, the display is automatically triggered. The span of the control enables the trigger point to be manually selected.

TIME-BASE (19): This selector switch selects sweep speeds from 50 ms/cm to 0.2 µs/cm in 11 steps. The position marked EXT is used when an external signal is to be applied to the Horizontal Input (24).

VERNIER (22): This control is a fine adjustment associated with the Time-base Sweep Selector Switch (19). It extends the range of sweep by a factor of 5. It should be turned fully clockwise to the CAL position for calibrated sweep speeds.

SYNC Selector (15, 16, 17): The INT/EXT switch (15) selects internal or external trigger signal. The +ve or −ve switch (16) selects whether the waveform is to be triggered on +ve or −ve step. NORM/TV switch (17) permits normal or TV (line frequency) frame.

STAB (20): It is a preset control on the panel. It should be adjusted so that you just get the base line in AUTO position of Trigger Level Control (18). In any other position of the Trigger Level Control, you should not get the base line.

EXT CAP (23): This pair of connectors enables the time-base range to be extended beyond 50 ms/cm by connecting a capacitor at these connectors.

HOR INPUT (24): It connects the external signal to Horizontal Amplifier.

EXT SYNC (25): It connects external signal to trigger circuit for synchronization.

14.4.5 Applications of CRO

The general-purpose CRO, used as a test-equipment in a laboratory, has following important applications.

Study of Waveforms To study the waveform of an ac voltage, sinusoidal or otherwise, it is fed to the Y-input. The size of the figure displayed on the screen can be adjusted suitably by adjusting the gain controls. The time-base frequency can be changed so as to accommodate one, two, or more cycles of the Y-input signal. For details, see Sec. 14.4.2. Some oscilloscopes have the provision of expanding only a part of the cycle of the signal, so as to examine this part in greater detail.

In modern oscilloscopes, automatic triggering is used. The beginning of each oscillation of the sawtooth-wave oscillator is controlled by the incoming Y-input signal. This makes the proper synchronization of the signal much easier and much more stable. In addition, it is also possible to start the horizontal oscillator on any part of the incoming signal.

The *dual beam* CRO uses a special CRT, which produces two completely independent beams. This CRT has two electron guns and two independent sets of vertical and horizontal deflection plates. With this CRO, we can display two signals simultaneously, and compare their waveforms. A similar function is achieved in a *dual trace* CRO. It uses the ordinary single-beam CRT, but has the capability to display two separate vertical input signals simultaneously. This is achieved by "time-sharing" of the electron beam by the two vertical input signals.

The *storage* CRO is very useful in the presentation of very slowly swept signals. It finds many applications in the mechanical and biomedical fields. It can store the events on the CRT.

Measurement of Voltages A dc voltage is measured by applying it between a pair of deflection (usually vertical) plates. The displacement of the spot on the screen is measured. Usually, the gain control (attenuator) of the vertical amplifier is calibrated in terms of deflection sensitivity. The *deflection sensitivity* of a CRO can be defined as the amount of displacement of the spot on the screen when a potential of one volt is applied to its deflection plates. In most CROs, the deflection sensitivity is expressed as the ratio of input voltage to the length of the trace. Accordingly, it is marked on the attenuator in V/cm, so that by multiplying the displacement of the spot by the deflection sensitivity directly gives the magnitude of the dc voltage.

An ac (sinusoidal) voltage is measured by applying it to the vertical deflection plates. A straight-line trace is obtained. Measuring the length of this straight-line trace and multiplying this length with the deflection sensitivity (given in V/cm) gives the peak-to-peak value of ac voltage. Half of this is the peak or maximum value of ac. Dividing it by $\sqrt{2}$, gives the rms value.

Example 14.4 The vertical gain control of a CRO is set at a deflection sensitivity of 5 V/cm. An unknown ac voltage (sinusoidal) is applied to the Y-input. A 10 cm long straight-line trace is observed on the screen. Determine the ac voltage.

Solution: The deflection sensitivity of the CRO is set at 5 V/cm. It means a potential of 5 V will displace the electron spot on the screen by 1 cm. Since the trace is a line of length 10 cm, the ac voltage must have a peak-to-peak variation of $10 \times 5 = 50$ V.

Therefore, the peak (or maximum) value of ac input voltage

$$= \frac{50}{2} = 25 \text{ V}$$

Dividing it by $\sqrt{2}$, we get rms value,

$$V = \frac{25}{\sqrt{2}} = 17.677 \text{ V}$$

Measurement of Currents A CRO with an electrostatic deflection system is basically a voltage indicating device. For measuring current, it is passed through a suitable, known resistor. Then the potential developed across this resistor is measured as explained above. The current can then easily be determined.

Measurement of Frequency One of the quickest and most accurate methods of determining frequencies is by using *Lissajous patterns*. A Lissajous pattern is produced on the screen when two sine-wave voltages are applied simultaneously to both pairs of deflection plates. A stable pattern is obtained when the ratio of the two frequencies is an integer, or a ratio of integers. The type of pattern observed depends upon this ratio (and also upon the relative phase of the two waves). Figure 14.28 illustrates how the Lissajous pattern is obtained for the two voltages applied to two sets of deflection plates, when the ratio of their frequencies is 1 : 2. Figure 14.29 shows some typical cases of Lissajous patterns for various frequency ratios and phase differences.

To measure the frequency of a sine-wave voltage, it is applied to one set of deflection plates (say Y-plates). To the other set (say, X-plates) we apply a sine-wave voltage obtained from a standard variable-frequency oscillator.

The frequency of this oscillator is varied till a suitable stationary pattern is obtained on the screen. Knowing this frequency, it is easy to determine the unknown frequency.

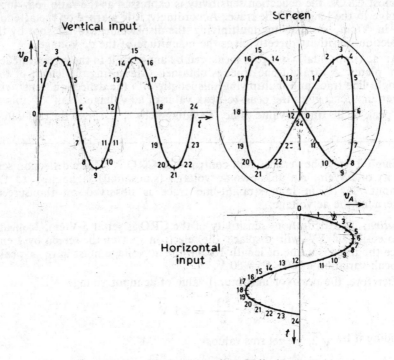

Fig. 14.28 Graphical construction of Lissajous pattern, when the ratio of frequencies is 1 : 2

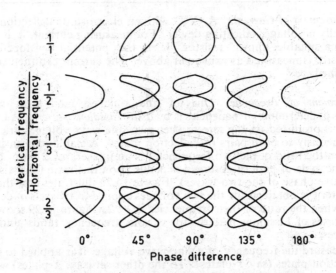

Fig. 14.29 Lissajous pattern for different frequency ratios and phase differences

In CROs available nowadays, the time base is calibrated. It is possible to display the unknown voltage wave and to read its frequency directly. This procedure is made clear in Example 14.5.

Example 14.5 A CRO with calibrated time-base is used for displaying a sine-wave voltage. For a convenient size of the display, the vertical amplifier attenuator is set at 2 V/cm, and the time-base control is set at 0.1 ms/cm. The display obtained is shown in Fig. 14.30. Determine the magnitude and the frequency of the sine-wave voltage fed to the Y-input.

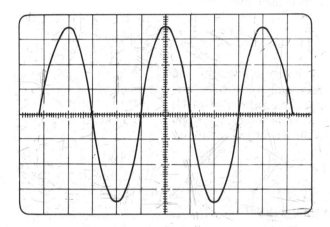

Fig. 14.30 Display of a sine-wave voltage on a CRO

Solution: The trace has been centred about the x-axis. The positive peak of the sine wave lies on the y-axis. The peaks of the sine wave extend by 3.5 cm on both sides of the x-axis. Therefore, the peak value is 3.5 cm. (In case the trace was not displayed symmetrically, we should have taken peak-to-peak value. Half of this p-p value gives the peak value of the sine wave.)

Since the sensitivity of the vertical amplifier attenuator is set at 2 V/cm, the length of 3.5 cm along the y-axis corresponds to

$$V_m = 3.5 \times 2 = 7.0 \text{ V}$$

Therefore, the rms value of the voltage is

$$V = \frac{7.0}{\sqrt{2}} = 4.95 \text{ V}$$

Now let us see how we read the frequency of the sine-wave voltage. It is observed from Fig. 14.30 that one cycle of the wave spreads over 4 cm along the x-axis. The time-base control is set at 0.1 ms/cm. Therefore, 4 cm along the x-axis corresponds to a time duration of

$$T = 4 \times 0.1 \text{ ms} = 0.4 \text{ ms}$$

Therefore, the frequency of the sine-wave voltage is

$$f = \frac{1}{T} = \frac{1}{0.4 \times 10^{-3}} = 2500 \text{ Hz}$$

$$= 2.5 \text{ kHz}$$

Measurement of Phase Difference A CRO can be used to determine the phase difference between two sine-wave voltages (of the same frequency). The two voltages are applied to the two sets of deflection plates simultaneously. The resultant pattern on the screen is an ellipse. The phase difference between the two waves is then given by

$$\sin \theta = \frac{Y_1}{Y_2} = \frac{X_1}{X_2} \tag{14.4}$$

where, Y_1 is the y-axis intercept, and Y_2 is the maximum vertical deflection, as illustrated in Fig. 14.31. (Similarly, X_1 is the x-axis intercept and X_2 is the maximum horizontal deflection.)

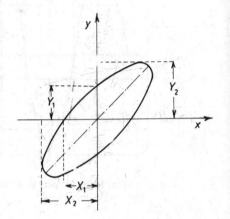

14.5 AUDIO SIGNAL GENERATORS

The signal generator is an instrument that generates an electrical signal in either the audio- or radio-frequency range. Audio signal generator produces audio frequencies (sine waves and/or square waves). It is a very popular instrument and is extensively used for testing amplifiers.

Fig. 14.31 Measurement of phase difference between two sine waves

At this stage, it is not possible to go into the details of the circuitry of an audio signal generator. However, an attempt will be made to explain how a signal generator works. We shall utilize a block diagram for this purpose.

14.5.1 Block Diagram of Audio Signal Generators

Figure 14.32 shows the block diagram of a sine-square audio oscillator (the signal generator is also called an *oscillator*). The heart of the generator is the Wein bridge oscillator. Though there are other types of oscillators, the Wein bridge oscillator serves best in the audio-frequency range. The frequency of oscillation can easily be changed by varying capacitors in the oscillator. The frequency can be changed in steps by switching in resistors of different values. (The details of this oscillator are discussed in Unit 13.)

The output of the Wein bridge oscillator goes to the "Function Switch". The function switch directs the oscillator output either to the sine-wave

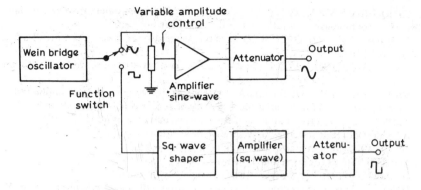

Fig. 14.32 Block diagram of an audio sine-square wave oscillator

amplifier or to the square-wave shaper. At the output, we get either a sine wave or a square wave. The output is varied by means of an attenuator.

The next section introduces the student to various controls on the front panel of a typical signal generator.

14.5.2 Front Panel Controls of a Typical Signal Generator

Figure. 14.33 shows the front panel of a portable sine-wave generator, Model 516-I. It is manufactured by Ruttonsha Simpson Pvt. Ltd. The instrument generates frequencies from 10 Hz to 1 MHz, continuously variable in 5 decades with overlapping ranges. The output sine-wave amplitude can be varied from 5 mV to 5 V (rms). The output is taken through a push-pull amplifier for low output impedance. The output impedance is 600 Ω. The square-wave amplitudes can be varied from 0 to 20 V (peak). It is possible to adjust the symmetry of the square wave from 36 % to 70 %. The instrument requires only 7 W of power at 220 V, 50 Hz.

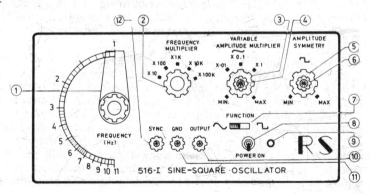

Fig. 14.33 Front panel controls of sine-square oscillator. (Model 516-I), manufactured by Ruttonsha Simpson Pvt. Ltd.

We give below the function of the various controls on the front panel of the signal generator, model 516-I. The numbers within brackets refer to the control number as given in Fig. 14.33.

FREQUENCY SELECTOR (1): It selects the frequencies in different ranges. It varies frequency continuously in a ratio of 1: 11. The scale is not linear.

FREQUENCY MULTIPLIER (2): It selects the frequency range over 5 decades from 10 Hz to 1 MHz.

AMPLITUDE MULTIPLIER (3): It attenuates sine-wave in 3 decades ×1, ×0.1, and ×0.01.

VARIABLE AMPLITUDE (sin) (4): It attenuates sine-wave amplitude continuously.

SYMMETRY (SQ) (5): It varies the symmetry of square-wave from 30 % to 70 %.

AMPLITUDE (SQ) (6): It attenuates square-wave output continuously.

FUNCTION (7): It selects either sine-wave or square-wave output.

ON POWER (8): It energizes the instrument in ON position.

NEON LAMP (9): It glows when the instrument is ON.

OUTPUT (10): This provides sine-wave or square-wave output.

GND (11): This provides ground reference for output (10) and also for Sync. Input (12).

SYNC Input (12): This terminal is provided to accept external sychronization signal.

14.6 STRAIN GAUGE

The strain gauge uses a passive* transducer which converts mechanical displacement into a change in resistance. The gauge is built of thin resistance wire, usually made of constantan (60 % Cu and 40 % Ni). This resistance wire can be used in two ways. In a *bonded strain gauge*, the wire is connected to a thin sheet of paper and covered with a protective coating of paper. It is then bonded with the structure in which strain is to be determined. The problem of bonding between the gauge and the structure is very difficult. The adhesive material must hold the gauge firmly to the structure, yet it must have sufficient elasticity to give way under strain without losing its adhesive properties. It should also be resistant to temperature, humidity, and other environmental conditions. When the structure is subjected to strain, the resistance of the gauge changes. This change in resistance is then measured using a Wheatstone bridge.

In an *unbonded strain gauge*, we use an armature which is supported in the centre of a stationary frame. The armature can move only in one direction. The movement in that direction is limited by four filaments of strain-sensitive wire. The filaments are of equal length and are arranged as shown in Fig. 14.34a.

When an external force is applied to the strain gauge, the armature moves in the direction indicated. The elements A and D increase in length, whereas the elements B and C decrease in length. As a result, the resistance R_1 and R_4 increase and R_2 and R_3 both decrease. These four resistances form the four arms of a Wheatstone bridge (see Fig. 14.34b). This arrangement of the four resistance elements (instead of using only one element) increases the sensitivity of the strain gauge. Due to change in the resistance values, the bridge gets unbalanced. The unbalanced current, indicated by the current meter, is calibrated to read the magnitude of the displacement of the armature. It is then an easy matter to calculate the strain produced in the structure.

---

*In contrast, a *self-generating* transducer, such as a thermocouple, produces its own voltage or current. It does not require external power.

Stationary frame

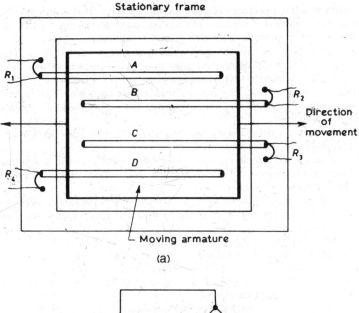

(a)

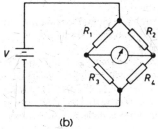

(b)

Fig. 14.34 Unbonded strain gauge: (*a*) Mode of construction;
(*b*) Measuring circuit (Wheatstone bridge)

REVIEW QUESTIONS

14.1 State the various electrical quantities that can be measured with a multimeter.

14.2 A multimeter uses the d'Arsonval movement along with associated circuitry. Explain in brief the construction and working principle of the d'Arsonval movement.

14.3 Explain what you understand by (*a*) current sensitivity (CS), and (*b*) voltage sensitivity (VS) of a d'Arsonval movement. Are these two quantities interrelated ?

14.4 The voltage sensitivity and current sensitivity of a movement (meter) is known. Derive a general expression for (*a*) the series resistance R_S, for extending its range to voltage V, (*b*) the shunt resistance R_{sh} for extending its current range to I_{max}.

14.5 The current sensitivity, or full-scale deflection current for a movement (meter) is known to be 0.5 mA. Can we extend its current measuring range to as low a value as 50 μA ? If yes, explain the procedure. If not, why ?

14.6 Explain why the resistance scale in a multimeter is (usually) inverted, i.e. its zero is at the right end of the scale and readings increase towards the left. Also, explain why the ohmmeter scale is not linear throughout its range.

14.7 A multimeter can be used to measure resistances. It is then called an ohmmeter. In an attempt to measure the value of a resistor connected in an electronic circuit, the ohmmeter terminals are connected across the resistor ends. The circuit remains energized (i.e. the power input to the circuit is not switched off). The ohmmeter will not only give wrong readings, but is also likely to get damaged. Explain why.

14.8 Explain in brief (in 5 to 7 lines) what you understand by ohms per volt rating of a multimeter. How is it related to its current sensitivity ?

14.9 The internal resistance of a meter is 100 kΩ in its 50-V range. If the meter is measuring a voltage of 25 V in this range, will its internal resistance remain the same ? Explain your answer.

14.10 What do you mean by loading effect of a meter ? Will the loading effect be more for greater ohms per volt rating of the meter ? Explain your answer in brief (in 8 to 10 lines).

14.11 In what respect does an electronic voltmeter differ from an ordinary voltmeter ?

14.12 What advantages does a VTVM have over an ordinary moving-coil meter? Does it have any disadvantages too ? If so, mention them.

14.13 Explain briefly the principle and working of a VTVM. State the reason why a VTVM has very high input impedance ?

14.14 Explain briefly the working principle of a digital multimeter (DMM).

14.15 A VTVM is basically a dc voltmeter. Explain how it can be used to measure (a) ac voltages, (b) resistances.

14.16 Sketch a CRT with electric focussing and deflection system. What are its main parts ? Give the function of each part. Why is the electron gun so called?

14.17 In a CRT, the vertical deflection plates are kept horizontally, whereas the horizontal deflection plates are kept vertically. Explain why.

14.18 Explain how an electron beam is deflected by a voltage applied to the deflection plates.

14.19 Explain briefly (in 8 to 10 lines) how an electric lens is formed when the focusing anode and accelerating anode are kept at different potentials. How does this lens help in focusing the electron beam ?

14.20 What is the purpose of the aquadag coating on the inside of the glass bulb in a CRT ?

14.21 Explain why you need a sawtooth voltage to display a waveform on a CRO (in 8 to 10 lines).

14.22 A sine-wave voltage of frequency 150 Hz is fed to the vertical input, and simultaneously a sawtooth/voltage of frequency 100 Hz is fed to the horizontal input of a CRO. Explain, with a neat diagram, what display you would get on the screen.

14.23 State what you would do for blanking off the retrace path on the CRO screen.

14.24 Draw the block diagram of a CRO and explain briefly the function of each block.

14.25 State the main applications of a CRO. Explain each of them briefly.

14.26 In some modern CROs two displays are obtained simultaneously. There are two methods to achieve this. State them.

14.27 Draw the Lissajous pattern you expect when the ratio of vertical frequency to horizontal frequency is 1 : 2.

14.28 What is the order of anode voltage in a general purpose oscilloscope ?

14.29 Draw the block diagram of an audio-signal generator. Explain the function of each block.

14.30 Give the order of output impedance of an audio signal generator.

14.31 Explain briefly the construction and working of a strain gauge.

OBJECTIVE-TYPE QUESTIONS

I. Here are some incomplete statements. Four alternatives are provided below each.
Tick the alternative that completes the statement correctly:

1. The voltage sensitivity of a multimeter is 20 mV. Its internal resistance is 40 Ω.
Then, its current sensitivity is

   (a) 1 mA   (c) 500 µA
   (b) 5 µA   (d) 0.8 A

2. A 1-mA, 50-Ω meter movement is used to make a voltmeter of range 50 V.
The series resistance to be connected is

   (a) 10 kΩ (c) 50 kΩ
   (b) 49.95 kΩ (d) 499.95 kΩ

3. A meter movement having full-scale deflection current of 50 µA and internal
resistance of 40 Ω is used to measure currents up to 50 mA. The value of the
shunt resistance should be

   (a) 0.040 04 Ω (c) 0.002 Ω
   (b) 0.5 Ω (d) 1.00 Ω

4. The current sensitivity of a meter is 100 µA and its internal resistance is 50 Ω.
Its ohms per voltage rating should be

   (a) 10 kΩ/V   (c) 20 kΩ/V
   (b) 50 kΩ/V   (d) 50 Ω/V

5. It is found that the internal resistance of a voltmeter is 100 kΩ when it is reading full-scale voltage of 100 V. On the same range, if it reads another voltage
of 40 V, its internal resistance will be

   (a) 400 kΩ (c) 40 kΩ
   (b) 4000 kΩ (d) 100 kΩ

6. In VTVMs, the ac voltage measurements correspond to

   (a) square-wave input   (c) sine-wave input
   (b) sawtooth-wave input   (d) any waveform at the input

7. The calibration signal, usually available from a CRO, is of the

   (a) sine-wave voltage   (c) sawtooth-wave voltage
   (b) square-wave voltage   (d) dc voltage

8. The colour of the bright spot on the screen of a CRO is the characteristic of

   (a) the signal being viewed
   (b) the primary electrons emitted from the cathode
   (c) the final speed with which the electrons strike the screen
   (d) the coating material of the screen

9. Lissajous patterns obtained on a CRO can be used to determine

   (a) phase shift   (c) amplitude distortion
   (b) voltage amplitude   (d) none of the above

10. The purpose of the SYNC control in a CRO is to

    (a) focus the spot on the screen
    (b) set the intensity of the spot on the screen
    (c) lock the display of signal
    (d) adjust the amplitude of the display

11. The length of the trace (sweep) on the CRT screen is controlled by

    (a) horizontal-gain control   (c) SYNC control
    (b) vertical-gain control   (d) trigger-level control

12. If the retrace is visible on the CRO screen, the trouble may be that

    (a) intensity is too high
    (b) there is loss of SYNC-control signal
    (c) blanking control is not set properly
    (d) the voltage of the accelerating anode is too high

13. The time-base of an oscilloscope is developed by

    (a) sine waveform   (c) square waveform
    (b) sawtooth waveform   (d) output from a built-in clock

14. To prevent loading of a circuit under test, the input impedance of a CRO must be

    (a) capacitive   (c) high
    (b) inductive   (d) low

15. The maximum frequency to be applied to a CRO is limited by

    (a) the vertical amplifier   (c) the SYNC signal frequency
    (b) the horizontal amplifier   (d) coating material of the screen

16. The attenuator in a signal generator is used to

    (a) provide an external shunt across the output terminals
    (b) vary the output impedance of the oscillator
    (c) increase the frequency of the output voltage
    (d) vary the output voltage amplitude in steps

17. The transducer used in a strain gauge is

    (a) an active transducer
    (b) a device that converts electrical voltage into mechanical displacement
    (c) a device that converts mechanical displacement into an electrical current
    (d) a device that converts mechanical displacement into a change in resistance

II. Indicate whether the following statements are TRUE or FALSE:

1. Most of the multimeters use iron-vane type movement. _____
2. The two spiral springs attached to the coil asssembly of a d'Arsonval movement provide a path for the current to reach the coil. _____
3. Higher the current sensitivity of a movement (moving-coil meter), the better will be the multimeter. _____
4. The resistance scale of a multimeter has zero on the right end, and it linearly increases towards the left. _____
5. An ohmmeter should never be applied to an energized circuit. _____
6. Whatever may be the deflection of the pointer, the internal resistance of a voltmeter is determined by the full-scale voltage. _____
7. The loading effect of an ordinary moving-coil multimeter is almost the same as that of an electronic multimeter. _____
8. If we give a negative potential to the upper vertical deflection plate with respect to the lower one, the spot on the screen will move upward. _____
9. The aquadag coating on the inside of the glass bulb is maintained at a high positive potential with respect to the ground, because it is required to collect the secondary emission electrons. _____
10. When we adjust the X-POS control on the front panel of a CRO, the dc voltage applied to vertically-kept plates inside the CRT changes. _____

Ans. I. 1. *c*; 2. *b*; 3. *a*; 4. *a*; 5. *d*; 6. *c*; 7. *b*; 8. *d*; 9. *a*; 10. *c*; 11. *a*; 12. *c*;
     13. *b*; 14. *c*; 15. *a*; 16. *d*; 17. *d*.

  II. 1. F; 2. T; 3. F; 4. F; 5. T; 6. T; 7. F; 8. F; 9. F; 10. T.

TUTORIAL SHEET 14.1

1. A d'Arsonval movement has a full-scale deflection current of 400 μA and internal resistance of 50 Ω. This is to be used as a voltmeter with ranges 1 V, 5 V, 10 V, 50 V and 100 V. Design the multiplier.

> [**Ans.** The series resistors are 2.45 kΩ, 12.45 kΩ, 24.95 kΩ,
> 124.95 kΩ, and 249.95 kΩ]

2. A moving-coil movement (meter) has a current sensitivity of 0.1 mA and internal resistance 100 Ω. This movement is to be converted into a multi-range ammeter with ranges 10 mA, 50 mA, 100 mA and 1 A. (*a*) Calculate the values of shunt resistors needed for each range; (*b*) design a universal shunt for this meter that requires no resistances less than 1 Ω.

> [**Ans.** (*a*) 1.010 101 Ω, 0.200 040 08 Ω,
> 0.100 100 1 Ω and 0.010 01 Ω;
>
> (*b*) the design is as given in Fig. T.14.1.1]

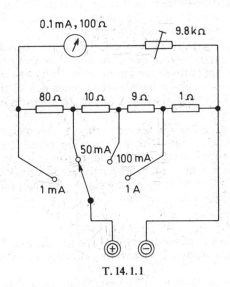

T. 14. 1. 1

TUTORIAL SHEET 14.2

1. The vertical amplifier of a CRO is set at a sensitivity of 0.5 V/cm. A sine-wave voltage is connected to the *Y*-input. A straight line trace of length 5.6 cm is obtained. Calculate the rms value of the voltage.   (Ans. 0.99 V)

2. A sine-wave voltage is displayed on a CRO. Its vertical amplifier sensitivity is set at 5 V/cm, and time base selector switch is set at a sweep speed of 50 μs/cm. The displayed sine wave has a peak amplitude of 2.7 cm, and its two complete cycles are accommodated in 8.4 cm along the *x*-axis. Determine the magnitude and frequency of the sine-wave input voltage.   [Ans. 9.546 V(rms), 4761.9 Hz]

3. Suppose two 2-MΩ resistors are connected in series across a 90-V source. A multimeter having a sensitivity of 40 000 Ω/V is used to measure the voltage across one of the resistors. The range used is 50 V. What will be the reading on the multimeter?
[Ans. 30 V]

EXPERIMENTAL EXERCISE 14.1

TITLE: Use of multimeter.

OBJECTIVES: To,

1. state the function of different knobs on the panel of a multimeter;
2. measure ac and dc voltages with the help of a multimeter;
3. measure resistances, and compare the measured value with the value indicated on the body of the resistor.

APPARATUS REQUIRED: Multimeter, signal generator, dry cells, power supply unit, resistors of assorted values.

BRIEF THEORY: A multimeter is an instrument that can measure ac and dc voltages and currents and resistances. It consists essentially of separate voltage, current and resistance measuring circuits. The meter movement is common to all the three circuits. A selector switch is provided to set up the required circuit for a desired measurement.

A multimeter is normally used to measure ac and dc voltages from 0 to 3000 V, ac and dc currents from 0 to 3 A, and resistances from 0 to 20 MΩ. In ac ranges, the measurements are possible from 40 Hz to 10 kHz. It is calibrated for a pure sine-wave signal. When used as a voltmeter, the resistance of the meter is determined by its sensitivity, expressed in ohms/volt and full-scale voltages.

While measuring resistance, never connect the meter terminals to an energized circuit. Also, ensure that there is no parallel branch across the component you are measuring. When in doubt, disconnect one terminal of the component from the circuit. The measured value of a resistor can be compared with the value written on its body. Very often, the value is written in the form of a colour code (see Unit 1).

PROCEDURE:

1. Sketch the front panel diagram of the multimeter. Clearly show the scales and other details. Indicate the functions of various knobs and controls.
2. Set the Function Switch for dc voltage measurement. Connect meter leads to the two terminals of a dry cell. The positive lead (usually red coloured) is connected to the positive side, and negative (or "common") lead (usually black coloured) to the negative side of voltage. Start with the highest range and then switch down to select the suitable range. Read the value of the voltage on a range in which the pointer deflection falls on the upper half of the meter scale. This gives better accuracy. Repeat the procedure to measure other dc voltages as obtained from a transistor (or IC) power supply.
3. Set the Function Switch for ac voltage measurement. Measure ac voltages of different values obtained from an audio-signal generator in a similar manner. You can also measure the voltage at the supply mains.

4. For the measurement of current, the meter should be connected in series. The conventional current enters the positive terminal and leaves the negative (or common) terminal.

5. Set the Function Switch for ohms measurement. Short-circuit the + and C leads. Does pointer indicate zero value ? If not, adjust the Zero Adjust control. Now connect the unknown resistor to the test leads and read the value of resistance. Select the range suitably so that the reading is in the upper half of the scale, where the markings are not very crowded. Similarly, read the value of resistance for other resistors. Using colour code, find the value of resistance of these resistors. Compare the measured value with the value found by using colour code. Do the measured values fall within the tolerance indicated?

OBSERVATIONS:

1. Voltage of the dry cell   = _____ V
2. Voltage of dc supply   = _____ V
3. Voltage of ac mains   = _____ V
4. Maximum voltage obtainable from signal generator   = _____ V
5. Resistance measurement

S. No.	Measured value	Value indicated in colour code	Tolerance indicated	Difference between measured and given value
1. 2. 3.				

EXPERIMENTAL EXERCISE 14.2

TITLE: To test a signal generator using a CRO.

OBJECTIVES: To,

1. explain the function of each knob on the front panel of the given signal generator and CRO;
2. measure the output voltage of the audio signal generator with the help of the CRO;
3. measure the frequency of a sine-wave voltage obtained from signal generator;
4. obtain Lissajous patterns on the CRO screen by feeding two sine-wave voltages from two signal generators;
5. measure phase shift produced by a phase-shift network.

CIRCUIT DIAGRAM: The circuit arrangement is shown in Fig. E.14.2.1.

BRIEF THEORY: Both the CRO and the signal generator are important test instruments. The signal generator contains an oscillator and produces sine- (and also square-) wave voltage of adjustable frequency and magnitude. This

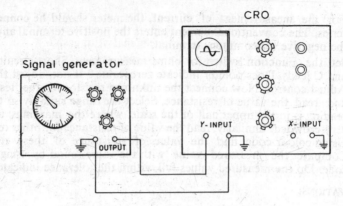

Fig. E.14.2.1

voltage can be used for testing the performance of electronic circuits. The main purpose of a CRO is to display waveshapes. The heart of a CRO is its cathode ray tube (CRT). To operate this CRT, the oscilloscope has a sweep (sawtooth) oscillator, deflection amplifiers (horizontal and vertical), power supply circuit and a number of controls, switches, and input terminals on the front panel.

An electron beam produced by the electron gun in the CRT strikes the fluorescent screen. As a result, a bright spot is observed on the screen of the CRT. By applying voltages to the horizontal and vertical deflection plates (in the CRT), the beam (and hence the bright spot) is deflected in any desired direction. To display a voltage wave, it is connected to the vertical input of the scope. To the horizontal deflection plates, a sawtooth-wave voltage is applied internally.

If we connect sine-wave voltages to both the vertical and horizontal inputs, we get a display called Lissajous pattern. The shape of this pattern depends upon the frequency ratio of the two sine waves.

PROCEDURE:

1. Sketch the front panel diagrams of CRO and signal generator. Mark the functions of each control.

2. Switch on the CRO. Rotate the intensity control clockwise. After some time, you will see either a bright spot or a line on the screen. If you see none, adjust X-POS and Y-POS controls to get the display in the centre of the screen.

3. Operate the INTEN and FOCUS controls and observe the effect on the spot (or line). Adjust them suitably.

4. Connect the output from the audio signal generator to the Y-INPUT terminals of the CRO. By adjusting the attenuator of the signal generator, adjust the output voltage at about 1 V. Adjust the frequency at 1 kHz. Now adjust the sensitivity of the vertical section to about 1 V/cm. The time-base may be adjusted to about 1 ms/cm. If a stationary display is not observed, adjust the TRIGG LEVEL.

5. To measure the voltage of the signal generator, adjust the vertical amplifier sensitivity suitably, so as to get a sufficiently large display. Read on the calibrated graticule, the vertical length of the display. This corresponds to peak-to-peak value of the signal. Multiply this

length by the sensitivity (in V/cm). Dividing this result by $2\sqrt{2}$ gives the rms value of the signal voltage. Repeat the measurement procedure for two or three other values of output signal voltages.

6. For measuring the frequency of the signal, adjust the TIME-BASE control suitably so as to get about 2-3 cycles of the signal displayed on the screen. Rotate the VERNIER control clockwise to CAL position. Read on the calibrated graticule, on the screen, the length of one cycle. Multiply this by the time-base setting (in ms/cm or μs/cm). This gives the time period of the signal. Taking inverse of this time period gives the frequency of the signal (i.e. $f = 1/T$ Hz).

7. Many CROs do not have a calibrated time-base. We can measure frequency of a signal using such a CRO. For this, feed the unknown signal (taken from the signal generator) to the Y-INPUT terminals. Take a standard signal generator, and connect its output to the X-INPUT terminals of the CRO. Put the TIME-BASE or HORIZON-TAL-AMPLIFIER knob at EXT position. Change the frequency of the standard signal generator till you get a stable Lissajous pattern. For the various frequency ratios, f_V/f_H, the Lissajous patterns are shown in Fig. E.14.2.2. The unknown frequency can thus be determined.

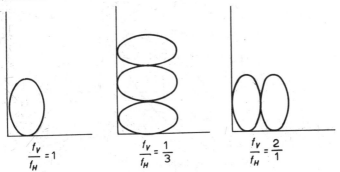

Fig. E.14.2.2 Lissajous patterns

8. To measure the phase shift introduced by an RC phase-shift nerwork, make connections as shown in Fig. E.14.2.3. Put the TIME-BASE control at EXT position. Adjust the vertical and horizontal amplifier gains (sensitivities) so as to get an ellipse of suitable size, as shown in Fig. E.14.2.4. Measure the lengths Y_1 and Y_2 (or X_1 and X_2). Calculate the phase difference between the two waves using the relation

$$\sin \theta = \frac{Y_1}{Y_2} = \frac{X_1}{X_2}$$

OBSERVATIONS:

1. Measurement of voltage:

S. No.	Signal generator output (measured by a voltmeter)	Measurement on CRO		
		(p-p) value in cm	Sensitivity in V/cm	rms value
1.				
2.				
3.				

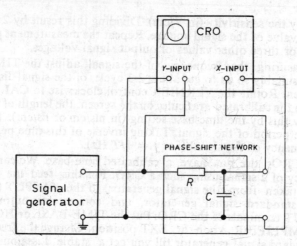

Fig. E.14.2.3 Arrangement for measuring phase difference

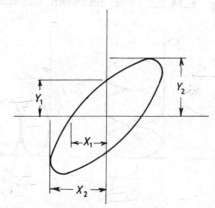

Fig. E.14.2.4 Lissajous pattern for measurement
of phase difference

2. Measurement of frequency:

S. No.	Frequency of the signal generator	Measurement on CRO			
		Length of one wave in cm	Sensitivity in ms/cm	Time period	Frequency in kHz
1.					
2.					
3.					

3. Measurement of phase angle:

$$Y_1 = \underline{\hspace{2cm}} \text{ cm}$$

$$Y_2 = \underline{\hspace{2cm}} \text{ cm}$$

$$\therefore \quad \theta = \sin^{-1} \frac{Y_1}{Y_2} = \sin^{-1} \underline{\hspace{1.5cm}} = \underline{\hspace{1.5cm}}$$

Index